INJECTION MOULDING MATERIALS

INJECTION MOULDING MATERIALS

A. WHELAN

National College of Rubber Technology,
The Polytechnic of North London,
Holloway Road, London N7 8DB, UK

APPLIED SCIENCE PUBLISHERS
LONDON and NEW JERSEY

APPLIED SCIENCE PUBLISHERS LTD
Ripple Road, Barking, Essex, England
APPLIED SCIENCE PUBLISHERS, INC.
Englewood, New Jersey 07631, USA

British Library Cataloguing in Publication Data

Whelan, A.
Injection moulding materials.
1. Injection moulding of plastics
2. Thermoplastics
I. Title
668.4'23 TP1150

ISBN-13:978-94-009-7360-2 e-ISBN-13:978-94-009-7358-9
DOI: 10.1007/978-94-009-7358-9

WITH 54 TABLES AND 44 ILLUSTRATIONS

Softcover reprint of the hardcover 1st edition 1982

Preface

In order to make the subject manageable the term 'injection moulding' has been restricted in its use so that only those processes which rely on thermal softening of the polymeric materials have been described and discussed in this book. It is intended to discuss the subject of reaction injection moulding in a separate book.

However, even with this omission, the subject is still a very large one as nowadays many sorts or types of polymers are injection moulded. For example, it is estimated that one-third of all plastics materials are injection moulded—the range of products produced is enormous and increases daily.

Because most moulding materials are based on plastics, in particular thermoplastics, the materials guides which form a large part of this book concentrate on the moulding of thermoplastics materials. Such guides should only be treated as general guidelines as each of the materials is normally available in a wide range of grades. These may differ in polymer molecular weight, molecular weight distribution, the additives used and their concentration, the physical form of the moulding compound, etc. A wide range of processing behaviours and end-use properties is therefore possible from any of the materials listed. This versatility is typified by the rubbery polymers which are compounded into an incredibly wide range of compounds. Because of this versatility only a very general guideline has been given for such materials.

It should also not be forgotten that the properties of a particular component can be drastically altered or modified by changes in processing conditions or processing procedure. Such changes can mask or disguise the differences between grades of the same material and can sometimes even mask the differences between different materials.

Many people have helped me during the writing of this book and to all such people I extend my sincere thanks. In particular I would like to thank my colleague J. P. Goff for helping me prepare many of the materials guides.

A. WHELAN

Contents

Preface v

Chapter 1 Moulding Materials and Materials Handling

1. Introduction 1
2. Delivery of plastics 2
3. In-plant materials handling 5
4. Material reclamation 9
5. Drying 18
6. Mixing and blending 21
7. Colouring systems 32

Chapter 2 Moulding Machines and Processes

1. Types of machine 37
2. Ram machines 38
3. Two-stage preplasticising systems 44
4. In-line preplasticising systems 51
5. Other machine types 64
6. Injection mould types 69
7. Alternative moulding techniques 73
8. Reaction injection moulding 78

Chapter 3 Machine Specifications

1. Facts included 80
2. The injection unit 80
3. The clamping unit 91
4. The hydraulic system 94

5. Control and setting 96
6. Optional extras 99
7. Safety 101
8. Machine purchase 108

Chapter 4 Component Finishing and Decoration

1. Flash removal and buffing 114
2. Machining of plastics 117
3. Metallisation 121
4. Printing on mouldings 139
5. Painting 141
6. Welding 143
7. Mechanical joining 150
8. Adhesive bonding 160

Chapter 5 Effect of Processing on Properties

1. Introduction 168
2. General effect of orientation and crystallinity on properties . 169
3. Mould filling, packing and stress relieving 172
4. Detection of orientation and crystallinity 179
5. Effect of moulding conditions on properties 185
6. Effect of changing gate size and position 193
7. Experiment design in injection moulding 198
8. Moulding quality and unknown changes 203

Chapter 6 Testing and Quality Control

1. The object of testing 205
2. Standardisation 207
3. Common short-term tests 207
4. Long-term testing 219
5. Injection moulding of plastics test specimens 223
6. Quality control and testing 238

Chapter 7 Material Recommendations for Injection Moulding

1. Polystyrene 251
2. Styrene acrylonitrile 254
3. Toughened polystyrene. 257
4. Acrylonitrile-butadiene styrene 260

5. Acrylic . . . 264
6. Unplasticised polyvinyl chloride . . . 268
7. Plasticised polyvinyl chloride . . . 272
8. Cellulosics . . . 275
9. Polycarbonate . . . 280
10. Polyethersulphone . . . 284
11. Polyphenylene oxide (modified) . . . 289
12. Low density polyethylene . . . 292
13. High density polyethylene . . . 297
14. Polypropylene . . . 301
15. Polybutylene terephthalate . . . 305
16. Nylon . . . 309
17. Acetals . . . 316
18. Polyvinylidene fluoride . . . 319
19. Polyphenylene sulphide . . . 323
20. Rubber reinforced polypropylene . . . 327
21. Thermoplastic polyurethane . . . 331
22. SBS block copolymers . . . 336
23. Thermoplastic polyetherester . . . 340
24. Fibre reinforced thermoplastics . . . 344
25. Structural foam . . . 349
26. Aminoplastics . . . 354
27. Phenolics . . . 359
28. Dough moulding compound . . . 363
29. Elastomers . . . 367

Appendix I Euromap documents . . . 384

Appendix II Troubleshooting . . . 387

Index . . . 391

CHAPTER 1

Moulding Materials and Materials Handling

1 INTRODUCTION

Injection moulding is a cyclic process whereby a heat-softened 'plastic' material is injected into a mould where it is set to the shape of the cavity. The mould is then opened, the part is ejected and the process is then repeated.

The term plastic is placed in inverted commas because in this case the word is used to indicate that the material is capable of flow. Thermoplastics are the most widely used materials handled by this moulding process but thermosetting plastics, rubbers and thermoplastic elastomers are also injection moulded. The products may be solid or cellular and by the introduction of a blowing stage, hollow enclosed parts (e.g. jars and bottles) may be made from thermoplastics materials.

The products of injection mouldings are many and varied and range from tiny gears and buttons to objects the size of hospital laundry baskets. Common mouldings include buckets, bowls, crates, toys and closures. This is purely a random list and not assembled in any particular order, as to try and list all the products made by injection moulding would be an exhausting and pointless task. Generally speaking such components have one thing in common: they are all required in large numbers. It is unusual, but not unknown, to find anybody moulding an item where only a few components are required. This is because the process demands expensive equipment (e.g. moulds) and machine setting can be difficult and therefore costly. However, once the mould and machine are set, then rapid production rates can be achieved even on complex components (such as the body of a camera). Such a component may be readily moulded complete with holes, slots, etc. and be virtually ready, as produced, for the next stage of assembly. A high degree of dimensional accuracy in the moulding can be attained provided that one makes due allowance for factors such as shrinkage.

It is probable that at least one-third of all the thermoplastics produced go through an injection moulding machine. In the UK about two-fifths of all money spent on equipment is spent on injection moulding equipment and therefore, from a cost point of view, the process is second only to extrusion in importance. One report (Bradley, 1981) stated that sales of equipment are expected to nearly double over the period 1980–90 within Western Europe. If 1978 dollars are used for comparison then the money spent is expected to rise from 475 million dollars to approximately 900 million dollars. West Germany's share of the market is expected to increase dramatically.

2 DELIVERY OF PLASTICS

Thermoplastics are normally fed to injection moulding machines in pellet or granule form, e.g. as a 3 mm cube. Although this is the most common raw material feed form, it is not the only one possible; PVC is often fed in powder form whereas compounds based on rubbers (see Section 6) are fed in the form of a strip. These two groups of material are often compounded 'in-house' whereas other thermoplastic and thermosetting materials are purchased already compounded from the raw material supplier.

The material may be supplied in sacks, tins, large bins or as a bulk delivery. Because of ease of handling, material packed in sacks is still popular, although the use of bulk handling is growing.

2.1 Sacks or Bags

Many plastics materials are packed into sacks which contain 25 kg (55 lbs) and apart from the supplier's name, and the material type, various other information is printed on the sack. Note should always be taken of this information and if in doubt, the manufacturer should be asked to explain the coded information so that the wrong grade of material is not used inadvertently.

Such sacks are commonly made from layers of paper, of which one layer or ply is polyethylene coated, and such sacks are stitched at each end. When the sacks are opened, care should be taken to ensure that no paper or thread is introduced into the material feed system (or hopper) with the plastics material; it is good practice to wipe the sacks clean before they are inverted and emptied, as this will prevent dust and dirt being introduced with the polymer.

In recent years sacks made from heavy gauge polyethylene have been introduced and after the sacks have been filled with the plastics material, they are partially evacuated and then heat sealed. This gives a very compact package which stacks easily as it lies flat; the ability to see the contents through the wall of the sack is an advantage as it helps to prevent the use of a differently coloured material. Foil-lined sacks are sometimes used for hygroscopic materials.

2.2 Tins or Cans

With certain materials, e.g. polycarbonates, such sacks are unable to stop the material picking up moisture. In this case (i.e. when an expensive, hygroscopic material must be kept as initially packed) the material may be packaged inside a metal tin or can. Such tins or cans should only be opened when required and only when the material has reached a specified temperature (after it has been brought from the raw material stores). With polyamides this specified temperature is usually room temperature but with polycarbonates the unopened tin must be heated for 3 h at a temperature of 115 °C before it is opened. Such precautions are necessary if moisture contamination is to be avoided.

2.3 Bulk Delivery

Portable bulk containers, or bins, are now used to transport plastics materials in either 500-kg or 1000-kg lots. Such lots may be supplied in fibre-board containers whose base is just over 1 m^2 (1 yd^2); a 1 tonne container would be approximately 2 m (2 yd) high. With some portable bulk containers a discharge valve is incorporated in the base so that the pelleted material flows under the influence of gravity from the container and into, for example, the blending machine.

A company which uses a large quantity of material should, in the interest of economy, elect to have its material delivered by a tanker. The material is discharged from this tanker into a storage silo which is either located inside the factory or outside; when outside, it is located as close to the factory as is conveniently possible so as to reduce the amount of conveying required. These silos are topped up periodically, e.g. once a week. The material, e.g. polystyrene, is conveyed from the silo, to the next stage in the process, either mechanically or pneumatically (Fig. 1).

Despite the advantages of pellets the use of plastics (e.g. HDPE, LDPE and ABS) in powder form is growing as such a material form is relatively cheap because the pelletising step is eliminated. The elimination of this step also means that the material has a comparatively low heat history. Powders

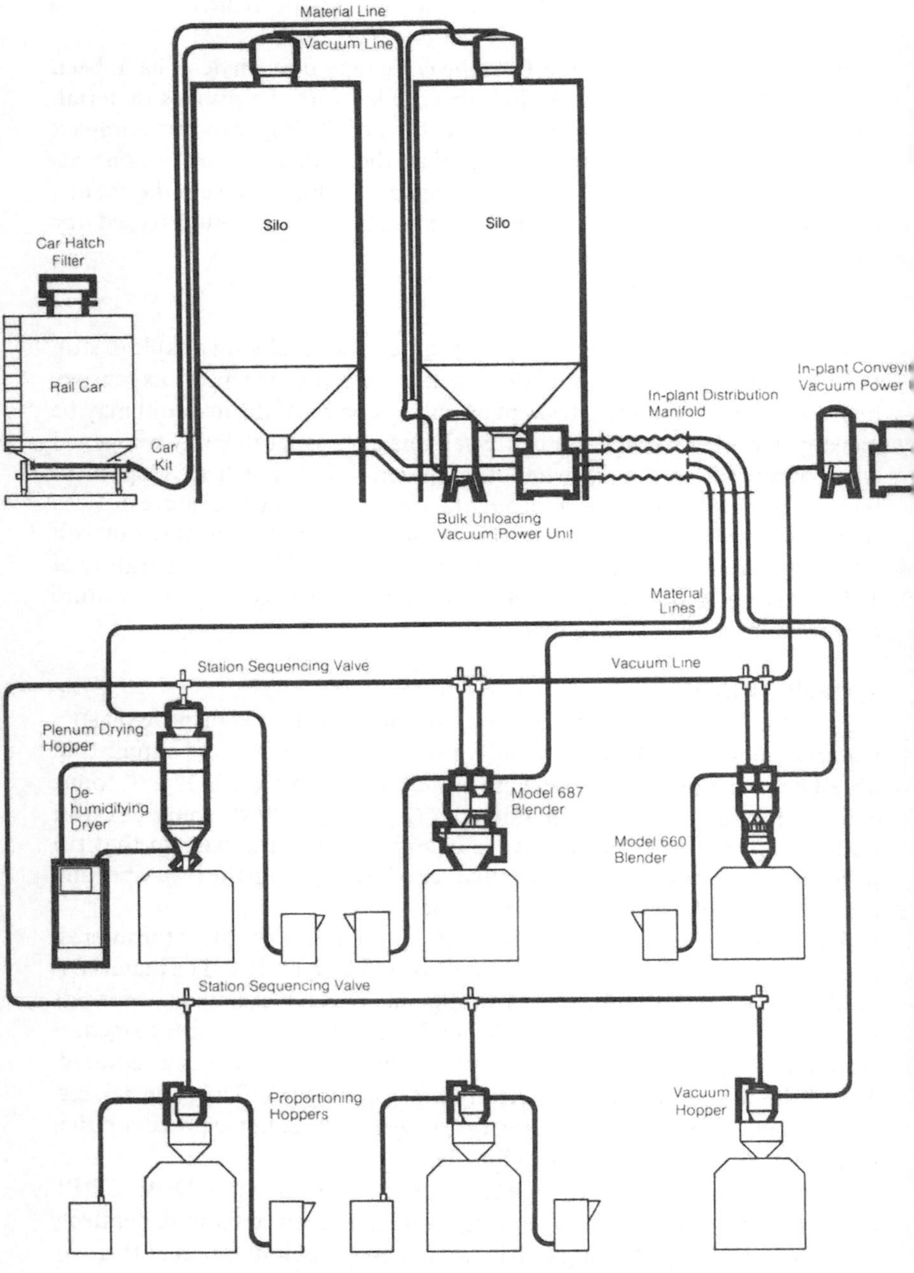

Fig. 1. Bulk storage system (after Whitlock, 1976a).

are, however, more difficult to handle as, for example, they may settle or pack in bulk storage systems. Increased moisture absorption and dusting problems may also arise and the risk of electrostatic charge generation (which may cause an explosion) should not be ignored. Because of such considerations a bulk handling system designed for pellets should not be used for powders. Powder handling systems cost more and this increased cost, together with the other problems, should be weighed against the advantages of powder.

2.4 Material Temperature

No matter which method of material delivery and storage is employed, it is important to ensure that the material is at the correct temperature when it reaches the moulding shop. One of the most common causes of material contamination is water contamination and this is often introduced when the moulding material is brought from an unheated store into the warm moulding area. When, for example, the bag containing the material is opened, the warm air comes into contact with the colder material and condensation (and therefore water contamination) occurs. If the whole storage area cannot be heated, then a storage space, which can hold sufficient material for one shift, should be made available. Cold material can be moved into this heated space at least 6 h before it is required so that it can reach moulding shop temperature before the containers are opened.

3 IN-PLANT MATERIALS HANDLING

Many moulding shops begin operation without any serious consideration having been given to materials handling. This is a mistake as it can lead to a cluttered and dangerous moulding shop.

3.1 Sack Handling

When the operation is small, and the amount of material which must be stored and moved about is also small, then this can easily be accomplished manually. For example, the sacks of material may be moved from the storage area on a sack-truck, opened at the machine and then either tipped into the machine hopper or lifted into the machine hopper by means of a pneumatic hopper loader. In some respects this is rather like a vacuum cleaner as it has a long probe which is inserted into the sack and through which the material is sucked. Where frequent colour changes are the rule, this simple material loading system has a lot to commend it because of the problems associated with cleaning of automated conveying lines.

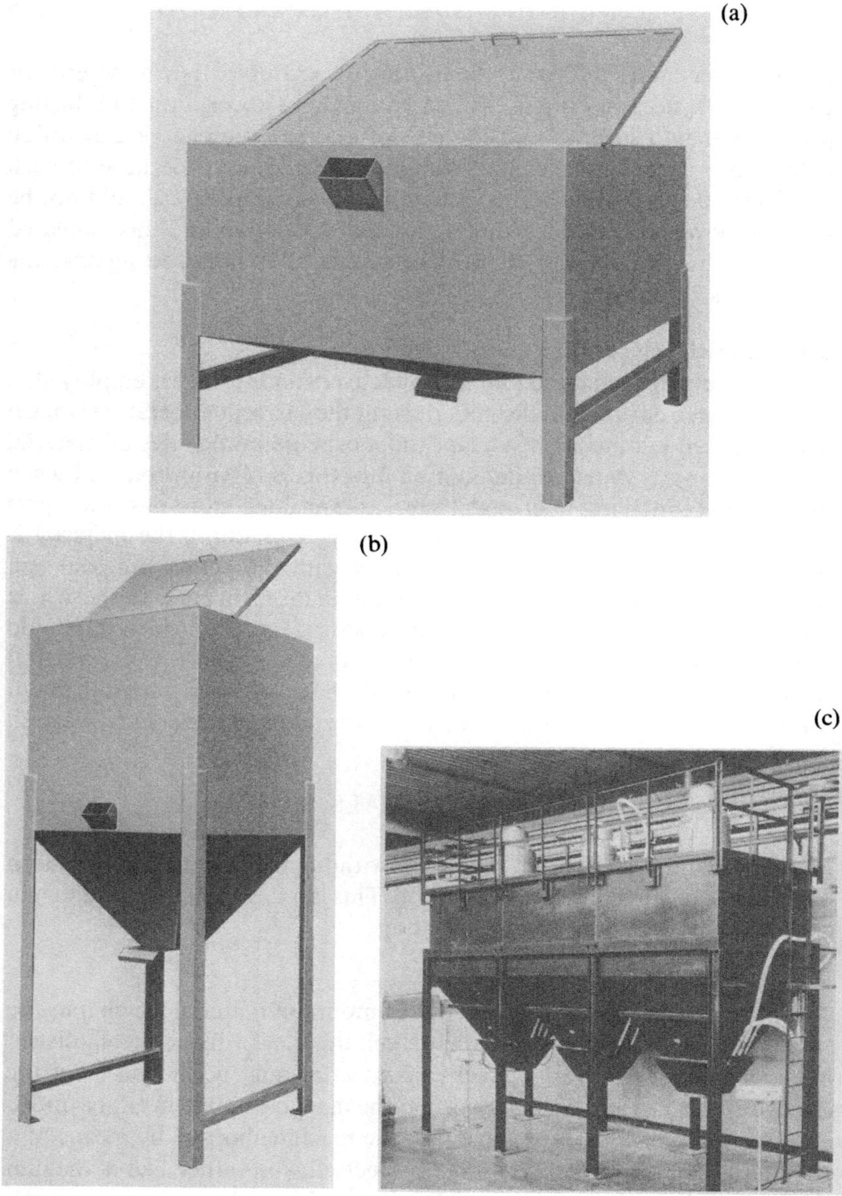

Fig. 2. Examples of material storage systems (after UPM, 1980). (a) Static material bin of 500 kg capacity. (b) Static material bin of 1000 kg capacity with discharge facility into smaller mobile bins. (c) Three intermediate bulk storage bins each with 5000 kg capacity fitted with con-evator vacuum hopper loaders conveying material from main storage silos.

As the size of the operation increases, then it becomes more convenient to handle the plastics materials on pallets. Each pallet may contain 1000 kg of material and such a load is handled by means of a fork-lift truck. It is usual to stack such pallet loads on top of each other but in order to avoid serious disruption of the storage system, no more than three pallets should be stacked. Suitable access must, of course, be provided for the fork-lift truck.

3.2 Bulk Containers

Many plastics materials are now delivered in portable, bulk-storage containers and such containers may be lifted on to a gantry or mezzanine floor, located above the machines, so that the material flows into the machine hoppers under the influence of gravity from a built-in discharge valve.

The material may also be stored within the factory in static, bulk-storage bins of intermediate capacity (for example, of 1000 or 5000 kg capacity) (Fig. 2; UPM, 1980) and then transferred to the moulding machine in a mobile storage bin. Such bins hold approximately 100 kg of material and are mounted on castors so that they may be wheeled to the appropriate machine; a probe may be inserted through the hinged lid so that the material can be transferred to the machine hopper by a vacuum loading device. Such a system reduces the amount of 'humping' that is required and makes for a neater, tidier moulding shop (Fig. 3; UPM, 1980).

Fig. 3. Mobile storage bin (after UPM, 1980).

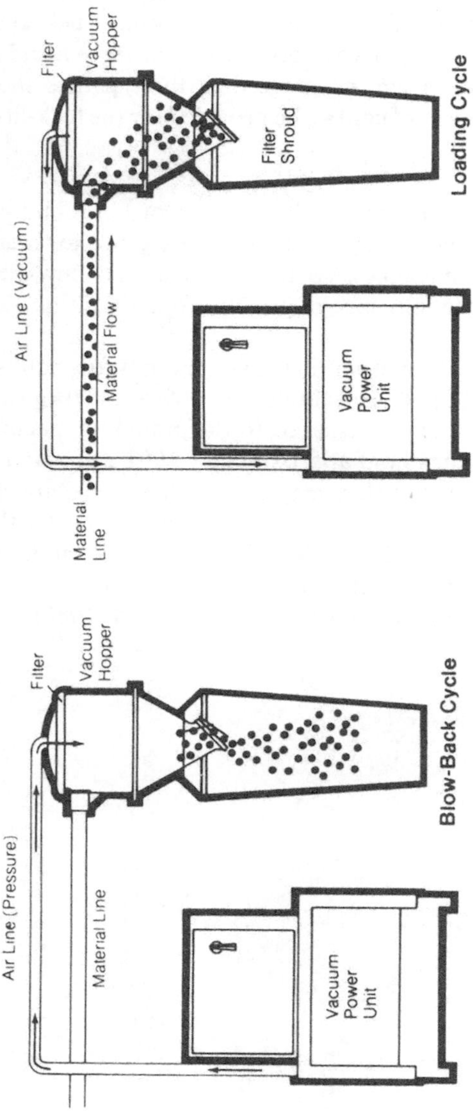

Fig. 4. A vacuum hopper system (after Whitlock, 1979).

3.3 Vacuum Hoppers

The material may be drawn up from the containers by means of, for example, a vacuum hopper and a telescopic pick-up tube. This tube is inserted into the storage bin and connected to the vacuum hopper by appropriate tubing (Fig. 4); a vacuum is induced in the conveying tubing (and the vacuum hopper) by connecting the hopper to an electric motor and blower. Free-flowing materials, such as plastics granules, are therefore transported until the electric motor is stopped; this may be done either manually or by means of a level detector.

When the vacuum hopper is full, a timer signals the air directional control valves to immediately shift from load cycle to blow-back cycle. The flapper valve on the bottom of the conveying hopper opens allowing the material to dump into the receiving station. The blow-back cycle forces air through the vacuum tubing and the protective filter located in the vacuum hopper cover or through a floor mounted filter chamber. The blow-back cycle cleans the filter after each loading cycle and aids in emptying the vacuum hopper. The power unit automatically cycles from load cycle to blow-back using air directional control valves without stopping and reversing the motor and blower (Whitlock, 1976a).

3.4 Automated Systems

For those moulders who use large amounts of materials completely automated systems are available. Material may be stored in bulk storage silos and conveyed by vacuum lines to individual machines. In many cases, material blending and/or drying is incorporated at an individual machine (Fig. 1). Such a materials handling system allows polymer to be purchased in bulk quantities (which is cheaper) and keeps the floor space around the moulding machinery free from material containers. With such systems, it is important to keep the conveying lines as short as possible, make bends as large and as few as possible and reduce the amount of flexible tubing in the system; such measures help to make materials conveying more efficient.

4 MATERIAL RECLAMATION

Despite the fact that moulding machines can be made to operate without producing any sprues or runners (i.e. by using a hot-runner mould), many machines operate, with both two- and three-plate moulds, so as to produce a feed system. In most applications the feed system must be recovered and re-used as this makes the process economically viable. This is possible with most thermoplastic materials but unfortunately, the same is not true for thermosetting moulding materials as the addition of even quite small quantities of reground, cured material can impair the properties of the final mouldings and can also impair the flow properties of the moulding material. This is because when reclaimed material is added to the thermosetting moulding powder one is, in effect, adding more filler.

4.1 Granulating Equipment

Ideally, the granulating equipment will only be used to reclaim the cold feed system (sprues and runners), however, it should also be capable of coping with reject mouldings. Most granulators used in the thermoplastics moulding industry consist of a rotor which carries a number of knives (e.g. three). These knives are rotated inside a housing which contains stationary or bed knives (Fig. 5). When the plastics material is introduced through the

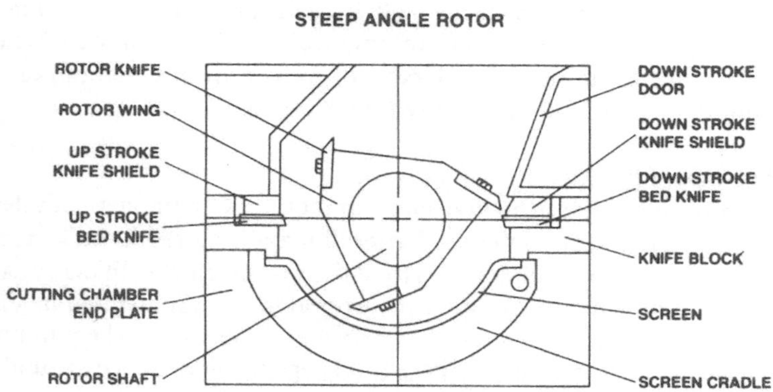

Fig. 5. A typical cutting chamber configuration (after Mills, 1979).

feed hopper at the top of the machine, it falls into the housing or cutting chamber where it is sheared and/or smashed by the knife action. The material escapes from the cutting chamber when its particle size falls below that of the holes in the screen which forms the base of the chamber. A coarse, irregular particle results and the particle size is by no means uniform; a great deal of fines is often produced by this process and care should be taken to ensure that this dust does not cause problems. It can be, for example, a fire, explosion and health risk and should be treated accordingly. Because of these problems, regranulation may be done in one location which has been designed for that purpose, e.g. the area is soundproofed, all machines are well earthed, a dust extraction system is incorporated and the operators may wear ear muffs and breathing masks.

4.2 A Manually Operated System

In many moulding shops the operator is required to separate the moulding from the feed system; this may be done using a pair of side cutters, a sharp knife, etc. The operator then places the moulding on a table or bench, gives it a quick inspection, and then feeds the separated feed system to a granulator which is located beside the press. Periodically, the granulator bin is emptied and the reclaimed material is fed back into the machine hopper after it has been blended with virgin material.

An easy way of achieving the blending required is to fit a hopper blender. A pneumatic conveying line is fitted to the granulator bin so that the reclaimed material is lifted automatically, in the correct proportion, by the hopper loader. The trimmed, inspected moulding is either packed (e.g. bag or box) or, if not of the required standard, it is regranulated.

4.3 Automatic Systems

In the system described previously, the operator must separate the feed system from the moulded components; however, this may be done mechanically. The object of doing this job mechanically is to save on labour but such savings can only be achieved if the ejection/sorting system works all the time.

4.3.1 Positive ejection

There are a number of ways available which help to ensure that the parts eject on every cycle, for example:

1. The moulds' ejection system must be designed so that the mouldings drop clear, i.e. there must be no hang-up points caused by flash around the ejector pins, machine marks on the snatch pin, etc.
2. Before closing, the mould may stop and start suddenly so as to shake any trapped mouldings free.
3. An air blast may be directed across the face of the mould so as to sweep any mouldings clear.
4. A pneumatically driven cleaning arm or sweep may move across the mould face so as to dislodge any trapped mouldings.
5. A robot take-off arm may move into the mould space and pick off the mouldings.

4.3.2 Feed separation

Various methods are also available to automatically separate the feed system from the components. These may be divided into two main classes, dependent on whether separation occurs within the mould or outside the mould.

Within the mould separation may be achieved by using a tunnel or submarine gate, a three-plate mould, a hot runner or insulated runner mould. Where the feed is ejected it may make subsequent sorting easier if two-stage ejection is employed; this means that the mouldings are ejected before the runners. After the mouldings have been conveyed away, a pneumatically operated blade can be used to divert the feed system to a granulator. With tunnel or submarine gating it is not usually possible to do this as both the feed system and the components fall at virtually the same time and a sorting operation is therefore necessary. The equipment that is used for this purpose relies on the fact that the feed system is of a different size and shape to the mouldings. In one system a rotating coil traps the runners whilst the mouldings drop through the coil into a collection shute; the runners are conveyed away by the coil and are collected at a separate chute.

If the mouldings are still attached to the feed system, then in many cases they may be separated by employing a rotary separator which is fed by a conveyor belt. This belt conveys the ejected mouldings, together with their attached feed system, to the rotary separator (consisting, for example, of a drum or belt) which has attached to it a series of pegs. As the drum or belt rotates, the pegs catch in the feed system and the lifting action cause the mouldings to break away from the feed system at the restricted gate. As the mouldings and runners have different shapes, matters can usually be arranged so that either the mouldings or the feed system drops through appropriately sized slots in the drum or belt.

4.3.3 Auger granulation

The mouldings may then be conveyed away (to be weighed, bagged or boxed) whilst the runner system can be transported to the granulator. Augers, or auger conveyors are commonly used for this purpose. Such devices consist of a large diameter, Archimedian screw, of appropriate length, which is rotated slowly inside an enclosed trough or barrel—this action conveys the sprue and runner system to the granulator. As the system is enclosed, it is less noisy and dust-producing than other alternative systems and there is less danger of accidental contamination, e.g. by ball-point pens, nuts and bolts, screwdrivers, etc. A conventional granulator would take up less space; however, such granulators are inclined to fling or spit back granules. These ejected granules make the moulding shop untidy and can make it unsafe, e.g. by making the floor slippery.

Alternatively a belt conveyor may be used.

After the feed system has been granulated, then it may be pneumatically fed back to the machine hoppers in the correct proportion and little or no operator supervision is required.

4.3.4 Heavy duty granulators

Large part size and high scrap volume may dictate the need for heavy duty machines with large throats and high horsepower ratings. In one system described (Mills, 1979) two 400 horsepower granulators were installed in order to recycle rejected polycarbonate dashboards. These machines were installed in a soundproofed room and the granulators were fed by a conveyor. The operator's only responsibilities were loading racks of reject parts into the system and taking away the empty racks. An automatic dumping system loaded the rejects on to the conveyor, which in turn fed the granulator. After granulation, the material was pneumatically conveyed to a silo for future use (Fig. 6).

4.4 Contamination of the Regrind

The grinding or granulating system can introduce material other than the desired polymer into the machine hopper, e.g. steel fragments from the

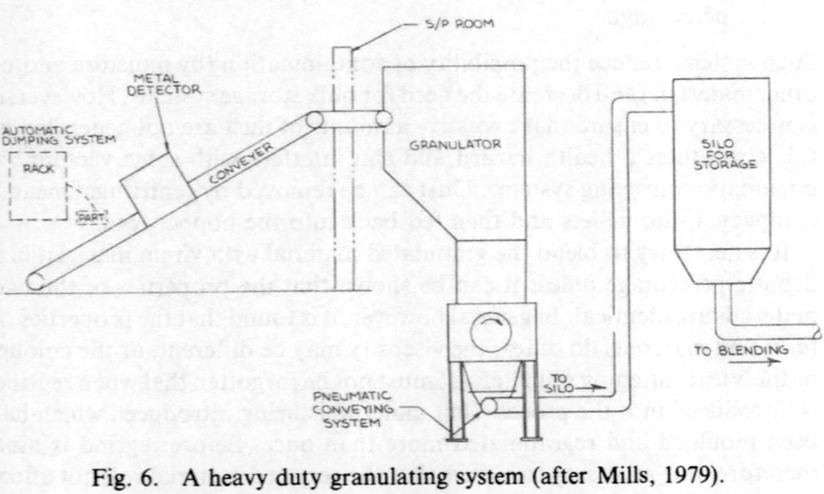

Fig. 6. A heavy duty granulating system (after Mills, 1979).

cutting blades may also be introduced. Because of this danger, it is recommended that, where a mixture of regrind and virgin material is used, that the moulding machine be fitted with a magnetic separator. Such devices consist of magnetic grids which fit in the base or throat of the machine hopper. Such a grid is only capable, however, of trapping ferrous metal fragments.

Care should be taken to ensure that the use of a pneumatic conveying system does not introduce moisture into the polymer. Unless the air has been specially treated, it will contain significant amounts of moisture.

Imperfect mouldings, sprues and runners should always be treated with great care as they are potentially very valuable. They should not be treated as scrap but should be carefully collected and stored if immediate granulation is not possible. To avoid contamination by dust, dirt or water, they should be stored in air-tight containers and they should be granulated as soon as possible to save wasting valuable storage space.

4.5 Level of Use

It is common practice in the plastics moulding industry to:

1. Automatically separate the feed system from the mouldings
2. Convey the mouldings to a finishing area
3. Drop the feed system into a besides-the-press auger granulator
4. Convey the granulated material to a proportioning hopper located on the injection machine
5. Blend the granulated material with virgin material in a definite percentage

Such systems reduce the possibility of contamination (by moisture and/or other materials) and decrease the need for bulk storage systems. However, it is necessary to ensure that excessive amounts of dust are not generated as this constitutes a health hazard and can interfere with some vacuum or pneumatic conveying systems. Dust can be removed by centrifugal means, compacted into pellets and then fed back into the hopper feed.

It is necessary to blend the granulated material with virgin material in a definite percentage unless it can be shown that the properties of the two materials are identical. In general however, it is found that the properties of reground materials do differ; the viscosity may be different, or the colour, or the Vicat softening point, etc. It must not be forgotten that when regrind is introduced into the process that material is being introduced which has been moulded and regranulated more than once. Before regrind is used therefore, it is necessary to ensure that the recycled material will not affect vital properties of the product (properties such as colour, stiffness, shrinkage, etc).

Ideally waste feed systems should be used as they are formed so that material is not accumulated. One way of seeing if this ideal state of affairs is possible is to:

1. Produce a set of mouldings and separate the feed system from the mouldings
2. Weigh both the mouldings and the feed system
3. Calculate the percentage of regrind, assuming that the feed system is to be reground
4. Consult Table 1 and ascertain how many times the feed system should be re-used
5. Granulate the feed system, remould and regranulate for the required number of times
6. Keep a sample from each pass through the granulator

For example, if the proportion of regrind is 20 % then the material should be moulded and reground three times. A mix may then be prepared which contains virgin material and reground material in the proportions suggested by Table 1. This may then be remoulded so as to determine if a vital property, or properties, have been retained.

If the properties of the mouldings are unsatisfactory then the amount of regrind will have to be reduced until acceptable results are obtained. For example, if the material has only been through the moulding and regranulation stage once then it is unlikely that its properties will change significantly. A 10 % level of use may therefore be acceptable for one job;

TABLE 1

MATERIAL RE-USE AND COMPOSITION OF REGRIND

Material type	*Percentage of regrind in hopper feed*					
	10	*20*	*30*	*40*	*50*	*60*
	Composition of hopper feed %					
Virgin material	90	80	70	60	50	40
Moulded and granulated—1	9	16	21	24	25	24
Moulded and granulated—2	1	3	6	10	13	15
Moulded and granulated—3	0	1	2	4	7	9
Moulded and granulated—4	0	0	1	2	3	6
Moulded and granulated—5	0	0	0	0	2	4
Moulded and granulated—6	0	0	0	0	0	2

accumulated material may then be worked off against other, less demanding moulding runs at a level of say 30 %.

Each formulation (i.e. 10 % and 30 %) should be given its own code number and its own formulation (Walker, 1980). This approach gives no problems over controlling the amount of scrap in hand and no 'local options' are allowed on the shop floor as to what constitutes a formulation.

4.6 Blending of Materials

It is very important that the ratio of virgin material to regrind be kept constant as otherwise mouldings with differing properties will result. One way of doing this (Whitlock, 1976a) uses a proportioning, vacuum hopper. This has a double acting, air operated valve mounted inside the hopper. The valve shifts from side to side so as to allow the alternative entry of regrind and virgin material in a set ratio into the vacuum hopper. At the end of the loading cycle, material flows and mixes in a cascade from the vacuum

TABLE 2

ACQUISITION AND SUPPORT COSTS FOR TYPICAL 300 TONNE MOULDING MACHINE (after Walker, 1980)

Purchase and installation including commissioning	£70 000
Support costs: (Based on a total operating life of 40 000 h)	
Energy costs: Avge 75 kW at 3·5p per kWh	105 000
Operator costs: 40 000 h at £2·50 h^{-1}	100 000
Maintenance and set-up costs:	
Labour 4 000 h at £4·00 h^{-1}	16 000
Materials and spares	16 000
Major overhauls: 2 at £10 000	20 000
Sub-total excluding material	257 000
Material costs: (Based on material at £750 $tonne^{-1}$ and 500 g per shot with 40 s cycle)	1 350 000
Overhead costs: (4 × direct operating costs)	400 000

$$\text{Whence static cost} = \frac{\text{overhead costs}}{\text{operating life}} = \frac{400\,000}{40\,000} = £10{\cdot}00\ h^{-1}$$

$$\text{dynamic cost} = \frac{\text{overhead costs} + \text{support costs (excluding material)}}{\text{operating life}}$$

$$= \frac{400\,000 + 257\,000 + 70\,000}{40\,000} = £18{\cdot}00\ h^{-1}$$

hopper into the machine hopper (Kadakawsky, 1979) (see Fig. 6 and Section 6.2.2).

4.7 Material Costs

An appreciation of the cost of purchasing and maintaining an in-line screw machine can be obtained from Table 2 (Walker, 1980). With a static cost of £10·00 h^{-1} the machine costs 16·7p min^{-1} to support and when working (dynamic costs) it costs 30p min^{-1} to support; this figure excludes material costs.

If the machine produces dirty or unusable reject mouldings then the cost of production must be added to the material price if a truer cost of those discarded mouldings is to be obtained. In the example shown in Table 2 it would take 1·3 min to produce 1 kg of material and at 30p min^{-1} the

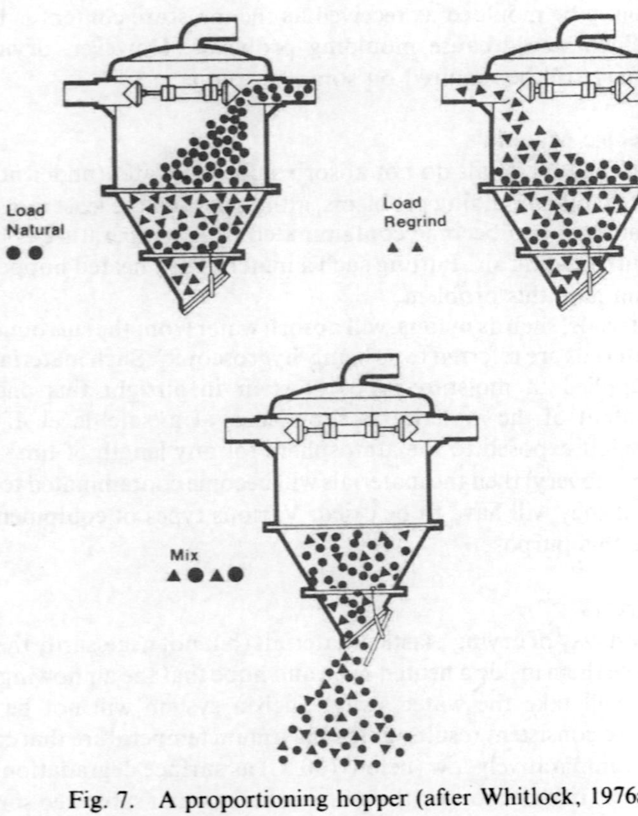

Fig. 7. A proportioning hopper (after Whitlock, 1976a).

processing costs are approximately 40p. This raises the material cost to 75p plus 40p, i.e. £1·15.

The cost of reclaiming clean, reject mouldings must also take account of the production costs. If the operator is used to feed the granulator and if he is granulating 60 kg h^{-1} then, ignoring power costs, the cost of reclaiming would be operator costs plus overheads divided by 60, i.e. 21 p kg^{-1}. If the production costs (40p) are added to this then it can be seen that the cost of producing rejects, even if they can be reused, is high. One way of using reclaim is shown in Fig. 7 (Whitlock, 1976a).

5 DRYING

When moulding materials are received from the raw material suppliers, they can usually be moulded as received as the moisture content is below that level which would cause moulding problems. However, drying or preheating may still be required on some occasions.

5.1 Hygroscopic Materials

Most moulding materials do not absorb sufficient water (under normal conditions) to cause moulding problems, although even the least moisture-absorbing material can become contaminated if its temperature is below that of the surrounding air. Putting such a material in a heated hopper will normally eliminate this problem.

Other materials, such as nylons, will absorb water from the surroundings and such materials are referred to as being 'hygroscopic'. Such materials are normally supplied in moisture-proof bags or in airtight tins and the moisture content of the material as supplied is at a 'safe' level. If such materials are left exposed to the atmosphere for any length of time (e.g., during scrap recovery) then the materials will become contaminated to such an extent that they will have to be dried. Various types of equipment are available for this purpose.

5.2 Oven Dryers

The simplest way of drying plastics materials (but not necessarily the best way) is to place them inside a heated oven and hope that the air flowing over the material will take the water away. Such a system will not be very efficient, or give consistent results, as the maximum temperature that can be employed is comparatively low (below 100 °C) as surface degradation may occur otherwise. In order to stop the air from being water saturated some of the heated air must be allowed to escape, or bled-off, into the atmosphere. The material should be spread out into thin layers, below 10 mm thick (0·4 in), and the trays should be covered with fine net or gauze so as to prevent contamination. Drying times are usually rated in hours and care should be taken to ensure that the properties of the material are not changed by processes such as oxidation and plasticiser loss.

More efficient drying can be achieved if a vacuum can be applied whilst the material is being heated. However, because of the practical difficulties involved in vacuum drying, this procedure is hardly ever employed in the injection moulding industry, rather hopper dryers are used.

5.3 Hopper Dryers

In its simplest form, a hopper dryer is similar in many respects to the simple, oven-drying equipment already described. It consists of a tall, circular hopper into which the plastics material is placed; preheated air is blown into the base of the hopper where it comes into contact with the plastics material. The air then passes through the column of material and some of this heated air is vented to the atmosphere; the remainder is reheated and recirculated.

Drying times are similar to those employed for oven drying unless dry air is used. Because of the limitations of a simple hopper dryer, some manufacturers prefer to refer to them as hopper heaters and advise against their use as drying systems for hygroscopic materials. Not all grades of some materials are hygroscopic; some black grades of ABS are more troublesome than, for example, white grades, because such black grades contain carbon black which absorbs water and causes problems.

In order to dry hygroscopic materials, it is necessary to pass hot, dry air over them at a specified rate and for a specified time. The air is dried by passing it over what is referred to as 'a molecular sieve desiccant'.

A molecular sieve desiccant consists of small beads of material (presumably based on a silicate) which has a high affinity for water vapour molecules after it has been treated by being heated to 290 °C. This treatment absorbs energy and so if a constant flow of air at 290 °C is passed through the desiccant bed at a constant rate, the temperature of the air will increase as the molecular sieve material becomes dry. This rise in temperature is used to judge when the sieve material is ready for use.

Filtered air is passed through a can which contains the molecular sieve desiccant and after treatment the air is very dry (e.g. it has a dew point of −40 °C). The air is then heated to the appropriate temperature and passed through the insulated, hopper dryer. After passing through the plastics

180° C	180° C	180° C	170° C	170° C	180° C	CAB
270° C	270° C	270° C	255° C	255° C	270° C	PBTP
230° C	240° C	240° C	235° C	230° C	240° C	ABS
250° C	260° C	260° C	255° C	240° C	260° C	SAN
240° C	240° C	240° C	240° C	220° C	240° C	PMMA
280° C	290° C	290° C	290° C	290° C	290° C	PPO
220° C	230° C	230° C	210° C	220° C	230° C	PA 6
280° C	280° C	275° C	260° C	270° C	280° C	PA 6.6
300° C	300° C	300° C	270° C	270° C	310° C	PC
350° C	350° C	340° C	330° C	350° C	330° C	PSU

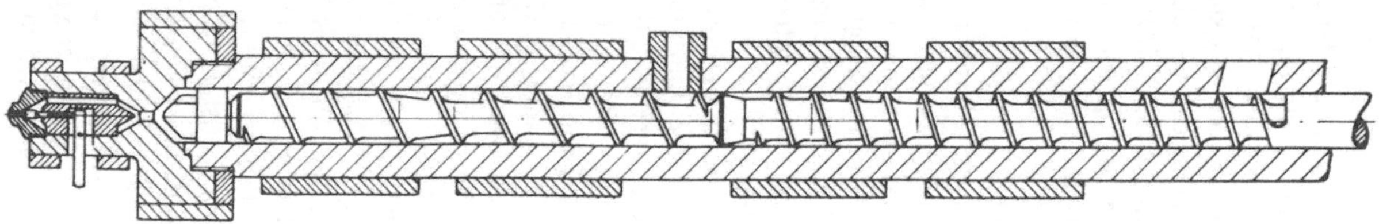

Fig. 8. Processing temperatures recommended for use in a vented unit (after Hausherr, 1977).

material, it is returned for further drying and recirculation. Drying systems usually have two beds of desiccant so that whilst one is being used for drying, the other is being regenerated as previously described.

The material for several machines may be dried in one large system and then conveyed to the individual machines; this is probably the cheapest system. Alternatively, each machine may be equipped with its own dehumidifying dryer and this, although expensive, gives maximum flexibility. Where the amount of material used at each machine is relatively small, then a third alternative exists—the material is dried in one large central dryer and then conveyed to the individual machines which are equipped with hopper heaters.

5.4 Vented Injection Units

'Wet' material may be fed directly to an injection moulding machine provided that the machine is equipped with a vented injection unit. The basic idea, behind a vented injection machine, is the same as that used for extrusion machines, i.e. the material is first changed into a melt and this is then decompressed so that the low molecular weight contaminant is flashed off. The melt is then recompressed and passed to the front of the machine ready for injection. When the contaminant is an organic substance then it may be burnt in a catalytic burner mounted above the vent; certain organic substances, e.g. flame retardants, may not be destroyed by such burners and due care should therefore be taken.

It is vital that the venting section of the screw be of sufficient length so that the vent only covers the parallel part of the screw root (Fig. 8; Hausherr, 1977). If this is not done then melt may escape through the vent or the vent may be blocked with granular material. The second stage of the screw should be operated, or constructed, so that it has a higher output than the first stage, e.g. by the use of deeper sections and a larger screw pitch. Such vented units dramatically decrease the chances of material contamination occurring, for example, in a drying oven.

6 MIXING AND BLENDING

Plastics and rubber mouldings are made from compounds; i.e. the mouldings contain other ingredients, or ingredient residues, besides the parent polymer. Such additives must be dispersed within the polymer.

6.1 Additives

Polymers are seldom used in the unmodified, or pure form as the unmodified materials lack desirable properties. For example, they may lack heat or light resistance, flexibility, cheapness, etc. Plastics and rubbers usually need to be protected against degradation and this is usually done by adding a protective system or additive. By using different types of additive the properties of the base polymer may be modified within wide limits so that the range of use of the base polymer is extended.

It must not be imagined that every polymer composition contains every additive listed; some compositions are relatively simple whereas others are very complex and contain a large number of additives. Some PVC formulations may contain, for example, six different heat stabilisers.

6.1.1 Accelerators and activators

These are commonly associated with crosslinking systems, i.e. rubbers or thermosetting plastics. When used in rubbers for example, accelerators reduce the time and/or temperature needed for vulcanisation; by using an accelerator it is possible to reduce the amount of sulphur needed. Mechanical properties and oxidation resistance are often improved by organic accelerators such as thiazoles and sulphenamides. Activators, e.g. zinc oxide, are often necessary to promote efficient functioning of the accelerator.

6.1.2 Blowing agents

Expanded or cellular products may be produced from polymer compositions by the use of what is known as 'blowing agents'. At processing temperatures a gas is formed and this gas usually dissolves in the polymer melt. During injection moulding the melt is expanded as the gas tries to escape from the melt during the mould filling operation.

6.1.3 Colourants

One of the big advantages of plastics is that they can be obtained in a very wide range of colours. This may improve the appearance of a product or it may make a component more easy to identify, e.g. a wire in a multicore cable. Colourants are usually added to the polymer, along with the other additives, and the mix is then melt compounded, e.g. in a continuous compounder. To save compounding costs and to achieve operational flexibility, colouring systems are often added at the moulding stage.

The colourants used may be dyes or pigments; dyes are soluble and pigments are insoluble colourants. Both are widely used, e.g. dyes give transparent colours and pigments opaque colours. One can thus obtain a complete spectrum of colours from clear to opaque black with a clear material such as polystyrene.

6.1.4 Crosslinking systems

A crosslinking system is added to a polymer so that a network of crosslinks is built up after shaping has been achieved; heat is commonly used to activate the crosslinking system. Probably the two most commonly known crosslinking systems are sulphur (used with rubbers) and hexamethylene tetramine (used with phenolics).

Some thermoplastics materials can be crosslinked and of those polyethylene is the most widely used material. For crosslinking this material, peroxides, or irradiation, or silicone-based materials have been used.

6.1.5 Fillers

Fillers are materials which are added to polymers in order to lower the cost of a compound and/or to modify the properties of that compound. They have been used for many years in the rubber and thermosetting plastics industry and are now being developed for widespread use in conjunction with thermoplastics. In the rubber industry the fillers that are used range from inert diluents (e.g. talc or whiting) to reinforcing fillers (e.g. fine particle size carbon blacks). This last group of materials improve the performance capabilities (tear resistance, abrasion resistance) of rubbery materials tremendously; it is only through the use of such materials that many rubber products can be made fit for service at an economic price.

Glass fibre is a filler that is being widely used with thermoplastics materials in order to improve the stiffness and/or heat distortion temperature of mouldings made from these materials. In many cases the surface of the filler may need to be modified, or treated, if significant property improvements are to be obtained and then maintained in service.

6.1.6 Protective agents or systems

Most plastics and rubber are organic materials which are degraded by heat and light; the extent and course of such degradation is influenced by the surrounding atmosphere, i.e. whether oxygen and/or ozone are present.

As polymers degrade by different reaction mechanisms (simply because they are different chemically) various types of stabilisers may be employed. For example, in the PVC industry, materials which react with hydrochloric acid, or which minimise the production of HCl, are widely used. This is because when PVC is heated, HCl is produced and colour changes occur in the polymer. As one of the big advantages of PVC is its transparency (and therefore a wide colour range) then any change is bad and must be prevented.

Other polymers degrade by oxidative degradation at their processing temperatures. For this reason antioxidants (usually phenol-based compounds) are added to materials such as PE in order to combat the effects of oxidation both during processing and during the service life of the component.

Many polymers are adversely affected by UV radiation; the major source of such radiation is sunlight. Polymer protection may be given by shielding the polymer or by adding UV absorbers. Shielding may be accomplished in various ways, e.g. by plating the plastic with metal or by adding carbon black to a polymer such as polyethylene. Some absorbers work by turning the harmful radiation into harmless radiation of a different wavelength; substituted 2-hydroxybenzophenones work in this way.

6.1.7 Impact modifiers

This is a term which is generally applied to a polymer which is added to another in order to improve the impact resistance of a product. Impact modifiers, based on acrylic polymers, are now used in conjunction with PVC in order to get clear rigid bottles with good impact strength.

High impact polystyrene (widely used for model construction kits) consists of polystyrene and up to 15% of an SBR rubber. The material is produced by polymerising styrene in the presence of the rubber and under these conditions the polystyrene is chemically bound or 'grafted' on to the rubber. This changes the polystyrene from being hard and very brittle to a softer tougher material suitable for making items such as refrigerator liners.

PP and PE are now blended with rubber (e.g. ethylene–propylene rubber) so as to produce materials which have certain rubbery properties and yet which can be processed like thermoplastics. Car bumpers are an example of such mixtures.

6.1.8 Lubricants

These additives are widely used in the polymer industries in order to facilitate processing. They are usually waxy materials which act by easing flow and/or reducing adhesion of the polymer to processing equipment. Stearic acid is a commonly used lubricant in the polymer industry.

6.1.9 Plasticisers

These are added to polymers in order to make the material more flexible. The most widely plasticised material is polyvinyl chloride (PVC) where, by the choice of the type and amount of plasticiser, a wide range of properties can be obtained. Plasticiser usage transforms this material from a hard, rigid material (suitable for pipes) into a soft, flexible material (suitable for upholstery).

The most widely used plasticisers are based on phthalates, e.g. dioctylphthalate (DOP) and diisooctylphthalate (DIOP); such materials are usually clear, colourless liquids.

6.2 Compound Blending

Once the ingredients have been selected they are weighed to a preset formulation; it is common practice in the plastics industry to blend the ingredients together before they are fed to a melt mixing stage. This preblending operation may be performed on the injection moulding machine or it may be performed as a separate operation.

6.2.1 Separate blending operations

In separate blending operations the solid polymer, in either pellet or powder form, is blended with the other compound ingredients and then the blend is fed to the injection moulding machine.

6.2.1.1 A simple blending system. For laboratory work on small scale production of thermoplastics a hand shaken, polythene bag provides a useful blender. The ingredients are carefully weighed into a clean polythene bag which is then slowly inflated from a clean, dry air supply. By twisting the end of the bag, it may be sealed so that the shaking operation can commence. It is best to use a large bag as it is important not to fill the bag more than half full; mixing will take approximately 5 min. The big advantage of this method is that it is cheap and quick and there is little risk of contamination.

6.2.1.2 Mechanical blending. Rotating blenders, which are usually based on either drums or conical containers, are widely employed in the plastics industry to blend granules with other granules or with additives. A system based on a steel drum is the simplest that can be imagined.

A steel drum is partially filled (not more than two-thirds full) with the required ingredients and then the sealed drum is slowly rotated, at say 25 rpm, until a uniform blend is obtained—this takes approximately 25 min. Simply rolling the drum along the workshop floor will not produce the required mixing action as there is no reason for the ingredients to become distributed along the length of the drum. A folding, spreading action is required and this is most easily achieved by rotating the drum end-over-end and at a slight angle. Therefore, as the drum turns the ingredients are folded in upon themselves when they fall into the 'corners' of the drum and are spread (or tumbled apart) as the drum rotation is continued.

Such a system is relatively cheap as the drum may be one that is already available—the material may have been supplied in it and in such a case, the chance of contamination is again reduced. However, it is probable that this system is not as effective as a conical blender. As the name implies, this blender uses cone-shape containers to improve the folding, spreading action previously described. Equipment for mixing small batches, e.g. 10 kg (22 lbs) is formed from two circular cones which are clamped across their base after the ingredients have been added. Such a construction means that the interior is available for cleaning. Conical blenders used for large batches are usually fabricated so that they are of one-piece construction and access is by means of hatches located at each end of the double-cone; machines are available which will mix 1 tonne of material.

6.2.1.3 High-speed mixers. PVC formulations (and to a lesser extent rubber formulations) are commonly mixed in high-speed mixers; such devices can generate considerable quantities of heat quickly and this is useful in the case of PVC as it aids plasticiser absorption and/or lubricant melting and subsequent coating of the PVC particles.

Figure 9 illustrates that these units are basically cylindrical in shape and usually, with respect to the mixing tank itself, the height is approximately equal to the diameter. The rotor or mixing impeller is situated as near to the base of the mixing tank as possible. The machine is driven from an electric motor which, on the smaller units, is usually a variable speed design. On most production-size units it is usually a two-speed motor which drives,

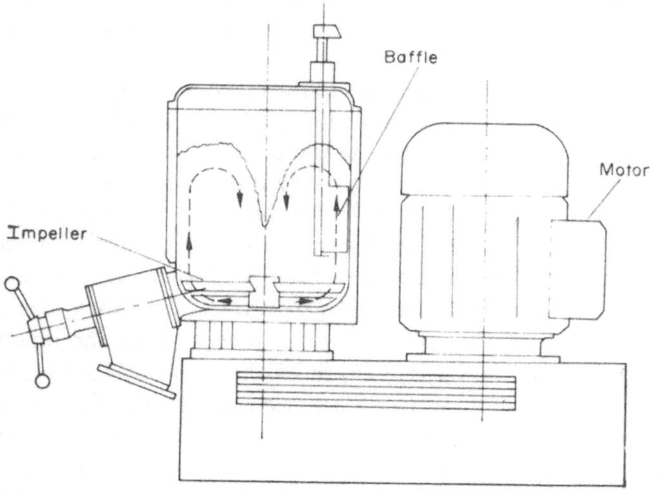

Fig. 9. Diagrammatic representation of a high-speed mixer (after Boulton, 1973).

through a pulley and vee-belt arrangement, the central shaft which enters through the base of the mixing tank. Most of the available machines are similar in basic design although there are small variations, these usually being associated with the form of the mixing tool or the impeller (Boulton, 1973).

This impeller is shaped rather like a ship's propeller but usually it has a relatively large number of blades e.g. 4 or 6.

Such machines may be operated in many different ways and the following procedure is only one of the many variations which could be imagined even for a simple UPVC formulation.

The impeller is set in motion at a relatively low speed and ingredients such as polymer, filler and impact modifier are added—the lubricant is added at a later stage. The speed is then increased and as a result of the friction generated and the heat supplied via the heated outer jacket, the temperature of the ingredients rapidly rises (on small mixers, impeller speeds of 3500 rpm are sometimes used).

Whilst the temperature is rising, blending is occurring as the impeller causes the material to be thrown outwards and upwards—it then falls back into the central vortex which has been created. The baffle breaks up streamlined flow and directs the material into the working area.

Material temperature is sensed (by means of a thermocouple inserted through the baffle) and when it has exceeded the melting temperature of the lubricant (e.g. 110 °C), this compounding ingredient is added. This melts and coats the particles. The blend is then dumped into a cooler-blender when a temperature of 130 °C has been reached. (The lubricant is added late in the mixing cycle so as to ensure that the heat build-up is rapid, e.g. 8 min.)

By cooling the material a more consistent feed to the injection moulding machine is obtained as otherwise, unless the hopper was heated, a variable feedstock temperature would result. Rapid cooling is usual so as to minimise the risk of thermal degradation.

Rapid cooling is achieved by using a large (about twice the capacity of the high-speed mixer), water-cooled vessel equipped with a slowly rotating sweep, e.g. at 30 rpm. After the temperature has fallen to an acceptable level the dry blend is conveyed to a storage silo so that any electrostatic charges can be dissipated. Such charges are introduced during the high-speed mixing operations and unless they are removed, e.g. by storage in a well-earthed silo, they can interfere with material handling and/or feeding.

6.2.2 Mixing and blending on the machine

There are obvious attractions in feeding the injection moulding machine with the required ingredients (which may be virgin polymer, reclaimed material, colouring system, etc.) and performing the blending operation on the moulding machine; this may simplify storage, materials handling and reduce the chances of contamination and thus lower costs.

6.2.2.1 Mixing hoppers. Mixing hoppers are available which can be mounted above a machine hopper and these hopper mixers will automatically proportion, blend and feed a continuous supply of, for example, virgin material, colour concentrate and reclaimed material to the moulding machine. Hoppers are available which can handle four components and give an output of up to 4 tonnes h^{-1}.

In one system (Whitlock, 1976b) materials are dropped through tubes on to a metering disc. During operation, materials are proportioned by volume on this disc in ratios determined by the height of the material metering tubes (which are mounted above the disc so that there is a gap between the tube and the disc). The higher the tube, the greater the material flow. Each tube has a vernier adjustment which is set to obtain the desired mix ratio. Proportioned materials drop through the centre of the metering disc where they are mixed by a cascade located in the blend supply hopper. Each individual ingredient (e.g. reclaimed material) is lifted (e.g. by vacuum) to the hopper assembly which contains the metering tubes. After dropping through the cascade, which consists of a series of flaps or baffles, the blended material is deposited in the hopper of the moulding machine. Melt mixing is then provided by the screw of the injection moulding machine.

6.3 Material Compounding

For many years, fully compounded material was widely used in the plastics industry as there was no real alternative until the advent of the in-line screw machine. Such materials were commonly produced using a single-screw extruder. Recent developments have resulted in compounders which combine the mixing and process control advantages of twin-screw extruders with the conveying advantages of single-screw machines.

6.3.1 Thermoplastics compound production

In the system described in Fig. 10 the mixing–compounding operation is separated from the pumping, conveying and pelletising function. Such machines are widely used for pelletising both rigid and flexible PVC formulations (Rice and Adam, 1977). Material degradation is avoided by heating the material in a short (length to diameter ratio of 15:1) twin-screw extruder (corotating at 300 rpm) under relatively high shear. It is then transported into the pressureless area of a slowly rotating (21 rpm) single-screw with a 6:1 L:D ratio. In operation, preblend is introduced into the

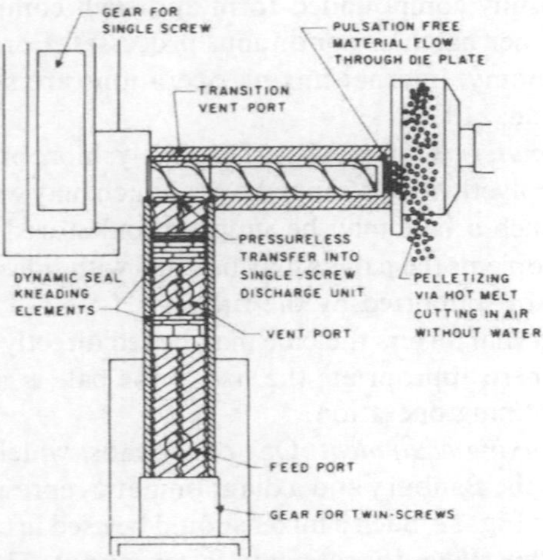

Fig. 10. Combination of a twin- and single-screw for processing heat sensitive material (after Rice and Adam, 1977).

feed section of the twin-screw compounder where high pitch screws maximise the intake and pump the feed, in a very positive manner, through three separate barrel zones where the compounding temperature is raised by externally supplied heat.

The heated PVC then passes through two kneading zones. By using interchangeable screw and kneading elements, the length and width of these sections can be varied so as to maintain the correct heat levels, necessary for dispersion of additives, without degradation of either the additives (such as blowing agents) or the polymer. The material then passes into the single-screw extruder where it is conveyed through a breaker plate for die-face pelletising. A vacuum can be applied to the transition chamber for removal of volatiles.

The single-screw section of the compounder need not raise the temperature of the material, and it can be equipped with barrel and screw cooling to lower the temperature if required. Since the temperature can be accurately controlled, the two-stage machine is suited to automatic, in-line, air-cooled pelletising, which is recommended for the production of moisture-free pellets.

6.3.2 Rubber compound preparation

Rubber compositions are nearly always supplied to the moulding machine in the fully compounded form and such compounding may be performed by either batch or continuous processes. For reasons of speed, output and economy, internal mixing operations are still the ones most widely used (Evans, 1978).

6.3.2.1 Raw material form. Apart from polychloroprene most rubbers are supplied in polyethylene-wrapped bales which may weigh, for example, 25 kg (55 lbs). Such a bale may be supplied and stored on pallets but to prevent creep problems the pallets may be fitted with sides so that the weight of the stack is not supported by the rubber.

With large internal mixers the bale may be fed directly into the throat of the machine. Where appropriate the size of the bale is reduced by a bale-cutting or guillotining operation.

6.3.2.2 The mixing equipment. One of the most widely known types of internal mixer is the Banbury and a diagrammatic representation of such a mixer is shown in Fig. 11. Such a mixer should be used in conjunction with a two-roll mill as this makes for safer and faster output. The curing system is added during the mill mixing stage, when the rubber has cooled down as this procedure keeps the internal mixer free from curing materials and helps eliminate scorch problems.

In some factories the output from the internal mixer (which is a large, intractable mass) is fed to two mills in succession. The first is used to remove the heat from the compound as quickly as possible and then the cooled material is passed to the second mill so that the curing system may be added.

6.3.2.3 Strip production. If natural rubber is being mixed, then the first stage in the mixing cycle is often molecular weight reduction (mastication) and when the viscosity of the polymer has dropped to the desired level, the ram is raised and the compounding ingredients are added.

Not all grades of natural rubber require mastication—constant viscosity grades are available which do not require this treatment. Synthetic rubbers also do not require mastication as they have a comparatively low viscosity.

The ram is then lowered so that mixing and dispersion can occur—this occurs very rapidly as the mix is held in the chamber by the air-operated ram. When the temperature reaches a preset value, the batch is then dumped on to the first mill where its temperature is reduced rapidly; it is passed through this mill several times and then transferred to the second mill where the curing system is added. At this stage the compound is banded around the front roll of the mill and the operator achieves the desired degree of mixing by cutting and folding the banded material into the nip formed by the two rolls. A skilful operator may, at the end of the mixing operation,

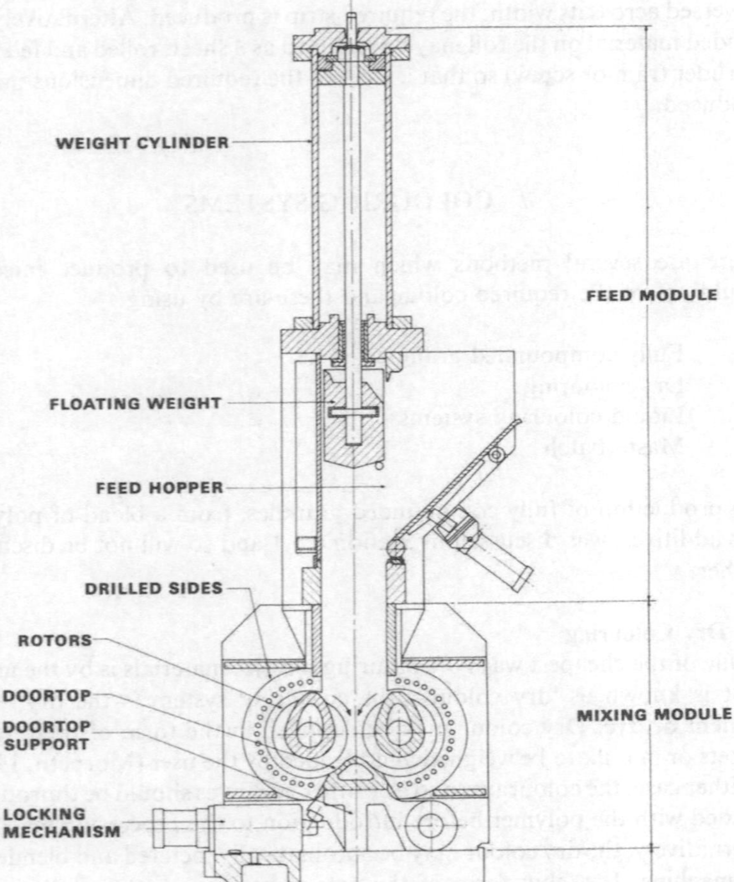

Fig. 11. A Banbury mixer (courtesy Farrel Bridge).

produce a strip of the required width by gradually traversing his knife across the roll face. To stop this strip sticking to itself, it is dusted with talc before being wound on a reel.

The strip, necessary to feed the injection moulding machine, may also be produced by using a roller knife which has two circular blades set at the required distance apart. When this knife is pressed against the roll face and traversed across its width, the required strip is produced. Alternatively, the banded material on the roll may be removed as a sheet, rolled and fed to an extruder (ram or screw) so that a strip of the required dimensions may be produced.

7 COLOURING SYSTEMS

There are several methods which may be used to produce injection mouldings of the required colour and these are by using:

1. Fully-compounded granules
2. Dry colouring
3. Liquid colouring systems
4. Masterbatch

The production of fully compounded granules, from a blend of polymer plus additives, was discussed in Section 6.3.1 and so will not be discussed further.

7.1 Dry Colouring

One of the cheapest ways of colouring plastics materials is by the use of what is known as 'dry colour'—the colouring system is the dry, solid pigment or dye. Dry colour may be supplied in the form of preweighed sachets or in bulk to be weighed and blended by the user (Morpeth, 1978). In either case, the colourants and any other additives should be thoroughly blended with the polymer before introduction to the processing machine. Alternatively, the dry colour may be automatically metered and blended at the machine. If a chip form of the colourant is used, then colour-dust contamination may be minimised; such dust-free grades also flow more easily than the original form and this suits automatic metering equipment better.

A very wide range of plastics may be coloured in this way at a comparatively low cost and with a relatively small colour inventory. However, it is difficult to meter the colour accurately and this can give rise to the development of inconsistent colours; if the material is pre-mixed, then colour contamination can occur as it is difficult to clean the mixers thoroughly. No matter which system of blending is used, thorough blending is necessary before the material reaches the screw.

7.2 Liquid Colouring

One of the cheapest ways of adding a colouring system is in liquid form. The pigments are dispersed in an inert, liquid carrier so that when they are added to the polymer, within the range of 0·1–1·5%, the desired colour results.

Such a colouring system may be tumble mixed with the polymer granules, although in practice, the preferred method of addition is at the moulding machine. This is because, as the system is liquid, automation is cheap and can give good results at low cost. A large range of colours can be produced from a small inventory of colouring systems and the colouring systems in turn, are compatible with a wide range of polymers. Although the colouring systems may be compatible with a wide range of polymers, it does not follow that the same colour will be produced with differing polymers. As the liquid carrier system may affect part properties, it is important that the effect on properties be investigated and that the level of addition be kept to as low a value as possible. The melt flow behaviour of the polymer is also affected—usually it becomes easier.

The colouring system is not the only additive that is incorporated in this way; when structural foam mouldings are being produced, the blowing agent is commonly incorporated into the liquid colouring system as this gives ease of addition and can result in a cleaner environment.

7.3 Masterbatch Colouring

Large concentrations of compounding additives are commonly dispersed in a carrier system, so as to form masterbatches, as this avoids the weighing of small quantities of powders and assists subsequent dispersion. The carrier system may be the parent polymer or it may be a carrier resin which has a low melting point.

When the carrier system is the parent polymer, then the masterbatch is referred to as a dilute masterbatch and such masterbatches may contain up to 20% of the colouring system. Such a dilute masterbatch would be prepared by extrusion, as this gives good dispersion, and would be supplied in chip or granule form. Up to 5% of this dilute masterbatch may need to be added to the natural polymer for colouring purposes.

Such systems can give very good results at comparatively low costs and are clean and easy to handle. However, to get the best results, it may be necessary to use a special mixing screw and it will probably be necessary to use a separate dilute masterbatch for each type of polymer. Where a large number of different materials are being run, it will be necessary to exercise strict control if mistakes are to be avoided.

Very much more concentrated masterbatches (which can contain up to 60 % colouring system) can be made by using carrier systems which are based on low molecular weight resins; such resins have a low melting point (e.g. 60 °C) and are compatible with a very wide range of polymers. Such 'universal' compatibility is a tremendous advantage, although at the same addition level, the same masterbatch may produce a slightly different colour in two different polymers. Concentrated masterbatches may be supplied in a range of granule sizes so that if, for example, powdered polymer is being run, then a fine-grain masterbatch could be selected so as to achieve improved pigment dispersion. Because of the concentrated nature of the masterbatch, the metering system must be capable of delivering a very accurately dosed feed consistently and the screw must be capable of achieving a high level of dispersion.

Despite these disadvantages, the use of concentrated masterbatches is increasing as the cost is low, they are easy to handle, the colour inventory is relatively small and as they are clean to handle, the amount of contamination produced is small.

7.4 Cost Considerations

There is no doubt that fully compounded material gives consistently good results but unfortunately it is expensive, particularly when only a relatively small lot is required. A moulder who only wishes to purchase such a small lot may find it difficult to locate a supplier who will supply the fully compounded material at an economic price. There is also a tendency for the size of minimum orders to be increased by the raw material suppliers and for the range of fully compounded materials to be decreased.

What this means is that many processors are finding it economically attractive to change over to in-house colouring using as the base feed-stock uncoloured or natural material; this natural material is then coloured at the machine. The number of grades of polymer that the processor must purchase obviously decreases and this means that bulk purchase of the natural polymer becomes attractive for several reasons. Apart from economic advantages (about 10 % saving), there are others such as lower book keeping costs and colouring flexibility. It should also not be forgotten that it is very difficult to predict exactly the quantity of material required for a particular moulding run and as it is politic to err on the positive side, it is easy to finish with some unused, coloured material. However, despite these disadvantages, some moulders still prefer fully compounded material because the raw material suppliers may have recommended the use of a particular grade for a specified application. This guarantee, or recommendation, is very comforting to injection moulders as they feel that there is someone else to share the responsibility if anything goes wrong, e.g. through a toxicity claim.

However, the cost of compounding must be paid for. According to Morpeth (1978) the relative costs of colouring would be:

1.	Fully compounded material	100
2.	Dilute masterbatch	30
3.	Concentrated masterbatch	20
4.	Liquid colouring	20
5.	Dry colour	15

7.5 Colour Addition

On many moulding machines the only addition made at the machine hopper is a colouring system. Controlled amounts of colourant may be added through a tube which is positioned just above the screw of the moulding machine. The colouring system is drawn into the flights of the machine screw, together with the material, where mixing occurs.

If the colouring system is a free flowing solid, then it may be dispensed, with reasonable accuracy, by means of a rotating feed screw which is powered by a DC variable speed drive. For example, from 0 to 1000 g min^{-1} of colour concentrate may be dispensed in this way.

If the colouring system is a liquid one, then dispensing is done by means of a pump (e.g. peristaltic) which operates by drawing colour from a storage container through silicone rubber tubing in the pump head. The liquid is propelled by rollers which compress the tube and thus cause the liquid to be moved along. A control unit ensures that the correct preset amount of colour is added to the polymer via flexible tubing which passes through a metal probe fitted in the machine throat. Pump operation is controlled by a contact in the screw circuit which allows material to be metered only during the screw return part of the cycle.

REFERENCES

Boulton, A. J. The preparation and properties of PVC dry blends, *Developments in PVC Technology*, Chap 6, Eds. Henson, J. H. L. and Whelan, A. Applied Science Publishers, London, 1973.

Bradley, D. Injection machinery sales predicted to double in a decade. *Plastics World*, p. 28, Feb 1981.

Evans, C. W. Powdered and Particulate Rubber Technology. Applied Science Publishers, London, 1978.

Hausherr, W. Venting units. *Netsal News*, p. 1, (10) Oct 1977.

Kadakawsky, R. Materials systems. Paper presented at a Demag Conference in Leighton Buzzard, 1979.

Mills, D. The importance of good scrap recovery. Paper presented at a Demag conference, Leighton Buzzard, 1979.

Morpeth, F. Colouring thermoplastics, *Plastics and Rubber Weekly*, p. 17, 2 June (1978).

Rice, P. and Adam, H. The continuous compounding of PVC, *Developments in PVC Production and Processing—1*, Chap 5, Eds. Whelan, A. and Craft, J. L. Applied Science Publishers, London, 1977.

UPM. *Examples of Material Storage Facilities Available from UPM*, trade literature of UPM Machinery Sales Ltd, Slough, 1980.

Walker, M. Scrap can damage your wealth, *British Plastics and Rubber*, p. 32, Dec 1980.

Whitlock. *Vacuum Conveying Systems*, trade literature published by Whitlock Inc., MI, USA. Form No. 9172, Oct 1976a.

Whitlock. *660 Series Blender*, trade literature published by Whitlock Inc., MI, USA. Form No. 9182 4 Dec 1976b.

Whitlock. *Material Handling Systems*, Whitlock Inc., MI, USA, 1979.

CHAPTER 2

Moulding Machines and Processes

1 TYPES OF MACHINE

From the previous definition given of injection moulding it should be obvious that the injection moulding machine must be capable of:

1. Softening the polymer
2. Injecting the softened material into a closed mould
3. Hardening the material so that it sets into the size or shape required
4. Ejecting the hardened material so that the cycle can begin again

Injection moulding machines are ranked or rated in various ways (see Chapter 3) but in Europe the Euromap system is becoming widely adopted.

Euromap is the European Committee of Machine Manufacturers for the plastics and rubber industries and this organisation issues various recommendations which, for example, may be used to describe and/or compare moulding machines. Euromap document No. 1 (1969) specifies injection machines in terms of the calculated swept volume during injection and in terms of the maximum clamping force that a particular machine can generate. The constructional features or design of the injection unit is also used to describe moulding machines.

There are, for example, plunger or ram machines, plunger-with-plunger preplasticising machines, plunger-with-screw preplasticising machines and extruder screw machines mentioned in that specification.

Each of these systems will be described in the sections which follow and where appropriate the operating sequence of a particular system will be described.

2 RAM MACHINES

Ram machines are also known as plunger machines; occasionally they are also referred to as piston machines.

2.1 The First Injection Machines

The first injection moulding machines used a heated barrel to thermally soften the material and a ram to shoot or inject the material into the mould. As the materials used were thermoplastics (e.g. cellulosics) they could be set into the shape of the mould cavity by keeping the mould cool.

Such ram, or plunger, machines were derived from die-casting machines which are used, of course, for moulding metals. These materials have properties which are drastically different to those possessed by organic polymers.

For example metals are good conductors of heat and they have low specific heats and low melt viscosities. In general, polymers have values for these properties which are very different and being organic they are thermally unstable. Some polymer compositions also set or cure at elevated temperatures. The range of materials which ram machines could handle was relatively limited and because of this they were rapidly replaced by screw machines which are much more versatile.

It must be pointed out that many plunger machines are still in use and that they are still made. The makers claim that their disadvantages are outweighed by their relative simplicity and low cost.

2.2 Operating Sequence of a Plunger Machine

A simple plunger injection moulding machine is shown in Fig. 1. If we assume that there is material in the hopper and barrel, and that the mould is clear of obstruction then the machine operating sequence is as follows when handling a thermoplastics material:

1. The mould closes. In order to save time, and therefore money, this is done as quickly as possible. As the mould is mounted on a platen, which in turn slides on polished steel tie bars (see Fig. 1), only low forces are needed to get the operating speeds required.

2. The mould clamps. Just before the two mould halves touch, the speed of the moving half is reduced so that mould damage is minimised. The preset clamping pressure is then applied to the mould only when the mould is together or touching. Forces of 30 $MN\,m^{-2}$ (2 tsi) are common (the area referred to in these figures is the projected area of the moulding) and as a

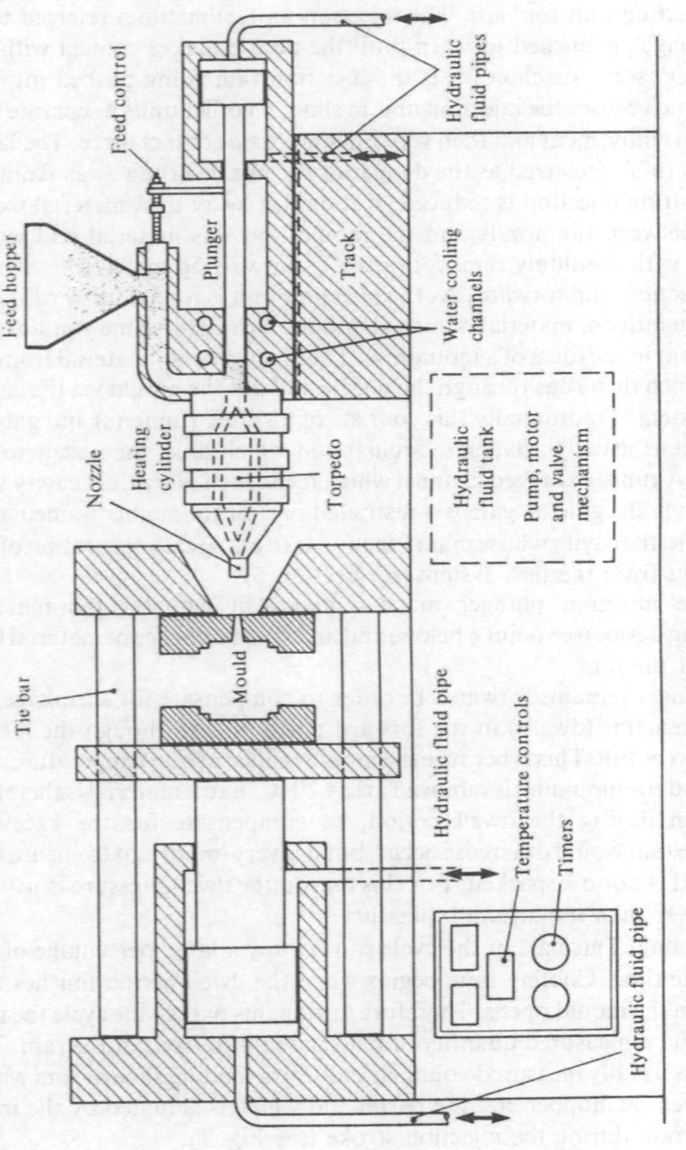

Fig. 1. A simple plunger machine (after ICI, 1978).

result of applying such high forces the tie bars stretch. Such stretching occurs no matter which clamping system is used.

3. Injection unit forward. The injection unit, sometimes referred to as 'the carriage', is pushed forward until the nozzle makes contact with the mould. On some machines it is the injection ram being pushed into the barrel which causes the injection unit to slide; on other units a separate ram causes this movement and then generates a nozzle contact force. The latter system is to be preferred as the danger of the nozzle lifting away from the mould during injection is reduced. If it did lift away then material would appear between the nozzle and the mould and this material and would interfere with moulding removal (apart from wasting material).

4. Injection ram forward. As the injection ram is forced forward then a known quantity of material is pushed into the barrel (the same amount that is taken out in the form of a moulding). This displaces hot material from the barrel which then runs through the nozzle and fills the mould via the cavity filling system. Traditionally this consists of a sprue, runner(s) and gate(s). The sprue is usually a tapered circular rod which links the nozzle to the runners. A runner is a feed channel which feeds each individual cavity with material via the gate. A gate is a restricted part of the runner immediately adjacent to the cavity whose main function is to allow easy separation of the mouldings from the feed system (see Section 6).

As the injection plunger moves forward it actuates the material measuring device (see point 6 below) and as a result a charge of material falls on top of the ram.

5. Plunger remains forward. In order to compensate for shrinkage the plunger remains (dwells) in the forward position even though the mould appears to be full. This is because the mould is filled at one temperature, e.g. 200 °C and the moulding is removed at say 20 °C. Extra material is therefore packed in during the dwell period, to compensate for the excessive shrinkage that would otherwise occur, but it is very important to ensure that the mould is not overpacked. For this reason the dwell pressure is usually much lower than the injection pressure.

6. Cooling. This part of the cycle can occupy a large percentage of the total cycle time. Cooling time begins when the dwell period finishes and ends when the mould opens. Therefore during this part of the cycle the ram returns and a measured quantity of material falls in front of the ram. The material is usually measured volumetrically by a loading shoe or ram which fits between the hopper and the barrel and which is actuated by the main injection ram during the injection stroke (see Fig. 1).

7. Mould opens. Initial mould separation may be slow (to avoid

scratching a deep walled part) followed by a fast mould opening movement (to save time and therefore money). The opening action then needs to be braked before the moving platen reaches its fully open position so as to save damaging the machine.

8. Moulding ejection. The opening action may be used to actuate the ejection mechanism (mechanical ejection) or the ejection system may be actuated by a hydraulic ram (hydraulic ejection). In both these systems the part is removed by means of pins which can be lifted from the base of the cavity by means of a striker bar.

9. Mould delay time. To allow time for the moulding to fall clear or, to allow a constant time for mould inspection, a mould delay timer is fitted. This keeps the mould open for a preset time, e.g. 1 s, and it is only when this time lapses that the cycle begins again.

The sequence described is a fairly common one but it must not be imagined that it is the only one possible. If holes or slots are required in the side-walls of a component then side cores may be used and this would mean a different operating sequence or cycle. Some people believe that it is better to keep the injection unit or carriage in constant contact with the mould as this gives a constant heat loss rather than a periodic one. If this procedure is adopted then a slightly different cycle will again result.

2.3 Disadvantages of a Plunger Machine

The operating sequence of a ram machine has been described in some detail because after studying it, and using a great deal of hindsight, one can appreciate why such machines are no longer the mainstay of the injection moulding industry. Their place has been taken by single-screw machines but before such machines are described it is proposed to explain why most of the machines now sold are not ram machines.

2.3.1 Injection unit requirements

The injection unit of a moulding machine is required to perform the following functions or operations:

1. Meter the material into the barrel. If this were not done then the amount of material in the cylinder would vary from one cycle to another and this would affect machine performance (e.g. pressure transmission) and moulding properties.
2. Raise the temperature of this material to its moulding temperature. This must be done quickly and perhaps most importantly, uniformly, i.e. the melt must be thermally homogenous.

3. Transport the thermally softened melt from the barrel into the mould. This should be done at a reproducible rate if uniform parts are to be produced.
4. Pack the mould with the desired quantity of melt. Again this must be done on a reproducible basis.

A ram machine does none of these jobs particularly well when it is used as described previously. Even the comparatively simple job of feeding the barrel with raw material can result in problems such as erratic feeding; this may be caused by a well lubricated material cascading through the volumetric feed device. Much more serious problems arise however.

2.3.2 Conflicting requirements

When the ram forces thermoplastic granules forward then a considerable amount of energy must be expended in overcoming resistance due to friction. This can mean that 50% of the energy supplied is lost, e.g. if 140 MN m^{-2} (20 000 psi) were applied to the ram then only 70 MN m^{-2} (10 000 psi) would be available at the nozzle. Such a loss is obviously undesirable but it could perhaps be tolerated if it was constant; the problem is that it is not. As the ram advances then the frictional pressure losses change and this results in different pressures in the mould. It must also not be forgotten that as the ram advances then the volume of material, which is being displaced, changes. This is because the bulk density of the granules is approximately half that of the melt (Munns, 1964); this is caused by the fact that the feedstock consists of a mixture of granules and air. Therefore, as the ram advances the material injection rate changes due to granules melting and fusing together.

This means that the barrel of a ram machine is not a good transmitter of pressure. Unfortunately it is not a good transmitter of heat either. Most machines are heated by means of heater bands which are clamped to the outside of the barrel. Heat is readily passed through the walls of the barrel but unfortunately this is not the object of the exercise. The object is to heat the polymer to a preset temperature throughout its entire mass.

The heater bands are therefore trying to heat a material which is separated from the heater bands by the thickness of the barrel and thermal lags therefore result.

Further complications result when the heat actually reaches the polymer as such materials normally have a low thermal conductivity (Table 1), a high specific heat and a comparatively low decomposition temperature. The machine must therefore try to put a lot of heat into a material which cannot conduct it away and which will degrade if heated too quickly.

TABLE 1
THERMAL CONDUCTIVITY VALUES FOR POLYMERS AND COMMON NONPOLYMER MATERIALS (after Majurey, 1977)

Material	*Thermal conductivity at 20°C W m⁻¹ °C*
Copper	400
Aluminium	230
Brass	110
Iron	80
Steel	46
Marble	3·5
Concrete	0·92
'Sindanyo' insulation	0·90
Glass	0·84–1·0
HD polyethylene (ρ0·982)	0·63
Brick	0·63
Water	0·60
Polyoxymethylenes (ρ1·441–1·432)	0·43–0·40
LD polyethylene (ρ0·918)	0·33
Nylon 6	0·31
PET (ρ1·409–1·337)	0·28–0·22
PTFE	0·25
Polypropylene (ρ0·911)	0·24–0·17 (atactic)
PMMA	0·20
Rubber	0·19
Cork	0·18
PVC	0·16
PS	0·15
PU	0·13
Paper	0·13
Wood (dry)	0·03
Asbestos	0·07
Foamed plastics (typically)	0·05

The situation is not improved by the fact that what is good for heat transmission is bad for pressure transmission. To improve the rate of thermal transfer, long, thin flow channels should be employed. Such flow channels impose restrictions which would obviously cause very high pressure losses and could therefore not be tolerated. The design of the cylinder is therefore a compromise between what is good for pressure transmission and what is good for heat transmission.

2.4 Torpedoes or Spreaders

Heat transfer, from the cylinder to the polymer, is improved by using a torpedo or spreader within the cylinder of a ram machine. A torpedo increases the surface area of hot metal in contact with the polymer and it also reduces the gap through which the polymer must flow.

A simple spreader would consist of a tapered, cylindrical core to which there are attached a number of fins or flights—these projections locate the spreader at the nozzle-end of the cylinder and provide additional hot-metal surfaces. The surface area can, for example, be increased by drilling holes through the torpedo.

Such systems resulted in large pressure losses, were difficult to clean and often introduced weld lines into the mouldings. Many of the improvements which were made to ram machines were concerned with torpedo design or modification. Heated torpedoes, rotating torpedoes and perforated torpedoes were all tried and used. Such devices did improve the output and/or the quality of the melt, but once the advantages of the single-screw machine were appreciated then the development of such devices lapsed. Improved melt quality by destroying laminar flow is now once again being used (see Section 5).

3 TWO-STAGE PREPLASTICISING SYSTEMS

3.1 Preplasticisation

The term preplasticising is widely used in the injection moulding industry to describe those machines in which heating of the material and transferring (or injecting) the melt can be considered as being done in two separate operations.

Those people familiar with the PVC industry may be confused by this term as in PVC technology the term 'plasticise' means that the polymer has been made soft by adding a plasticiser. In injection moulding the term means that the material has been made soft by the use of heat. The plasticising capacity of a machine is therefore the amount of thermally softened polymer that the injection unit can prepare in unit time, e.g. $kg\,h^{-1}$. (Unless a standard method of assessment is used then misleading information may be obtained and for this reason the term is not as widely used as it once was.)

To avoid ambiguity alternatives to preplasticising have been suggested, e.g. preplasticating and preplastifying, but so far these terms have not been widely adopted.

3.2 Early Preplasticising Systems

These were originally developed before the Second World War but their usage did not become widespread until the 1960s. By this time the single-screw machine had been refined and developed and was a commercial reality. It is that type of machine which is the most widely used type of machine at the present time although in some respects a two-stage system has certain advantages.

In this context the term 'two-stage' implies that the material is softened in one barrel and then transferred to another barrel—it is from the second barrel that the mould is filled. The second barrel is sometimes called the 'shooting cylinder'. The heating or plasticising barrel was initially based on a plunger unit and many ingenious attempts were made to overcome the fundamental disadvantages of a ram machine. However, because of the superior plasticising performance of a screw most two-stage machines are now based on screw units.

Probably the first system using a screw in what may be described as a two-stage system was that revealed in Celluloid Corporation's British Patent No. 422432 granted in January 1938 (Munns, 1964). This machine used an extrusion screw to thermally soften the material (first stage) and this unit pumped the melt into the injection chamber (second stage) which was fitted with its own injection plunger (Fig. 2). The system so described is very

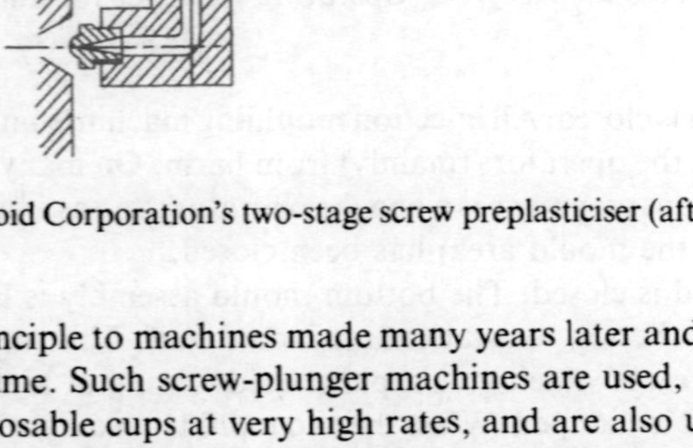

Fig. 2. Celluloid Corporation's two-stage screw preplasticiser (after Munns, 1964).

similar in principle to machines made many years later and is still in use at the present time. Such screw-plunger machines are used, for example, to produce disposable cups at very high rates, and are also used to produce structural foam components (see Chapter 7). In such applications use is made of the high plasticising capacity of a screw and the fast injection possibilities of a ram.

3.3 Screw/Plunger Combinations

Because the plasticising unit is commonly mounted above the ram injection unit such machines are sometimes referred to as 'piggy-back' machines. Such an arrangement is not the only one possible. Co-axial plunger and screw units were offered by some companies but such systems never seemed to become really popular. In such units a plunger operated inside a hollow screw so that as the screw rotated the ram came back to allow the polymer melt to accumulate at the front of the barrel. This gave a compact injection unit but apparently the problems associated with such a relatively complex system hindered its widespread application. Such an injection unit could not handle difficult materials, e.g. UPVC, as material stagnation and subsequent degradation occurred.

For rubber-based compositions a different sort of two-stage machine seems to have become established as the most widely used machine type; the injection unit is mounted at right angles to the extruder and both are connected by a V-shaped head. This assembly is usually mounted above the machine and mould closing and clamping is accomplished with a vertical press. Single and double station machines are possible and the maximum shot size is 12 000 cm^3 (Fig. 3).

At present the most popular size of machine in the UK has a clamping capacity of 220 tonnes. Such machines can deliver up to 1000, 1250 or 2000 cm^3 of compound, depending on the injection barrel that has been fitted; the screw diameter is of the order of 50 mm (2 in).

3.4 Operating Sequence of a Rubber Injection Moulding Machine

A two-stage machine is shown in Fig. 3 and if we assume that a cycle has just been completed then a typical operating sequence for that machine will be as follows:

1. The guard is closed. All injection moulding machines must be guarded so as to protect the operators (mainly) from harm. On many machines the cycle of operations commences when the operator presses a button after the guard (around the mould area) has been closed.
2. The mould is closed. The bottom mould assembly is lifted upwards until it touches the top, or stationary, mould half. This may be done by using two jack rams which give rapid mould closing at low cost.
3. The mould is clamped. When the two mould halves touch a steel block is moved (hydraulically) beneath the bottom platen and the full clamping pressure is applied to the closed mould via this by means of a large area, small stroke hydraulic ram. This ram may only have a stroke length of

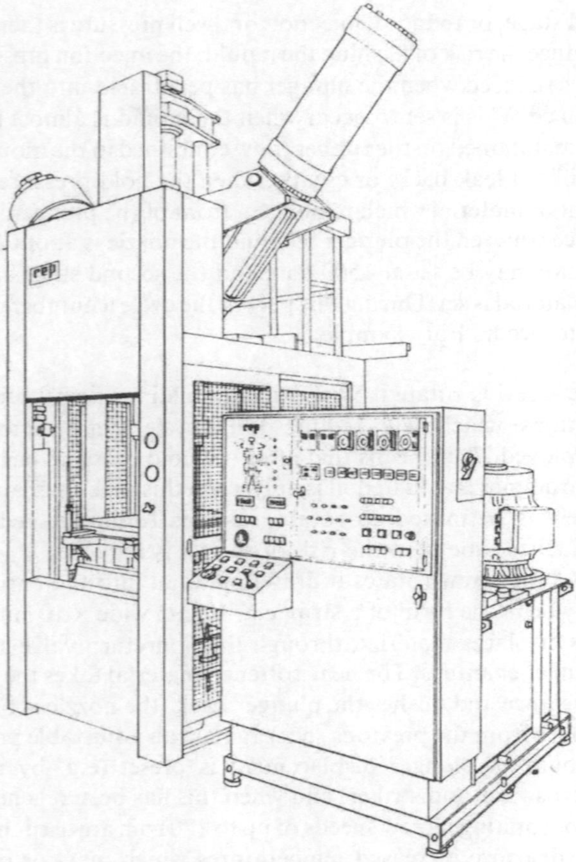

Fig. 3. A rubber injection moulding machine (courtesy REP).

10 mm, i.e. such a machine has a 'lock and block' clamping unit (other systems could of course be used).

4. Mould filling commences. On machines of this type it is common for the nozzle to remain in contact with the mould, although sprue break can be employed if desired. The pressure and speed controls are adjusted to give the mould filling rate required (this depends on the compound, the feed system, surface finish required, etc.). As the pressure builds up inside the system it causes a nonreturn valve (contained in the nozzle) to close; this stops injection pressure from being transmitted to the extrusion unit as the thrust bearings, which support the screw, must be protected.

A second stage, or reduced, injection or dwell pressure is then applied. In order to reduce the risk of flashing the mould, the injection pressure (and/or the speed) is reduced when the plunger has penetrated into the barrel for a preset distance. This is set to occur when the mould is almost full and this pressure is maintained on the rubber (now contained in the mould) until the material will not leak back, or escape, when the hold pressure is reduced. The amount of material which remains in front of the plunger is small (e.g. the clearance between the plunger face and the nozzle is about 1 mm). First stage pressure may be set at 1500 bars and the second stage at 1200 bars.

5. The material is set. During this part of the cycle a number of events are scheduled to occur. For example:

(a) The screw is rotated. With some mouldings (i.e. those with thick sections which will require long cycle times) screw delay is employed. This means that after the hold pressure has dropped to zero a timer is actuated; it is only when this timer has lapsed that the screw is permitted to rotate. As this technique reduces barrel residence time then the risk of scorch is reduced.

    As the screw rotates it draws material into the barrel (the feed may be in the form of a strip, e.g. 40 mm wide × 10 mm thick) and this displaces material, through the nonreturn valve and into the plunger chamber. The heat-softened material takes the line of least resistance and pushes the plunger back (the nozzle is blocked with rubber from the previous shot) against an adjustable pressure. The amount of plunger displacement is preset (e.g. by means of a microswitch and striker) and when this has been reached the screw stops rotating. Screw speeds of up to 170 rpm are used; higher screw speeds cause increased temperatures which may, or may not, be desirable.

(b) The plunger is pulled back. With some 'scorchy' compounds (i.e. those which may set prematurely) the plunger may be withdrawn before the screw starts rotating; this lowers the heat generated by the screw and thus reduces the risk of scorch. If the plunger is not withdrawn then the heat buildup within the compound can be considerable, e.g. 25 °C.

(c) The injection unit is withdrawn from the mould. This may be done if, for example, the compound has set in the nozzle. To clear this it is usually sufficient to inject a small amount of material through the nozzle.

(d) Whilst all this is happening the material is being supplied with heat from the electrically heated mould/platen assembly, e.g. set at 185 °C. During this part of the operation the mould may be 'burped' or 'bumped' (i.e. quickly opened and closed) in an effort to remove any trapped air. An alternative which is growing in popularity is to remove the air from the mould by vacuum, before the injection stroke.

(e) Mould clamping pressure is removed. After the required curing time has elapsed the clamping pressure is removed and the block is retracted so that the mould can open.

6. The mould is opened. Gravity assists the mould in opening quickly but if this high-speed opening causes problems (e.g. tearing) then the speed may be reduced. This downwards acting, mould opening stroke may be used to actuate the ejector system if desired.

If the now projecting ejector pins interfere with the next operation (e.g. insert loading) then the bottom platen may be set so that it lifts, and thus allows the ejector pins to drop.

7. The guard opens. This is usually done automatically, e.g. by using an air actuated system connected to a timer or to a microswitch.

8. The moulding is removed. Because of the nature of the mouldings and of their associated runners (e.g. soft, flexible intricate mouldings are common) part removal is commonly performed by hand; the operator may also inspect the mould, trim, inspect, and pack the components.

On some machines the bottom platen can be moved out of the press (if required) at the end of each cycle so as to make loading and unloading (e.g. of inserts) easier.

The guard is then closed and the cycle begins again.

3.5 Attractions of a Two-stage System

It is the rubber industry which makes extensive use of two-stage machines and therefore the comments which follow apply to the type of V-machine which has been described in the previous section. The company which is probably best known for this type of machine is REP.

Unlike thermoplastics which are seldom compression moulded (except for items such as gramophone records, large sheets and test specimens) rubber compounds are still widely moulded by compression techniques. However, injection moulding does offer the following advantages:

1. No cutting, shaping and weighing of blanks is required
2. Manual handling of moulds is reduced, i.e. there is a greater level of automation

3. Cure times are substantially reduced, especially for thick sections
4. By suitable mould design and manufacture (e.g. the use of moulds with self-trim grooves) flash can be virtually eliminated from the ejected mouldings
5. When properly set up, extremely reproducible cycles are obtained and this results in mouldings with more uniform physical properties
6. Injection into a closed mould gives better control over dimensions and therefore fewer rejects
7. Trimming and inspection costs are reduced

However, the equipment (the injection moulding machine, moulds and ancillary equipment) is more expensive to buy and maintain, it is more difficult to set up and tighter control over the compounds must be exercised. The feed must be available in the right form (e.g. strip or granules) and if something goes wrong (e.g. if scorch occurs) it is more difficult to put right. The higher temperatures employed also give rise to more fumes and more mould fouling problems.

Two-stage injection machines are used because of the control that separation of the plasticisation and injection function gives. Shot control is precise and consistent; on in-line screw machines, wear can cause leakage and reworking of the polymer. As scorch (premature vulcanisation or setting) is to be avoided this reworking is obviously undesirable.

On two-stage machines the extrusion unit is set at a lower temperature (e.g. 80 °C) than the injection unit (e.g. 90 °C). Because of the mechanical working that the material receives, the melt temperature is higher than the set extrusion temperature. On transfer to the injection unit this high, consistent temperature is maintained; the temperature is consistent because the L:D ratio remains constant—on an ILSM (in-line screw machine) it decreases as the screw moves back.

The screw used on such a machine could have a L:D ratio of 17:1 and a compression ratio of 1·24:1—such a screw could generate considerable quantities of work heat (for a definition of these terms see Chapter 3, Section 2). The temperature of the rubber compound could be raised by 25 °C by this means. A further 25 °C rise could easily be obtained by forcing the material through an appropriately sized hole in the nozzle (e.g. 3 mm) or through a small gate. Such techniques of heat generation are particularly useful for thick-sectioned parts as they allow comparatively short cycles to be used.

This type of two-stage machine is vertically arranged and as such it takes up less floor space than a 'traditional' horizontal in-line screw system.

Large shot weights from a relatively small, simple and cheap machine are decided advantages for this type of unit.

4 IN-LINE PREPLASTICISING SYSTEMS

The machines that became popular for plastics were based on in-line units; units in which screw rotation and reciprocation were possible.

4.1 A Twin-screw System

A twin-screw machine, marketed by R. H. Windsor (UK), was the first in-line machine to be generally available. Such a unit is called in-line because the melt is not diverted from a straight flow path; the same barrel is used to heat and to transfer the material. At first glance it would appear that the preplasticising system has been lost but this is not so.

As the screws rotated then the material was fed forward and the screws were pushed back by the pressure developed. A reservoir of melt then accumulated in front of the screws and this charge was then shot into the mould by the screws coming forward and acting as rams. What is surprising is that such a machine was available in 1953, i.e. before single-screw systems became readily available. However, when single-screw machines were introduced they rapidly gained widespread acceptance (as they are comparatively simple) and single-screw machines are those which are now used to handle the widest range of materials.

4.2 A Single-screw, In-line Unit

Most machines now sold use one screw and this one screw is used to plasticise the material and, after it has done this, the screw is then driven forward so that it acts as a ram. The screw must therefore be capable of being rotated and reciprocated, and such actions must be performed very precisely, and repetitively, if consistent products are to be obtained.

Basically, the machines are similar to the ram or plunger machines previously described but the injection unit differs and the machine has additional controls, so that screw stroke and rotational speed can be controlled to the desired accuracy. A separate material measuring device is not normally necessary. A diagrammatic representation of an in-line reciprocating, single-screw machine is shown in Fig. 4.

4.3 Operating Sequence of a Single-screw Machine

If we assume that the machine has just completed a shot then the

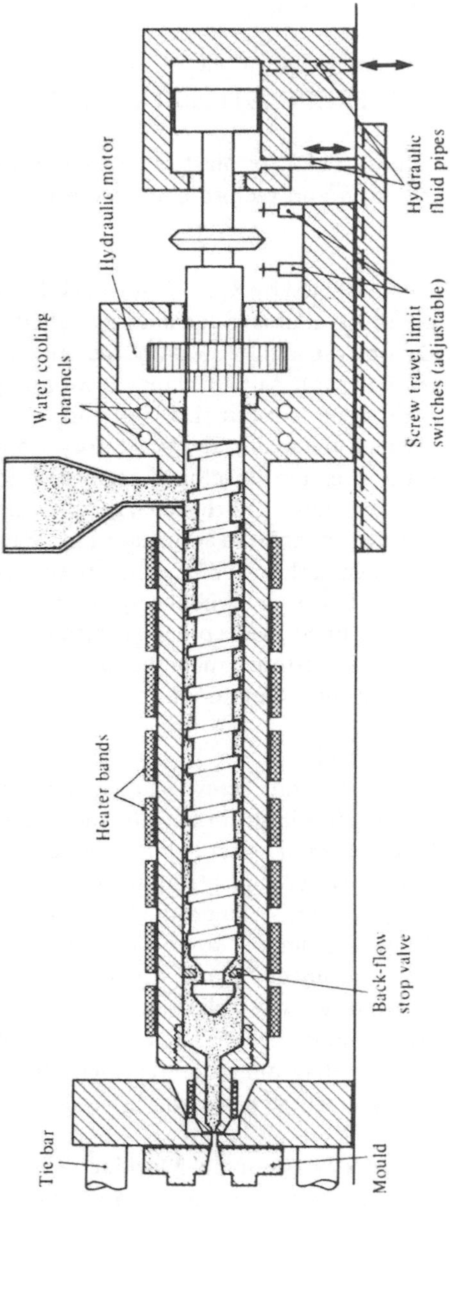

Fig. 4. An in-line reciprocating screw machine (after ICI, 1978).

operating sequence will be as follows (if a thermoplastic material is being moulded) once the safety guards have been closed:

1. The mould closes. Again this is usually done by a hydraulic system, either direct hydraulic, toggle or lock and block.
2. The mould is clamped. It is found that lower clamping pressures are needed on screw machines, compared to ram systems, as lower injection pressures are needed.
3. The injection unit moves forward. This implies that sprue break has been employed.
4. The screw is pushed forward. Most commonly the screw is not rotated as it is pushed forward by the hydraulic ram(s). As there is only melt in front of the screw, pressure losses, compared to a plunger system, are much reduced and it would be reasonable to expect that the pressure in the mould would be higher. However, because of the improved plasticisation of the melt in the screw machine it is found that moulds can be filled with significantly lower set pressures than those necessary when the same moulds are fitted to a ram machine. Because there are no granules in front of the screw then mould filling is more rapid. Melt is prevented from flowing back down the screw, by means of a valve located at the screw tip (Fig. 5).
5. Dwell under pressure. After the screw has penetrated a preset distance into the barrel then the pressure is dropped to the second stage pressure by means of a microswitch and striker or the drop is triggered by a signal which

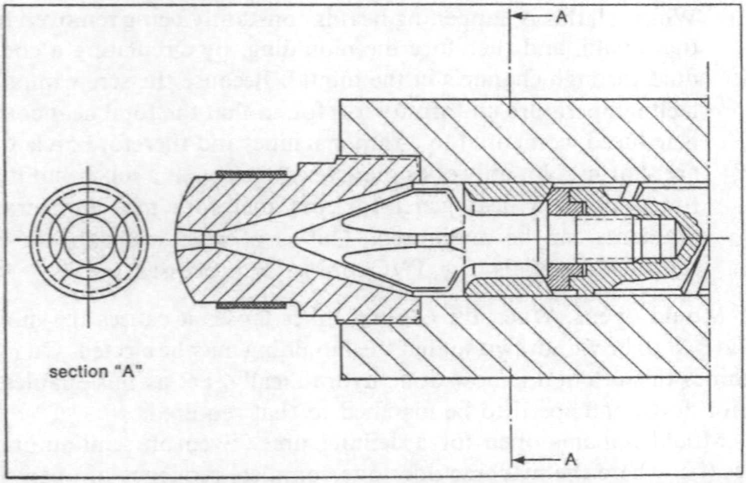

Fig. 5. Open nozzle with screw tip and back flow valve (after Bayer, 1974).

originates from a transducer located behind the mould. This drop is usually set to occur just as the melt fills the mould. Apart from energy savings, other benefits result, e.g. lower moulding weight, more consistent moulding weight and lower stress levels in the mouldings (which results, in part, in superior properties). The amount of melt which remains in front of the screw after injection is completed (screw cushion) should be adjusted to give the minimum consistent with good pressure transmission, e.g. 3 mm on small machines.

6. Cooling. During this part of the cycle a number of events are scheduled to occur, for example:

(a) The screw rotates and this causes material to be pumped forward. The pressure developed pushes the screw back and the amount of screw travel is used to regulate the feed by making the injection assembly strike a limit switch which turns off the screw motor. By throttling the oil flow from the injection ram assembly the pressure which the screw must overcome before it can return (back pressure), can be increased so as to improve mixing.
(b) At the end of the screw rotation the screw is pulled back hydraulically so as to decompress the melt. This action is called suck-back, or decompression, and it helps to prevent the melt from drooling from the nozzle. In this way the need for a shut-off valve may be eliminated.
(c) The injection unit is withdrawn so as to effect sprue break.
(d) Whilst all this is happening heat is constantly being removed from the mould, and therefore the moulding, by circulating a cooling fluid through channels in the mould. Because the screw improves melt temperature uniformity it is found that the total heat content is reduced, compared to a ram machine, and therefore cycle times are shorter. Not only does the screw function as a mixer but it also functions as a heat generator and therefore melt temperature gradients can be minimised. This is of great assistance if heat sensitive materials (e.g. PVC) are to be processed.

7. Mould opens. When the cooling timer lapses it causes the moving mould half to be withdrawn so that the moulding may be ejected. On many machines this ejection is now done hydraulically, i.e. as this enables the ejection force and speed to be matched to that required.

8. Mould remains open for a definite time. Even on semi-automatic cycles (i.e. where the machine does one complete sequence of operations and then stops) it is usual to automatically time the mould open stage so

that consistent cycles result. When this timer lapses then the cycle starts again.

It should be noted that the feed to such injection moulding machines is very often a mixture of virgin material and regranulated scrap (regrind). Such material must be kept clean and dry; a regular ratio of the two materials must be used if machine operation and part consistency is to be maintained. Regranulation (of sprues, runners and faulty mouldings) is usually achieved by feeding such scrap through a grinder which may be located by the side of the machine.

Cycle time can be extremely fast, for example, disposable pots are now being moulded on a total cycle time of 1 s.

4.4 Advantages of an ILSM

The in-line screw machine (ILSM), or more correctly the in-line, reciprocating, single-screw machine, is the most popular injection moulding machine. This is despite the fact that it is more expensive than a ram machine to buy and to maintain—this is because it is a more complex system. Amongst its advantages (over a simple plunger machine) are the following:

1. Reduction in total cycle time
2. Lower injection pressure required to fill the mould
3. Reduced clamp required per unit moulding area
4. Smaller machine capacity required to produce a given size of moulding (as measured by shot weight rating)
5. Improved dry colour or masterbatch dispersion
6. Ability to handle heat sensitive polymers or mould at lower stock temperatures
7. Ability to mould harder flow materials
8. Less packing in the mould and easier ejection
9. Ability to handle powdered resins.
10. Rapid change-over from one type of resin or colour
11. Improved physical properties in the moulded part
12. Improved surface appearance
13. Reduction in warping, and internal stress and better dimensional stability

The above results were obtained (Weir and Zimmermann, 1962) by mounting moulds first on a ram machine and then mounting the same moulds on a screw type machine; machine and moulding performance were then compared.

It must not be imagined that the single-screw machine is the ultimate in moulding machines; it may be the best at the present time but there is still great room for improvement. Screw tip or valve wear can be excessive and in many cases the temperature distribution within the melt is excessive due to incorrect design or incorrect operating procedures. Compared to a plunger machine, the screw machine is expensive to buy and its complexity makes it difficult to service.

4.5 Injection Moulding of Thermosetting Plastics

Thermosetting materials, or thermosets as they are commonly called, are complex mixtures of resins (PF or UF, etc.), fillers, accelerators, pigments, etc. They are supplied as partially cured materials and are usually in powder form. Once they have been set to the shape of the moulding cavity by heat and pressure, the moulding will not subsequently soften when reheated. They can be decomposed by excessive heating and the above definition implies that these materials cannot be heat softened and reprocessed like thermoplastics materials.

4.5.1 Moulding techniques

For many years these materials were processed by compression moulding and transfer moulding techniques and such methods are still widely employed. It is found, for example, that high frequency preheating of the moulding powder followed by automatic compression moulding gives fast production of mouldings with a very high surface gloss. However, when the in-line screw machine was developed the two well established techniques faced competition because of the advantages that injection moulding offered.

The chief advantage of injection moulding is that very fast rates of production are possible; this is due to the preheating efficiency of the screw and it is particularly important for the slower curing materials, e.g. PF and MF. Such preheating efficiency is also particularly advantageous when thick wall parts are being moulded. Another major advantage is that labour costs are generally lower and this may be attributed to the reduced number of steps or stages involved in the injection moulding of thermosetting materials.

These two advantages help offset higher machine costs and material wastages. It has been said the advantages of injection moulding are economic and the disadvantages are technical. This means that mouldings may cost less to produce but they are not necessarily of high quality. Flow marks and weld lines may mar the surface and impair the components' performance. Fibre orientation can introduce problems such as warping; filler degradation may occur as a result of the screw action and high residual stresses may be present within the component. These high residual stresses may be caused by excessive packing pressures.

The high capital cost of injection moulding equipment is also a serious deterrent to many companies.

4.5.2 Machine differences

The basic injection moulding machine is similar for both thermosetting plastics materials and for thermoplastics; what is different is the design of the injection unit and the operating temperatures employed.

1. The barrel is run, or operated, at a low temperature, e.g. 80 °C. Such a low temperature is necessary if precure in the barrel is to be avoided. Instead of electrical heating bands the cylinder is jacketed for water circulation as this system gives accurate temperature control. It must not be forgotten that once the machine is operating the screw will generate heat and any excess heat must be removed from the system. Many machines have provision for more than one zone on the barrel (e.g. two). The rear zone is usually run at a comparatively low temperature. With UFs, for example, the rear zone is often cold.

2. The mould is run at a high temperature, e.g. 160 °C. This high temperature is achieved by electrical resistance elements and the mould temperature is controlled by thermocouple actuated instruments. To prevent excessive heat loss, the mould is usually insulated from the platens and in order to allow volatile matter to escape from the cavity (e.g. air and condensation products) the mould must be well vented.

3. The machine is fitted with pressure reduction. At the barrel temperatures employed, thermosetting materials have very high viscosities and in order to force the material into the mould at the desired rate the machine should be capable of generating high injection pressures, e.g. 150 MN m^{-2} ($\simeq$20 000 psi). However, as the temperature of the material rises (when it enters the hot mould) its viscosity falls, and the material could escape from the mould forming 'flash'. For this reason the machine should be fitted with a pressure system which allows the pressure to be dropped just before the mould fills.

4. A screw delay timer is commonly used. Many injection moulding parts have relatively thick wall sections and these can take a long time to cure. If the screw returned immediately after completing the injection and hold pressure part of the cycle, then the plasticised material (which forms the next shot) would be kept in front of the screw for a long time. This could

set (or crosslink) but the screw delay timer helps eliminate this problem. It is only when this timer lapses that the screw starts to rotate.

5. The screw flights are made thicker in order to reduce damage, which could be caused through forcing the screw forward against hard material, the flights of the screw are made thicker and therefore stronger than those used for processing thermoplastics. In order to prevent excessive heat build-up the screw is built with a low compression ratio of, for example, 1:1 or even 1:0·8. This also means that if the material cures in the barrel then by removing the nozzle (which should be designed for easy and quick removal) the cured material may be extruded out.

The tip of the screw can be of the smear head type (similar to that used for PVC) and in order to prevent the material slipping back along the screw during injection the screw pitch may be made finer than that used for thermoplastics, e.g. three-quarters of the screw diameter. On many machines the direction of screw rotation may be reversed if required as it is found that this is useful in the event that the material cures in the barrel.

It is held by some authors (e.g. Butler, 1975) that a direct hydraulic clamp system is better for thermosets than a toggle system, as the thermoset 'dust' may cause wear on the toggle pins and thus result in a loss of clamping force. As cycle times are generally quite long for thermosets there is less reason to provide a high speed clamp unit as the clamp closing speed is only a small part of the cycle.

4.6. Machine Lay-out

4.6.1 Machine components

The modern injection moulding machine is an extremely complex and expensive piece of equipment which nowadays may be attached to a microprocessor or a computer. Despite this sophistication the modern ILSM machine bears a remarkable resemblance to its predecessors in many respects. The machine may still be considered to be made up of a number of units or separate systems. These are as follows:

1. The injection unit
2. The clamping unit
3. The mould
4. The power system
5. The control system
6. Ancillary equipment

The way in which the injection unit and the clamping unit are arranged may be used to classify the machine.

4.6.2 The horizontal machine

A conventional single-screw machine layout is shown in Fig. 4. In such a layout the injection unit and the clamping unit are both horizontal and the power system is housed in the base of the machine. The mould usually moves along the tie-bars to generate the mould opening and closing actions, and the operating sequences are controlled or supervised, by the control unit which is located in a separate cabinet immediately adjacent to the moulding machine. Ancillary equipment, such as waste grinders, hopper loaders, etc., are located either close to or on the machine.

Such a layout gives a machine which can be easy to operate and to service as machine parts or components are generally at a convenient working height. Both the mould and the injection unit can be readily moved, using relatively low forces, as they are only required to slide along the tie bars which are parallel to the ground. Gravity assists in clearing the mould as once it opens then the moulding tends to drop clear thus freeing the machine for its next cycle. However, this type of machine occupies a large amount of floor space and the accommodation of large moulds which require long strokes and large daylights, can be difficult.

Some manufacturers offer machines in which the injection unit and the clamp unit are at right angles to each other but both are still horizontal with respect to the ground. Such an arrangement is useful if the moulding requires a side gate as the mould may be filled via a feed point which is located on one side of the cavity.

On horizontal machines it is possible to prevent melt drooling from the nozzle during the cooling part of the cycle by employing a technique known as 'suck-back'. Alternatively, a valve may be fitted in the nozzle.

More than one barrel may be mounted on the same machine bed, e.g. if multicoloured or sandwich mouldings are being produced.

4.6.3 Alternative arrangements

Although the horizontal machine is the most popular machine type it is not the only one; it seems that every combination of, or orientation of, injection unit and clamping unit has been tried in an effort to improve performance or convenience and/or to reduce running costs. In the main these two units are arranged to be either horizontal or vertical and if only these two orientations are considered then one can have:

1. Horizontal locking and horizontal injection
2. Vertical locking and vertical injection
3. Vertical locking and horizontal injection
4. Horizontal locking and vertical injection

The first has already been discussed (see Section 4.6.2) and the second can be derived from the first if one imagines that both units are slewed or moved from the horizontal to the vertical plane (Fig. 6). Machines have been made which do precisely this; both units are mounted on common tie bars and this total assembly is mounted on a pivot so that it may be swung into either position. This is useful if inserts are being used as with the machine in the vertical position gravity helps to keep such inserts in place. However, the

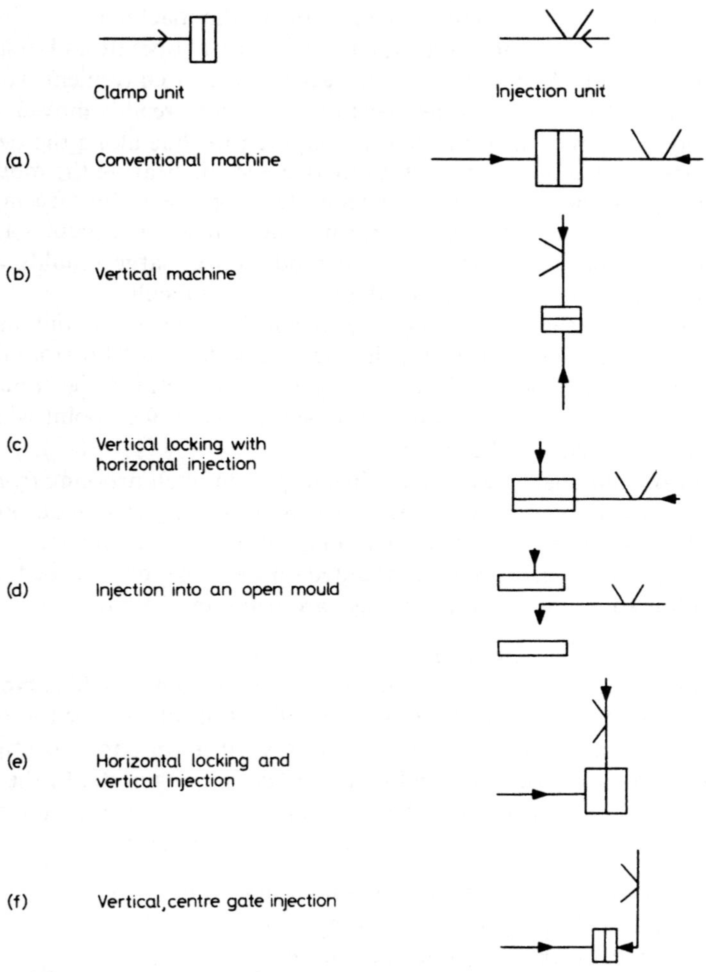

Fig. 6. Diagrammatic arrangements of machine components.

amount of floor space which these machines occupy is virtually the same as a horizontal machine.

If a machine is designed as a vertical system then the floor space required is approximately half of that necessary for a horizontal type and more machines per unit area can be accommodated. On such machines the injection unit is usually mounted above the clamping system as this means a convenient operating height at the mould face. If the injection unit is mounted below the clamp then this could be advantageous in certain situations, e.g. if the mould is being operated at a very low temperature as then the up-draught of hot air from the barrel could be used to stop condensation occurring on the mould.

The large amount of head-room required and the loss of operating convenience are probably the main reasons why vertical machines are not more widely employed. Vertical locking with horizontal injection is usually accomplished by using a downstroking clamp unit as this gives a fixed convenient height and fast closing speeds (due to the influence of gravity). Such machines are commonly employed when side gating is desired and the nozzle of the injection unit is set to coincide with the parting line of the mould (Munns, 1964). The material is injected into the parting line, along the runners and through the gates and into the mould. If the moulding cannot be edge gated then it may be necessary to use a hot runner mould due to the length of runner employed. It is relatively easy to design the machine so that long strokes and large moulds can be accommodated using a vertical clamp. Automatic operation can be achieved using techniques similar to those employed for automatic compression moulding.

Machines have been built where the thermally softened material is shot into the open mould, after it has been turned through 90° by a suitable extension to the nozzle. The injection unit then moves away, so that the nozzle is clear of the press, and the melt is then moulded just as in ordinary compression moulding, i.e. the injection unit is functioning as a dosing and preheating device for thermosetting materials.

Horizontal locking and vertical injection is sometimes employed on small machines as this gives a compact arrangement and yet the machine components are at a convenient height. With such systems injection takes place into the top parting line of the mould. Shut-off nozzles are usually used so that melt loss is minimised. If centre gating is desired then the mould design must be more complex or the machine must be designed so that the melt can be turned through 90° by a special nozzle. Injection then takes place through the centre of the mould in the usual way, i.e. as for a horizontal machine.

4.6.4 Multistation machines

Most injection moulding machines only have one mould assembly. This mould assembly may be capable of producing one impression at a time or it may be capable of producing more than one impression every cycle. However, there is only one complete mould set associated with a particular injection unit at any one time. When the setting time is very long, the use of one mould set can result in excessively long setting times. If more than one mould set is used, then the machine is referred to as a multistation machine. Such multistation machines are most commonly used for thick sectioned thermoplastics materials or for rubber components which are required in very large numbers, e.g. boot or shoe soles, seals, washers, etc.

The simplest multistation system to imagine is one which consists of a number of vertically mounted presses; the press platens therefore move up or down as the occasion demands. Each press or station contains a mould or moulds and the cavities within the moulds are fed by a side-feed, sprue and runner assembly. The presses are located around the arc of a circle and each can be fed in turn by the single injection unit. The injection unit therefore pivots to feed each mould station in turn. Machines have been made which are also capable of a vertical movement so that, for example, two shots may be delivered to each press station. This assumes of course that each press station contains more than one mould assembly, i.e. it is a multidaylight press.

The alternative to this type of system is to keep the injection unit in one place and to bring the mould assemblies to the injection unit. The moulds may be mounted on a table or carousel and after one mould has been filled it is moved away so as to enable another mould to be brought into position.

Carousel machines are called rotary machines and they have been made with either horizontal or vertical rotating mould carriers. Normally the mould carrier or carousel stops at a number of predetermined positions so that injection, cooling, mould opening, injection mould closing, etc. are progressively carried out as the carousel is indexed around.

Multistation machines are now most commonly associated with the rubber injection moulding industry where significant reductions in cure times and production costs (of relatively thick products, e.g. rubber boots) have been accomplished. Large machines with eight, ten or twelve stations have been used for the production of such boots. Some of the factors which affect the number of mould stations are mould costs, machine costs, order numbers, vulcanisation time, stripping time, etc.

The firm of Desma (Italy) is closely associated with the production of rotary multistation machines and Schleuter (1970) made a production rate and cost comparison of a four-station Desma machine with single-station alternatives. It has been stated (Anon., 1974) that three or four separate 500 tonne presses and three or four operators would be necessary to match the output of a single four-station (500 tonne multistation) machine which would need only one operator.

Some multistation machines have been built with two injection units, e.g. one feeds the mould with a soft rubber compound suitable for boot uppers and the second unit feeds the mould with a hard sole compound formulated for abrasion resistance, etc.

The injection unit of a multistation machine need not be confined to a reciprocating screw unit. Two-stage machines have also been used for this purpose and again the best known examples are found in the rubber injection moulding industry.

4.6.5 Multibarrel machines

The use of machines with more than one barrel has been described earlier. For example, one type of preplasticiser (see Section 3) has two barrels—the first is used to thermally soften the material and the material is transferred to the second where it is stored until the injection process is ready to begin.

However, the use of multibarrel machines is not confined to such machine types. In the plastics industry multibarrel machines are commonly used for the production of (a) multiple-colour mouldings and (b) sandwich mouldings.

4.6.5.1 Multiple-colour mouldings. The multiple-colour mouldings may consist of, for example, a clear plastics piece with a coloured portion moulded on to it so as to produce a lens or light-fitting assembly. Such products may be made on machines in which the moulding assembly rotates or indexes from one barrel unit to another.

At least one company makes machines suitable for the production of, for example, tortoiseshell effects. More than one colour is present but in these cases the two melt streams flow into the mould at the same time so as to produce markings or patterns which are considered suitable for certain products, e.g. boxes, combs, etc. The pattern so produced is capable of being reproduced exactly from one moulding to another.

4.6.5.2 Sandwich moulding. The other major use mentioned for multibarrel machines was sandwich moulding and in this context the term means that the finished moulding consists of two components; these two components are a skin and a core. Various combinations are possible, for example the skin and core may be based on the same material with the core being an extended version of that material. Alternatively, the skin and core may be of different materials, for example, the core may be made from a heavily filled polyethylene material and the skin may consist of a polyamide. A moulding which has the required properties could therefore be obtained using only a small amount of an expensive polymer. It has been suggested that this would be a good way of using, or losing, plastics scrap materials.

An example of one system is where a semipositive mould is used and the machine has two barrels—one injects a skin material and the other injects the core material. The core material contains a blowing agent and the gas which is liberated dissolves in the plastics melt under the pressures employed during processing. Each injection unit may be connected to the other injection unit by means of a valve.

The sequence of operations commences with both injection units charged with polymer, both screws retracted and the mould closed. The first material injected is the 'skin' material, this injection stroke only partially fills the mould. The valve is then shifted so that the injection unit which contains the polymer plus gas (formed by the decomposition of the blowing agent) is connected to the mould. This material is then injected. As the second material flows into the mould it stretches the material which is already there into a uniform thin skin. Strict control over variables such as melt temperature, mould temperature, injection rate, etc. is required if this stretching operation is to be successful.

When the mould has been filled (with a plastics material which is still capable of expansion) the valve then shifts so that the first injection unit is connected again to the mould. A small charge of 'skin' material is then injected into the cavity in order to clear the sprue of expandable plastics material. Unless this small additional charge is injected, then the foam or cellular structure would be revealed when the sprue was cut from the moulding; residual foamable polymer would also appear in the skin of the next moulding thus marring this moulding's surface finish.

Once the mould has been filled and the sprue has been cleared, the mould is then allowed to open a small amount. The gas trapped inside the core material tries to expand and in so doing causes the plastics material to expand also. A product with a dense outer skin (which has a good surface finish) and a lightweight cellular core results on cooling.

5 OTHER MACHINE TYPES

Single-screw preplasticisers are now firmly established as the major types of injection moulding machines—the two types which are widely used are the two-stage system and the single-stage system. Of these two types it is the single-stage machine which is the most widely used at present. Periodically, other systems are introduced but it is difficult to imagine that the in-line screw machine will be displaced from its present position of popularity in the foreseeable future. Despite several disadvantages, it has the supreme virtue of being versatile—it can handle or mould a very wide range of polymer compositions. However, single-screw machines do have limitations.

The range of use of plastics materials is being extended at the present time by adding fillers and/or other polymers to the base thermoplastics material. When fillers are added to thermoplastics, usually the objective is to reduce cost but occasionally, other reasons predominate. For example, some fillers improve the resistance of the moulding to burning. Ideally, such fillers should be capable of being incorporated during the moulding operation but a single-screw machine finds it difficult to convey and mix heavily filled blends.

5.1 A Rotor and Ram System

One alternative to a conventional single-screw machine is based on an injection unit which contains a rotor and a ram (Fig. 7). The rotor passes through the centre of the ram and the rotor tip operates inside a tapered chamber. As the ram comes forward, then the material is sheared between the rotor tip and the barrel wall, thus giving a good mixing action. When the melt has filled the chamber in front of the rotor, then injection takes place (Anon., 1978). It would be expected that such a system could handle heavily filled compositions as the material is held captive under pressure whilst shear is being applied.

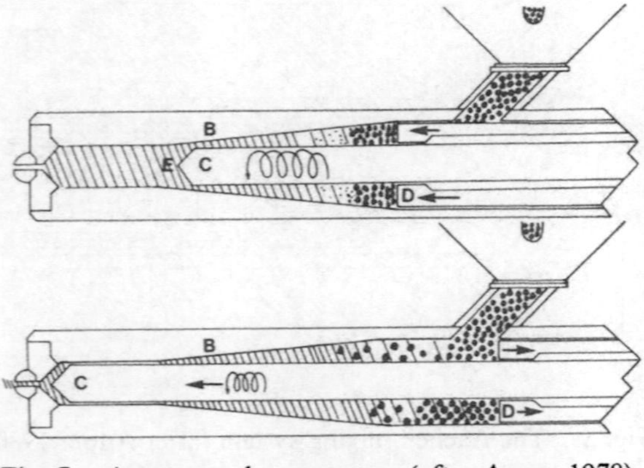

Fig. 7. A rotor and ram system (after Anon, 1978).

5.2 The Aachen Mixing System

In the field of extrusion, improved mixing has been obtained by incorporating pins into the melt flow channel. These pins improve mixing as they break up the laminar flow which occurs within the screw flight. Workers at the West German IKV Institute (Anon., 1980) have applied this principle to injection moulding by incorporating a screw which has segmented spiral mixing blades; these intermesh with fixed pins on the inner surface of the barrel. These pins are arranged radially around the barrel and each line of pins occupies the space between the spiral segments. As the screw turns, then the material is sheared between the pins and the spiral mixing blades (Fig. 8).

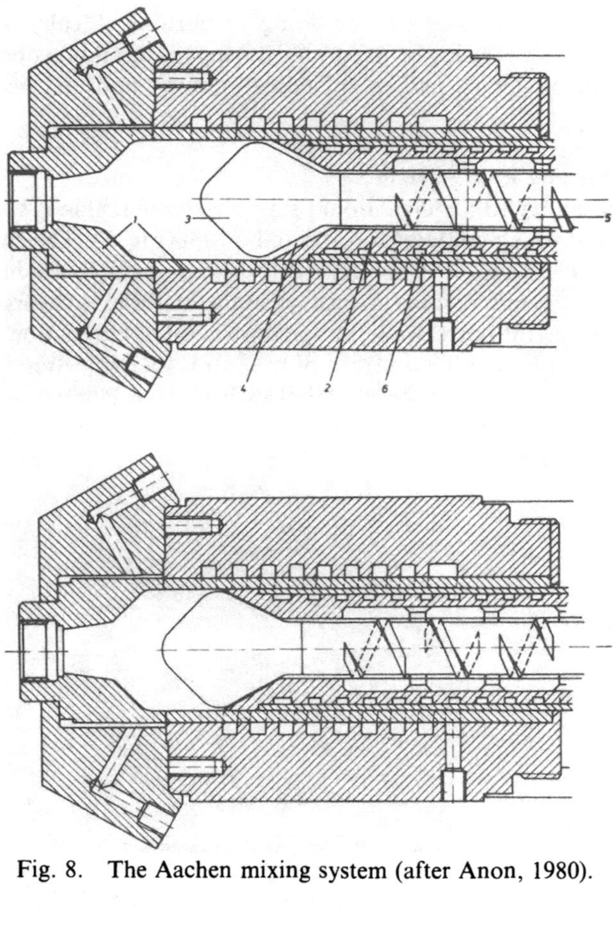

Fig. 8. The Aachen mixing system (after Anon, 1980).

The screw and the mixing barrel together make up 'the injection ram' as this assembly fits inside another larger cylinder. To achieve injection 'the injection ram' moves forward thus displacing material into the mould. During the preplasticising stage the melt is pumped along the screw in the usual way and undergoes extensive mixing. The melt passes through an annular gap which is formed between a torpedo at the end of the screw and the end of the mixing barrel. As the screw pumps the melt forward, the screw and the mixing barrel (the injection ram) move backwards together.

This type of unit has been used to mould such difficult combinations of materials as 80% aluminium hydroxide and 20% polyethylene. It has also been claimed that it can handle polymer mixtures which contain polymers of widely differing melt flow indices.

A major disadvantage of this system would seem to be that it is relatively complex compared to a conventional single-screw machine and it would therefore be expensive to purchase and to maintain.

5.3 Adiabatic Machines

It has been known for many years in the extrusion industry that adiabatic heating of plastics is possible and occurs on many production machines, either accidentally or deliberately. The term adiabatic implies that all of the heat necessary to cause thermal softening of the material is generated by shearing the material with the screw. Adiabatic extruders seemed to go out of favour as it was found that better control over the process could be obtained by using a combination of adiabatic heat-input and conducted heat-input (Fig. 9).

Probably the most well-known adiabatic injection moulding machine is the Dynameltor which was originated by the Meiki Company Ltd of Japan. The barrel is divided into zones but only the front zone is maintained at melt temperature, the centre zone being, for example, 5 °C below melt temperature and the rear zone approximately 30 °C below melt temperature. The main screw, which conveys and shears the material, is not equipped with a check valve at its tip as the packing under back pressure of semi-plasticised material immediately behind the melt zone prevents back flow during injection.

Material is fed into the barrel by a feed screw which is situated at the base of the hopper and matters are arranged so that the amount of material present in the barrel is comparatively small, e.g. less than twice the total shot weight. This small amount of material and the absence of a check valve helps to minimise degradation and improves colour or material changing. It is claimed by the manufacturers that because the material is heated from

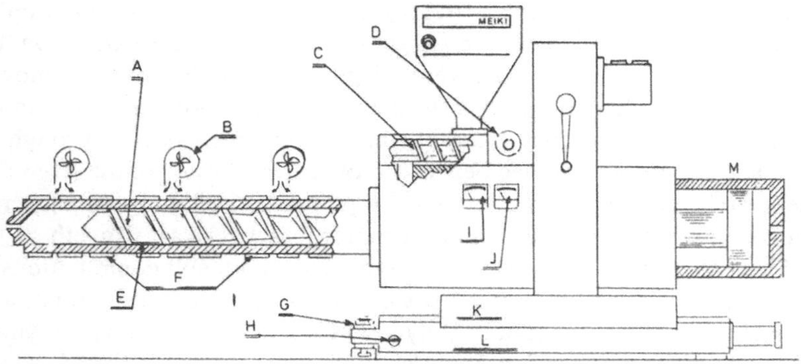

Fig. 9. Adiabatic injection unit (courtesy Eastern Plastics Machinery Ltd). A, main screw; B, air cooling device; C, feed screw; D, feed screw speed adjuster; E, shallow flight; F, heater; G, nozzle height adjuster; H, carriage swing device; I, main screw rpm; J, feed screw rpm; K, carriage; L, slide base; M, injection cylinder.

'within' a more homogeneous melt at a low temperature results. If this is true, then substantial savings in cycle time should result.

5.4 The Weissenberg Effect

It has been noted that if a rod is rapidly rotated on its own axis when immersed in certain viscoelastic materials, that instead of the vortex forming (as occurs with viscous fluids), the liquid climbs up the rod. This rod climbing is sometimes known as the Weissenberg effect (Brydson, 1970). When the rod rotates, then viscous drag forces will cause the material to rotate around the rod and the long-chain polymer molecules will initially be deformed in an arc-like shape. Because of the natural tendency of the molecules, which wish to take up a random configuration, tension effects will transform this arc-like shape into a straight line with a net inward movement of the molecules. In order to accommodate this movement, the viscoelastic material flows up the rod.

Unusual movement of a polymer melt also occurs, due to the same effect, when the material is sheared between a rotor and a plate. If there is a hole in the centre of the plate, and material is introduced between the outside edge of the rotor and the plate, then material will emerge from the central hole. The rotor-and-plate arrangement has been used to feed a short screw on an injection machine so as to obtain a very compact unit which is significantly shorter than a conventional in-line screw machine. Another attraction of this type of system is that the residence time of the polymer within the injection unit is comparatively short and this helps to minimise polymer degradation.

Despite these advantages, this type of system does not appear to have become widely adopted as, presumably, it lacks the operational flexibility of a conventional single-screw machine.

5.5 Extrusion Moulding

Most injection mouldings are of comparatively low wall thickness and this means that there is considerable resistance to flow during the mould filling operation. If setting or hardening occurs during this mould filling stage, for example, by using a thermoplastics material and a cold mould, then the resistance to flow increases dramatically and this is why high injection speeds and high injection pressures are often required.

If the wall thickness of the component is large, then high filling rates may not be required, or even desirable, and in such cases the pumping action of a rotating extruder screw may be used for mould filling. Machines are available in which heavy sectioned components such as toys, shoe soles, etc. are moulded by using an extruder unit connected to a cast metal mould. Such a system is relatively cheap as the 'injection unit' is simple, and because the pressure involved is low, mould costs can be kept down by using cast metal moulds.

6 INJECTION MOULD TYPES

The moulds used in injection moulding vary enormously in their size, complexity and cost; in fact the range of moulds employed is as great as the range of injection mouldings themselves. Moulds used for prototypes may be relatively cheap, as only a few mouldings may need to be produced from them; such moulds have been produced from a range of materials, e.g. other plastics and dental plaster. Mass production moulds on the other hand may be required to produce tens of millions of parts at very high speed and so their cost and complexity is many orders of magnitude higher.

Normally the injection mould consists of two or more pieces of metal which contain an impression of the component which it is desired to produce. If the mould contains only one impression or cavity, then it is referred to as a single impression mould and if it contains more than one impression then it is referred to as a multi-impression mould.

6.1 Two-plate Moulds

Many injection moulds can be considered to be made from two halves with each half being bolted or attached to one of the machine platens (Fig. 10). Mould closing, opening and clamping are achieved by the hydraulic system and in order to withstand the large forces involved, the mould is usually manufactured from tool steel. Apart from the cavities themselves, the mould contains heating/cooling channels, and a feed system; provision

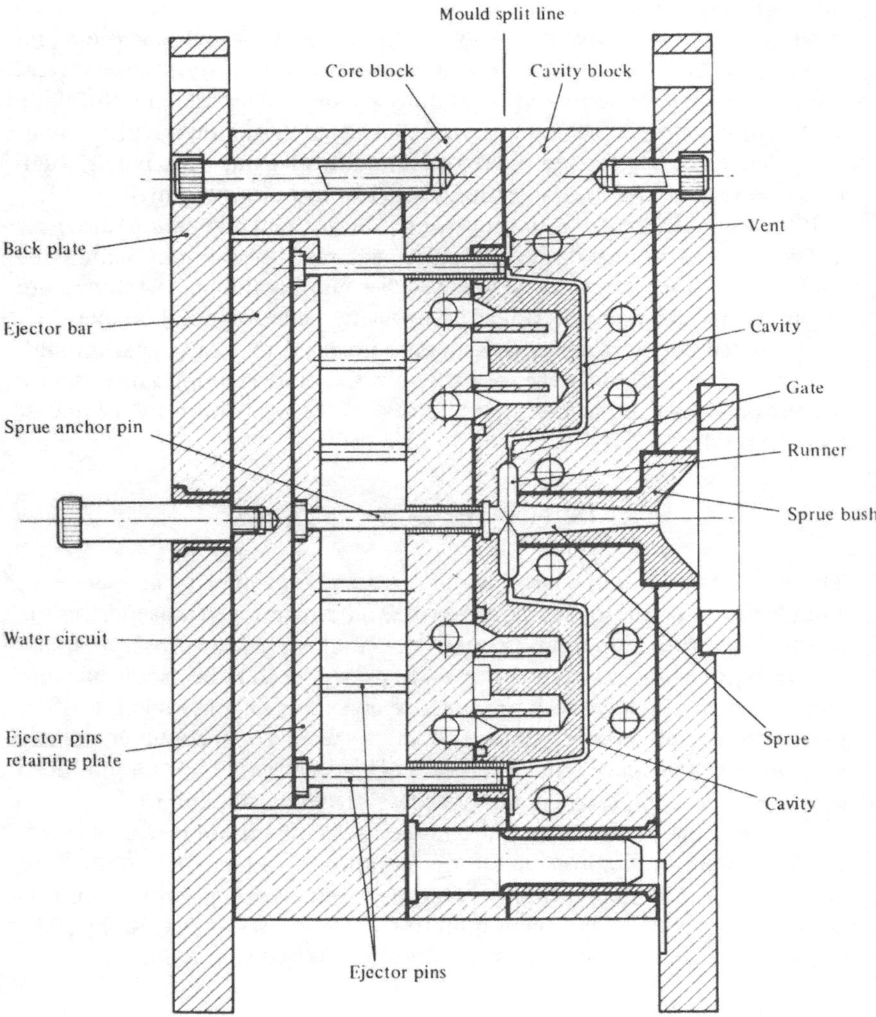

Fig. 10. Layout of a typical two-plate mould (after ICI, 1978).

is also usually made for part and runner ejection, and the mould may also contain vent channels. Each mould half is usually made from a number of pieces or subassemblies. Such two-plate moulds are normally specified because of their relative simplicity and therefore low, relative cost. For many moulding jobs, moulds other than the two-plate mould described are preferred because of considerations such as ease of finishing, feeding and feed system elimination; the moulding technique may also be different to that conventionally used with a two-plate mould.

6.2 Three-plate Moulds

As the name implies, this type of mould has three main parts or plates which when separated give rise to two main openings or daylights. When the mould opens, the feed system drops from one daylight and the moulding or mouldings drop from the other; segregation of the moulded spray or shot is therefore possible. The use of this type of mould construction means that centre pinpoint gates can be used in a multi-impression tool. Alternatively, more than one gate may be used to feed the cavity of a single impression mould or a three-plate mould may be used when it is required to feed a cavity from a point that is not at the centre of balance of that moulding.

6.3 Runnerless Moulds

In order to eliminate the ejection of the feed system from the mould at the end of each moulding cycle, modified three-plate moulds are now in common use. These are modified so that the material in the feed system does not set or harden during normal running; in the case of thermoplastics materials, this means keeping the polymer hot whereas in the case of thermosets and rubbers, this means keeping the material cool.

When only the sprue is prevented from setting then the mould may be referred to as a 'sprueless mould'. If all the feed system is prevented from setting then the mould may be referred to as an 'insulated runner' or as a 'hot runner mould' or, a 'cold runner mould'. Feed system setting can be prevented by (a) insulation and (b) temperature regulation.

If the feed system is made larger than normal (e.g. three times normal diameter) and if long delays are avoided then insulated runner mouldings may be performed on a three-plate mould as the outer layers of material, in the feed system, will insulate the central core.

By heating the runner system a hot runner mould (as used for thermoplastics) results whereas if the runner is kept cool a cold runner mould (as used for thermosets) is obtained.

6.4 Stack Moulds

It is believed that stack or stacking moulds (sometimes called 'double-deck moulds') were first used to mould gramophone records but in any event this process is not as new as many people imagine. Such moulds consist, in effect, of a number of cavities one mounted behind the other and fed by a common sprue. As the mould cavities are superimposed one above the other the mould opening force is much reduced and the clamping force can therefore also be correspondingly reduced. Stacked moulding can be used to increase the output from a moulding machine and/or to reduce the clamp pressure requirements.

6.5 Family Moulds

Family moulds are used to produce a number of different sized plastics components which together form one item, e.g. an assembly kit or a child's toy. In general it is not good practice to have different sized cavities within the same mould as problems are usually experienced in production and/or in use. The moulding conditions chosen may suit one particular cavity but it is unlikely that they will suit all the cavities and so some of the mouldings may be, for example, overpacked.

6.6 Vertical Flash Moulds

On a conventional two-plate mould the parting line (the flash line) is normal to the injection direction as this simplifies mould construction. Large mouldings and/or structural form mouldings have been made in moulds of a different construction, i.e. part of one mould half fits inside the other mould half. Because of the fit between the two mould halves, material finds it difficult to escape down the parting or flash line. Such moulds have been used for the production of sandwich mouldings and for the production of injection–compression mouldings.

6.7 Undercut Moulds

Many mouldings contain holes or undercuts and these are commonly formed in the sides of mouldings by sliding cores or cams. The cores or cams are withdrawn before ejection takes place so that normal ejection is possible. Cam movement may be achieved by utilising the mould opening/closing movement or the cams may be moved by pneumatic or hydraulic cylinders. Cylinder operation is sometimes more convenient as the movement can be actuated at any point in the moulding cycle. Once the cams are in position they are usually locked into position so as to prevent the polymer flow from moving them out of position. Moulds which contain such devices are relatively expensive as they are more difficult to design, build, operate and maintain. In many cases it is cheaper to machine holes or slots into a moulding in a postmoulding operation.

6.8 Unscrewing Moulds

Such moulds are used so as to allow threaded components to be ejected from the machine; a cavity or core is rotated and this action frees the threaded section. It may be possible to eliminate such a relatively complex system by:

1. Producing the component in two parts and then assembling
2. Moulding over inserts
3. Pushing an insert into the moulding in a separate operation

As far as possible threads should not be cut or tapped into polymers as undesirable notches may be formed.

7 ALTERNATIVE MOULDING TECHNIQUES

Modern injection units, based on screws, can plasticise large quantities of polymer melt; such units may also be used as metering or dosing systems and so save on materials handling.

7.1 Intrusion or Flow Moulding

Flow moulding or intrusion is an extrusion moulding process which utilises a conventional single-screw machine so as to allow that machine to produce very large components. Additional controls are provided so that after the screw has injected material into the mould, the screw immediately starts to rotate and pumps additional material into the large mould cavity. This is commonly done against a back pressure which is adjusted so as to obtain the required quantity of material within the limits set by the screw stroke. When the screw has reached its fully back position, it is then moved forward again, without being rotated, so as to pack or consolidate the moulding under pressure whilst the moulding cools.

Other variations on this theme are possible. For example, on a conventional ILSM additional controls may be provided so that the screw may be rotated during the injection process; this gives a comparatively large shot weight but it does make additional demands on the hydraulic system and can cause excessive wear. Because of this, and because of the erratic shot weights which may result, injection with screw rotation is not normally employed.

7.2. Injection–Compression Moulding

Injection–compression moulding is used, for example, in the rubber industry for the production of footwear. Such footwear consists of canvas or fabric uppers and rubber soling.

The upper material (canvas or fabric) needs to be kept below 150 °C and the use of this comparatively low temperature means that the rubber viscosity will be high; large injection pressures will therefore be needed. Such high forces can distort or tear the fabric.

One technique which solves this problem uses multistation machines. The rubber is injected by a screw-ram injection unit into a gap between the shoe upper and the sole plate which is held 3 or 4 mm (0·118 or 0·157 in) wider than it would be in its final position (Wheelans, 1974). After injection of rubber (under the heel at the first mould station) and passage to the next mould station the gap between the last and the sole plate is closed by pressure in the manner of compression moulding. Rubber flows from under the heel of the upper along the foot to the toe and over the toe cap. Some rubber may be pushed backwards through the sprue bush and this is wasted as flash.

Injection–compression moulding is also used in the rubber industry to produce items such as small O-rings. The injection unit on such a machine functions as a metering unit and also as a preheating unit; it feeds a charge of preheated material to what is, in effect, an open compression mould. The mould then closes and forces the rubber to conform to the shape of the cavities and after curing the many small components (e.g. 1000) are removed as one piece, i.e. each component or item is joined to its neighbours by a thin film of flash and it is this which is the main attraction of the process. When the mould opens the operator simply removes one piece and, providing the flash is kept thin, part separation is easy and material wastage is low.

The plastics industry also uses injection compression moulding, e.g. for the production of items made from thermosetting materials. Once again the injection unit functions as a metering unit and as a preheating unit; in-line injection units may be used for this purpose. The compression moulding process is therefore simplified in some respects and mouldings with a reasonable surface finish can be obtained. During the feeding of the thermally softened material into the open mould volatile matter can be eliminated and as the process is virtually a compression moulding process, moulding pressures are much reduced. By using this technique unsightly flow marks (caused, for example, by the melt being divided by a pin in a conventional mould) can be effectively avoided (Fig. 11; Parker, 1975).

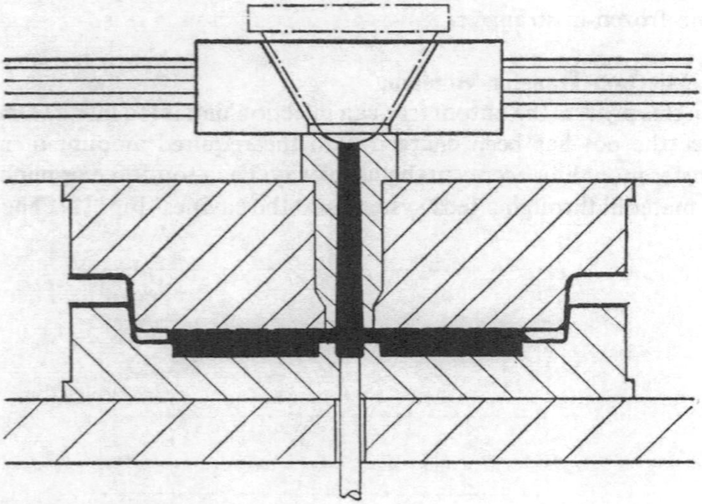

(a) Mould in injection position

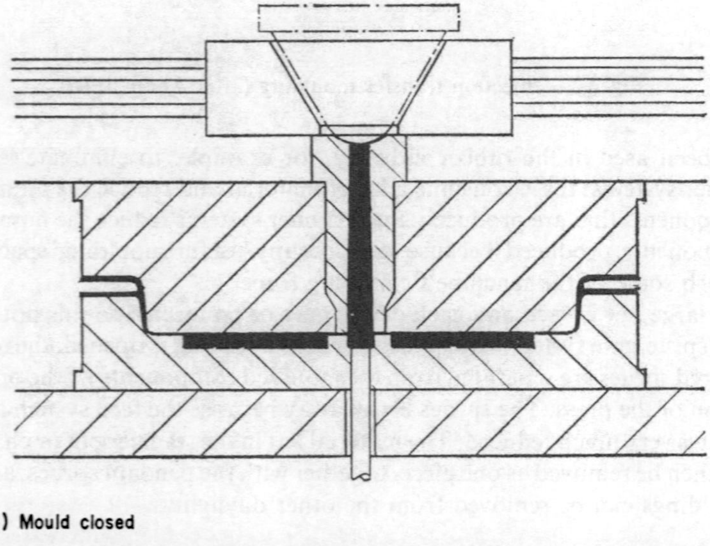

(b) Mould closed

Fig. 11. Direct screw transfer (DST) injection compression method (after Parker, 1975).

(Acrylic lenses are also moulded by this technique as the products have little or no frozen-in strain.)

7.3 Injection Transfer Moulding

In this process the output from an injection unit is fed into a transfer pot. Once the pot has been charged with the required amount of material, transfer moulding occurs in the usual way, i.e. a top-force or punch forces the material through a feed system into the cavities (Fig. 12). The process

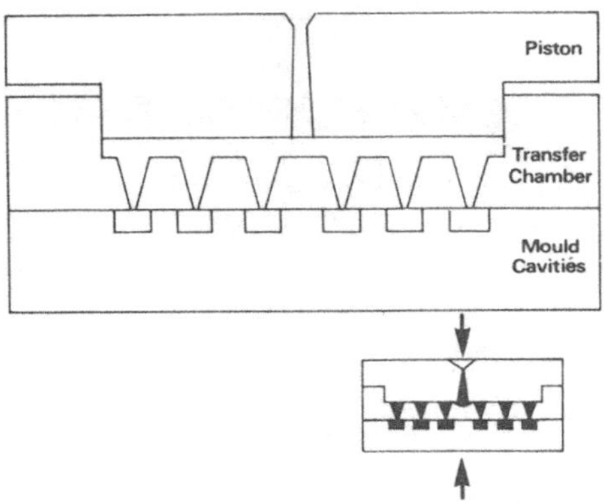

Fig. 12. Injection transfer moulding (after Anon, 1980).

has been used in the rubber industry, for example, to eliminate lengthy runner systems; these consume a lot of material and reduce the number of components that are produced. Such runner systems reduce the number of components produced because they occupy useful moulding space and absorb some of the machine's clamping force.

A large pot is used and each cavity may be connected to this pot by its own sprue/gate system. After curing, when the mould is opened, the reverse tapered sprues are separated from the moulded components by the opening action of the press. The sprues break away because the feed system at that point is very much reduced. The material left in the transfer pot or chamber can then be removed as one piece, together with the pendant sprues, and the mouldings can be removed from the other daylight.

7.4 Injection Thermoforming

Thermoforming is a technique whereby a thermoplastics sheet material is deformed to a desired shape using pressure after it has been made soft and pliable by the application of heat. The required shape is normally obtained by deforming the sheet material over, or into, a mould using an air pressure difference. One of the big advantages of thermoforming is that components with extremely thin wall thicknesses can be obtained but unfortunately the thermoplastics material must be available in sheet form. Sheet extrusion is widely used for this purpose but the equipment is relatively expensive and can be difficult to operate. Because of this, systems have been devised which eliminate the need for the extrusion equipment.

The injection unit of an in-line screw machine is used to mould a disc of a thermoplastics material and whilst this is still hot, it is shaped by means of air pressure into, for example, a disposable cup.

7.5 Injection Blow Moulding

Blow moulding is used to produce hollow enclosed parts such as bottles and many early machines used an injection moulding unit to produce the tube or parison from which the bottle was made. The early machines used plunger machines but machines in current production use an in-line screw unit. In injection blow moulding a blowing stick or mandrel is placed inside the heated mould. A plastics material is then forced into the mould and flows around the mandrel thus forming a tube with a closed end. Whilst the material is still hot, the mould is opened and the mandrel is transferred to a blowing stage or station. Air is then introduced into the plastics tube, via the mandrel, and this causes the plastics tube to be expanded to the shape of a mould. When the product (e.g. a bottle) is rigid enough to be ejected, the mould is opened and the mandrel plus bottle is removed.

In recent years, this process has enjoyed a revival as it is relatively easy to arrange for biaxial orientation to occur. The mandrel is made of telescopic sections and these are extended before the air is introduced. Mandrel extension causes orientation in one direction and the blowing process causes orientation in the other.

Such a process can, because of the orientation, give strong products with improved resistance to gas diffusion. If a material such as polyethylene terephthalate (PET) is used then the bottles may be suitable for the packaging of carbonated drinks, e.g. lemonade.

In injection blow moulding, the blowing pressure is of the order of $0{\cdot}7\,MN\,m^{-2}$ (100 psi) and this has obvious advantages. However, an injection mould must be constructed to withstand the forces involved in that process. The need for two moulds (one of them an expensive injection mould) and the comparatively low rate of production are the main disadvantages of this particular process. However, as the moulds used are of the hot runner variety, there is little or no wastage and as a closed, preformed tube is used, the product requires no finishing, i.e. finished-form production is possible.

8 REACTION INJECTION MOULDING

This process is associated with polyurethanes and is commonly known as the RIM process or the RSG process (RSG is the German equivalent to RIM). It is a self-skinning process, i.e. one which produces mouldings which have a cellular core and higher density outer layers.

Such complex mouldings are produced in one step or stage by mixing together a polyol (a material which contains hydroxyl groups), poly-isocyanate (a material which contains isocyanate groups (NCO)) a blowing agent, e.g. methylene chloride, and catalysts. The relatively cold mould (e.g. 50 °C) is only partially filled with this mixture which reacts to form the polymer (a polyurethane). Heat is liberated and this causes the blowing agent to boil and thus gas is produced.

The reaction mixture expands to fill the mould cavity but because the temperature is lower at the mould wall, and because of the internal pressure, the blowing agent condenses at the surface thus forming the higher density outer layers.

By varying the raw materials used it is possible to produce a range of mouldings whose properties range from rigid to flexible; fillers may be incorporated and reinforcing sections (e.g. based on steel) are easily incorporated. Because of such considerations, and because of the low clamping pressures (e.g. 1 $MN\,m^{-2}$ (140 psi)) the use of this process is growing rapidly. Systems other than polyurethanes can be used, e.g. based on polyesters and epoxides.

REFERENCES

Anon. *Rubber World*, p. 43, **171**(3), 1974.

Anon. Rotor cuts output risks. *Plastics and Rubber Weekly*, p. 12, 4 August 1978.

Anon. Aachens IKV redesign the injection machine. *British Plastics and Rubber*, April, 1980.

Bayer. Technical literature on processing of Durethan, Bayer, W. Germany, 1 Oct 1974.

Brydson, J. A. *Flow Properties of Polymer Melts*. Published for The Plastics Institute by Iliffe, London, 1970.

Butler, R. A. Machinery for thermoset injection mouldings, Chap 10, *Developments with Thermosetting Plastics*, Eds Whelan A. and Brydson, J. A. Applied Science Publishers, London, 1975.

Euromap document No. 1. *Description of Injection Moulding Machines*. Available from The British Plastics Federation, London, Jan 1969.

ICI. *The Principles of Injection Moulding*. Technical Service Note G 103 (3rd edition), ICI, Herts, 1978.

Marjurey, M. J. Demands for the availability of thermal conductivity data for polymers. *Plastics and Rubber International*, p. 111, **2**(3), 1977.

Munns, M. G. *Plastics Moulding Plant, Vol. 2, Injection Moulding Equipment*. Published for the Plastics Institute by Iliffe, London, 1964.

Parker, F. J. Injection moulding—material and process requirements, Chap 11, *Developments with Thermosetting Plastics*, Eds Whelan, A. and Brydson, J. A. Applied Science Publishers, London, 1975.

Schleuter, R. H. Multi-station injection machines with simple or dual injection units. *SPE J.*, p. 95, 26 April 1970.

Weir, C. L. and Zimmermann, P. T. Facts and figures on ram versus screw injection. *Modern Plastics*, p. 122, Nov and Dec 1962.

Wheelans, M. A. *Injection Moulding of Rubber*, Newnes Butterworths, London, 1974.

CHAPTER 3

Machine Specifications

1 FACTS INCLUDED

A major machine manufacturer offers each machine in his range, in a number of different variations, so that it can handle a wide range of materials and/or different jobs. Such a wide range is achieved by fitting on to a standard or basic machine, different items or control circuits whose function is specified in that company's trade literature or machine specification.

The specification contains basic facts about the machine such as the machines rating. Injection moulding machines are commonly classified according to the clamp force that they can generate and the amount of material that they can inject. A machine specification normally contains numerical details of the injection unit, the clamping unit, the drive and the overall machine dimensions. A brief description is also given of the basic or standard equipment which is normally supplied with the machine and then further details of alternative equipment available are listed. As specifications are normally consulted when machines are being purchased a brief section on machine purchase has been included.

2 THE INJECTION UNIT

When a screw machine is purchased it is usual for a screw to be included in the machine price. The buyer is asked to select one screw from a range of screws and this is then supplied together with an appropriately sized barrel.

2.1 Injection Capacity

The injection capacity is the maximum volume of material (and therefore given as cubic inches, cubic millimetres, etc.) that can be injected each time the screw or ram moves forward during a cycle of operations. The amount that is injected depends on a number of factors. For example, the diameter of the screw and the amount of movement that it makes during the plasticisation stage—this can be varied so that injection capacity will in turn obviously vary. What is not so obvious is that leakage occurs during injection. As the screw moves forward to inject the material into the mould, the shut-off ring or valve at the end of the screw leaks and allows material to flow back down the screw channel. Machine manufacturers know this and apply safety factors to compensate. However, different machine manufacturers may not use the same safety factor and so, unless injection capacity has been determined according to a standard method, it is important to ascertain what safety factor has been used. It could, for example, have been 10 or 20%.

By tradition, in the plastics industry it is customary to rate injection moulding machines in terms of the weight of polystyrene that that machine can inject with one stroke of the screw or ram. For many years, machines were rated in terms of the number of ounces that they could inject but, then with the growing spread of metrication, the units changed to grammes and the figures were adjusted to make them more easily comprehensible. For example, a 1 oz machine became a 25 g machine and a 2 oz machine became a 50 g machine.

This method of rating machines is useful as long as one appreciates that because of the different properties of polymers (e.g. density) different shot weights will be obtained from the same injection unit when one changes materials. To introduce a degree of uniformity into classification procedures, various bodies now offer specifications which can be used to rate the capacity of injection units. One such standard, issued by Euromap, rates injection machines in terms of the calculated swept volume (in cubic centimetres) that the injection unit can deliver in one stroke when set at an injection pressure of 100 MN m^{-2} (14 500 psi) (Euromap document No. 1, 1969).

2.2 Length: Diameter Ratio

This is always quoted on screw machines and is given as a ratio, e.g. 18,1 or 18:1. It may be listed as L:D or effective screw length.

It is common practice to define a screw in terms of its length to diameter ratio and so if a screw is described as being an 18:1 screw, it means that its length is 18 times its diameter. In actual fact, the screw may be longer than this as the length quoted is only the effective length. In the case of injection moulding machines this means the distance from the rear of the machine hopper (when the screw is in the fully forward position) to the end of the screw flights, i.e. it does not include the shut-off valve.

For plastics materials L:D ratios are commonly 20:1 for normal moulding jobs but the L:D ratio of vented screw units may reach 30:1. In general, as the L:D ratio increases, melt uniformity and plasticising capacity also increase but unfortunately so does unit cost and drive power requirements.

If the same machine is offered with a range of screws, then if a screw of larger diameter is chosen, it will be found that the L:D ratio drops. For example, a 70 mm screw may have an effective screw length of 18:1 whereas the corresponding 90 mm screw would have an effective screw length of only 14:1.

2.3 Screw Stroke

The figure that is quoted is given in either millimetres or inches and it refers to the maximum amount of linear movement that the screw can make whilst plasticising. As the screw turns and pumps material forward, the effective screw length decreases and as a result a poorer quality melt can easily result. This effect becomes most marked if the L:D ratio of the screw is initially small and helps to explain why L:D ratios have steadily increased since in-line screw machines were first introduced.

Normally the maximum screw stroke is not more than 3·5 screw diameters, as if it exceeds this figure, melt quality suffers. When selecting a machine, it is wise to use a lower figure than the maximum, but on the other hand, one should not use too low a proportion of the screw stroke (e.g. less than 10 %) as otherwise metering errors will become excessive. It is probably best to use approximately 60 % of the machine's rated shot capacity.

2.4 Plasticising Capacity

The units of this parameter are in pounds per hour or kilogrammes per hour and it is commonly expressed in terms of polystyrene (PS). The figure quoted is the maximum one possible for that particular unit and care should be taken when comparing figures from different sources as different methods of measurement may have been used by the different manufacturers.

2.4.1 Factors influencing capacity

The figure quoted is dependent upon many factors, for example:

1. The material. The particle size, shape and size distribution, the coefficient of friction between the material and the barrel surface, the thermal properties of the material, etc.

2. The barrel. The temperature of the barrel and the temperature distribution, the surface treatment of the barrel, etc.
3. The screw. Screw length, diameter, pitch, etc.

If a mould is being used in the method of determining the plasticising capacity, then obviously the mould used will, in turn, influence the result obtained. Methods which do not rely upon a mould being used exist and of these the one suggested by the SPI (1968) appears to be most useful, although this method of measurement is not widely practised in Europe (see Section 2.4.2).

Normally screws that are fitted to injection moulding machines are general purpose screws which have been designed to handle the widest possible range of materials reasonably well. However, if it is known that a particular machine is going to spend most of its time moulding one particular material, then it will be better to order a screw which has been designed for that particular material. It has been calculated that if a machine is to spend 65% of its working time on one material, it will pay to have a special screw made for it (Ireland and Smith, 1977). The value will come in more efficient plasticisation, better melt quality, faster screw retraction (without any need for back pressure) and higher part quality.

When determining plasticising capacity, please note that the object is not simply to produce large quantities of molten polymer. The object is to produce melt of the desired quality at the highest possible rate. The quality is commonly defined in terms of temperature, temperature distribution and/or dispersion of ingredients. It is possible to produce good quality melt on short L:D screws but in general it will only be available in smaller quantities than if a higher L:D screw was used.

2.4.2 Society of the Plastics Industry Inc. (SPI) Comparison Test

This test became effective on 1 January 1968 and as yet does not seem to have become widely accepted in Europe. This is despite the fact that it gives to the moulder a means of comparing the performance of one screw injection machine with another under certain specified situations and using standard materials (SPI, 1968).

To perform the test one must have available the following:

1. A nozzle whose dimensions have been specified by the SPI (Fig. 1)
2. A melt thermocouple whose hot junction is located at the centre of the transducer body, e.g. West Instrument Company type TCA-0052-2 (Fig. 1)
3. A recording instrument to record stock or melt temperature from the thermocouple specified in 2 with a minimum chart speed of

1 mm s^{-1} (0·039 in s^{-1}) and a marker response time of 1/10 of a second per 6 °C

4. Mutually agreed material, e.g. polystyrene (PS) such as ASTM D703 type 1

For each material a target melt temperature is either specified or must be agreed upon by the interested parties. For example, in the case of the polystyrene mentioned above, the target melt temperature is 216 °C. At the agreed or specified melt target temperature, the melt must be acceptable, i.e. it must show no signs of gassing, degradation, etc.

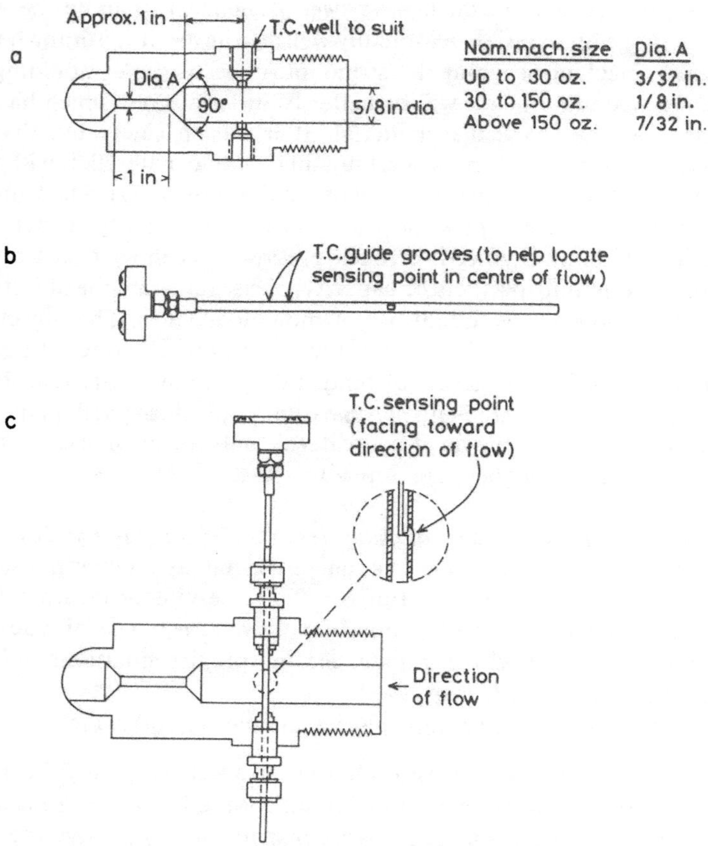

Fig. 1. (a) Nozzle for use in new plasticating test. (b) Melt thermocouple for use in new plasticating test. (c) Thermocouple installed in nozzle. (After SPI, 1968.)

The rated injection capacity is used to determine the following:

1. Shot size. For a given test the shot size shall be 25, 50, 75 and 100 % of injection capacity (unless otherwise agreed)
2. Injection rate. For example, the injection rate shall be between $1\frac{1}{2}$ and 2 oz s^{-1} (43–57 g s^{-1}) for machines with a nominal rating of up to 30 oz (850 g).

At a particular shot size setting, the screw running time is adjusted to give an output of good quality melt. This means that the melt must be free of gassing, degradation, etc. and also free from thermal inhomogeneities—the allowable recorded variation during 10 successive shots shall not exceed 6 °C. Once the screw running time has been decided, then the cycle is adjusted so that it is twice this time. Ten successive shots are individually collected and weighed; the melt temperature variations that occur within each shot are also recorded together with the total screw running time.

A recovery rate is calculated by dividing the average weight of a shot by the average screw running time at a particular shot size setting. The results are plotted as recovery rate against per cent of injection capacity.

2.4.3 The Euromap recommendation

The European Committee of Machinery Manufacturers for the Plastics and Rubber Industries (Euromap) issues various recommendations relating to injection moulding machines. One such recommendation (Euromap document No. 5, 1972) is relevant to this discussion on injection units but unfortunately it seems to ignore temperature variations that can and do occur in the melt if an injection unit is operated incorrectly. Plasticising flow (screw recovery time), rate of injection and specific energy consumption are the important production data measured in this recommendation. Such data is determined by means of any injection moulded part provided that the conditions of manufacture of such parts are clearly set out in the report. If the recommendation is to be of use, therefore, in comparing one injection moulding machine with another, then the same moulding material must be used and the same injection mould. It does not appear, therefore, to be as useful, nor as stringent, a test as the SPI one mentioned previously.

2.5 Injection Rate

The injection rate or injection velocity is the maximum rate at which the screw or ram can eject or shoot material from the barrel during a single shot. It is normally quoted in terms of weight transferred per unit time (e.g. oz min^{-1}) or volume displaced in unit time (e.g. mm^3 s^{-1}).

2.5.1 Maximum injection rate

The maximum injection rate will be obtained when there is least resistance to flow and so the maximum figure is always obtained by shooting the material into the air, i.e. no mould is used in the measurement. In addition, the injection rate valve is set to maximum and the injection line pressure is also set to maximum.

The injection rate should always be capable of being adjusted, and adjusted finely as the injection rate influences many properties of the moulding. For example, its setting has a decisive influence on the surface appearance of the mouldings. In general, as part thickness increases, injection rate decreases. As small, thin-walled parts are commonly moulded on the smaller injection moulding machines, this means that these machines usually have the highest injection rates. Machines used to injection-mould disposable items (made from polypropylene or toughened polystyrene) have the fastest injection capacity at the present time, e.g. a 250 g flower pot can be moulded with a mould filling time of 0·1 s.

2.5.2 Velocity control

On many moulding machines the injection speed is obtained at the beginning of the moulding run by adjusting a valve—this adjusts the rate at which oil flows behind the injection piston. On many modern machines the injection rate is capable of being altered or programmed during an individual cycle so that parts with a good surface finish can be produced at the fastest possible rate.

To do this the hydraulic system must be capable of delivering the correct volume of oil at the right time and the system must be capable of developing sufficient pressure to maintain the required flow rate despite the load changes that occur during mould filling. The hydraulic system must be also capable of decelerating the screw forward movement and must not rely only on the resistance to flow of the material in that particular machine and mould. The systems that are used may be divided into two main types, these are (a) those systems capable of giving virtually any speed and (b) systems which given a restricted number of speeds. Examples of (a) include (i) servo-valve based units and, (ii) variable volume pump machines. Type (b) systems include (i) multipump combinations and (ii) machines based on digital hydraulics or cartridge valves.

At the present time servo-valve and cartridge valve systems are the most popular but it is probable that variable volume, pump machines will grow in popularity as machines equipped with such pumps use less energy than other types.

2.5.3 Pump and accumulator size

High injection rates can only be achieved by supplying oil to the injection ram (which in turn drives the screw forward) at the required rate. When a machine is being planned, the designer knows approximately the largest moulding, for a given screw diameter, that his machine will be capable of making. He then assumes that the screw will not travel more than four screw diameters during injection (as otherwise poor quality melt will result) and knowing this he can easily calculate the amount of oil that will have to be supplied to the screw/ram assembly. The filling time of such mouldings is (approximately) known by previous experience and so the amount of oil which must be supplied in unit time can be calculated. If an accumulator is not being used, then an appropriately sized, hydraulic pump must be installed. If an accumulator is being used (and they often are in order to get high speed injection capability from a relatively small pump) then it is important that the accumulator size is adequate. The accumulator should be capable of completing at least one injection stroke before it needs to be topped up or reconnected to the pump. When an accumulator runs out of fluid during the injection stroke, then not only does injection rate suffer (as the pump must share its output between the accumulator and the injection ram) but part quality is also affected.

2.6 Injection Pressure

Until comparatively recently, it was generally accepted that the maximum pressure which the screw or ram was capable of exerting on the melt should be approximately 3000 $MN\,m^{-2}$ (20 700 psi). Such pressures are necessary in order to maintain the specified rate of mould filling and/or in order to pack additional material into the mould so as to compensate for shrinkage. As the resistance to flow increases (caused by the use of higher viscosity polymers, filled polymer compositions, thinner wall thicknesses and longer flow paths), then higher pressure will be required. Some machines are supplied with injection units that can generate pressures of up to 5000 $MN\,m^{-2}$ ($\simeq$35 000 psi).

The injection pressure (which is directly related to injection line pressure) should be capable of being adjusted and in order to save excessive wear on the hydraulic system, it should be set to the lowest possible value consistent with the injection rate required. If the relief valve which controls the line pressure is set to an unnecessarily high value, then a lot of noise will be produced and the pump life will be considerably shortened. If injection pressure and injection rate are capable of being independently set, then there is no need to set the injection pressure at an unnecessarily high value.

2.7 Nozzle Contact Force

When pressure is applied to the plastics material, to force it from the barrel and into the mould, then some of that pressure works in such a way as to force, or to lift, the nozzle away from the sprue bush. To combat this separating force, a counter-force must be applied, e.g. by means of the small hydraulic rams which are used to move the injection carriage backwards and forwards (i.e. to achieve sprue break).

The nozzle contact force is usually expressed in tons, tonnes, or, kilonewtons. Ten kilonewtons are equivalent to 1 tonne and on an injection machine with a screw diameter of 80 mm the nozzle contact force would typically be 75 kN or 7·5 tonnes.

2.8 Sprue Break

This is often not mentioned on machine specifications but this does not mean to say that it is not important. Sprue break is measured in inches or millimetres and specifies how far the injection unit can be withdrawn, or moved away, from the mould. On many machines there are two sprue breaks—an operating sprue break and a maintenance sprue break. The operating sprue break is used during machine operation and is normally set to a minimum value so that time is not lost by making the injection carriage move more than absolutely necessary. The setting of this small sprue break may be over-ridden so that, for example, the injection carriage may be withdrawn further away from the mould so as to assist in nozzle cleaning or a heater band change. A large sprue break facility is always useful as it can ease maintenance and simplify the installation of extended nozzles.

2.9 Screw Rotation

On most injection moulding machines the screw rotational speed (the rpm) is capable of being varied so as to suit a particular material/machine combination. The specification usually indicates the speed range (e.g. from 0 to 200 rpm) and it should also indicate if a stepless speed range is possible. With many electric drive systems the number of speeds which can be selected is limited (e.g. to 5) whereas with hydraulic screw drive, any speed within the range specified may be obtained.

As it is the surface speed of the screw which is important it should not be forgotten that if the screw size is changed, then the speed range will change also. At the time of purchase, the torque, or driving power available should also be checked as the torque required will be different from one material to another. It is also worthwhile checking that the screw is adequately protected against excessively high torques. The hydraulic motor is inherently safe, whereas an electric motor will probably require a separate safety device.

2.10 Screw Decompression

After the screw has been forced back by the melt which it has produced, there will be some residual pressure remaining in the melt. By drawing the screw back a controlled distance, this residual pressure can be relieved and the need for complex shut-off nozzles may be eliminated. The ability to decompress the melt is extremely useful and is now considered by many moulders to be an essential feature.

2.11 The Heating System

The number of zones, into which the barrel has been divided for heating purposes, is usually given on machine specifications. As each zone has associated with it a temperature controller, then it is important to check, when comparing different machine specifications, that each machine has the same number of zones as temperature control systems can be expensive.

The temperature of the barrel is raised by means of heater bands and as different types are available, it is again worthwhile checking that one is comparing like with like. Ceramic heater bands will probably be the most expensive but with any heater band, the cost also depends on the heating capacity of that particular band.

The type of temperature controller which the manufacturer will supply, as part of the standard price, should also be determined as there are significant price differences between one type of temperature controller and another. On some machines it may be possible to check if the heater bands have failed, simply by pressing a button, whereas on another machine this useful facility is classed as 'additional equipment' and will therefore have to be paid for.

In the case of the 70 mm machine mentioned earlier (see Section 2.7) the number of temperature controllers on the barrel could very well be five and the installed heating capacity would be of the order of 30 kW. When checking the heating system, ensure that the nozzle is capable of being heated and controlled separately from the remainder of the barrel. A simple energy regulator is not good enough for this important part of the machine. The temperature control instrument specified for the nozzle should be at least as good as those supplied for controlling the barrel. As it is the tip of the nozzle which is important, the controlling thermocouple should be located in that part of the machine.

2.12 Material Considerations and Unit Design

No polymer is infinitely stable at the processing temperatures which are employed in injection moulding and changes in material properties are to be expected as a result of heating. As time as well as temperature plays a part in the thermal degradation process, it is important to ensure that no stagnation points exist within the injection unit assembly, for example, within the back-flow valve, the nozzle assembly and the feed throat. Thermal degradation is more immediate for some materials than others and if in doubt the advice of the material suppliers should always be sought so that, for example, acid-resistant screws and barrels may be specified if necessary. With heavily filled materials and/or glass-filled materials, the use of wear resistant screws and barrels (e.g. bimetallic) may also be necessary. Ideally, the screw should have a separate tip, which is capable of being easily removed, if material characteristics or wear dictate that a change is necessary.

The ring or sleeve type of shut-off valve is now the most common type in use, although it cannot be used for some materials, e.g. UPVC. When processing this material, a plain screw tip may be necessary and, provided that the screw will withstand acid attack, an exchange of screw tips will allow a range of materials to be processed on one screw. If, however, extensive use is going to be made of one material, then it is best to order a screw assembly which has been designed for that material.

The nozzle should again be ordered to suit the material as some valved or sealing nozzles are positively dangerous when used with certain materials. One type of nozzle is opened when the nozzle contacts the sprue bush; this action forces back a pin and a series of holes is thus exposed. When the injection unit moves away from the sprue bush, residual pressure in the barrel moves the pin forward again, thus blocking the holes. Unfortunately, if material degradation occurs within the barrel the pressure cannot be relieved and if the pressure becomes excessive, then serious damage and/or danger may result (e.g. with acetals).

On many injection moulding machines, the injection unit is not supported adequately and after a period of time, the cylinder assembly may drop so that the nozzle no longer lines up properly with the sprue bush. If this occurs, then incorrect seating or partial blockage of the sprue may occur and a material stagnation point will be created; injection unit support should therefore be checked before purchase.

2.13 Alternative Processes

Most machine manufacturers offer alternative screws and barrels so that by employing an appropriately designed unit, thermosetting plastics materials or rubber compounds may be moulded on the same basic machine. If it is thought that either of these materials will be run on the machine, then it is worthwhile asking the supplier what additional controls he would recommend. The cost of fitting such additional controls during the manufacturing stage may be very small compared to the cost of retro-fitting the same controls in the buyer's factory. For example, it may prove useful to have fitted a screw delay timer and a clamp delay timer. The first timer only allows the screw to rotate after a preset time has lapsed, this is useful as it means that the material is not 'cooking' at the front of the barrel whilst a thick-section part is curing in the mould. The second timer mentioned may be fitted to direct hydraulic clamp machines and permits the clamp pressure to be applied after the mould has been filled. By its use, and with special moulds, semipositive moulding is possible.

By fitting additional controls (e.g. a timer and a selector switch) the process of intrusion, or flow moulding, may be performed.

3 THE CLAMPING UNIT

An injection moulding machine is commonly classified according to the type of clamp that it possesses and also according to the force that the clamping system can apply to the mould. Thus a machine may be referred to as a '150-tonne, toggle machine'.

3.1 Type of Clamping System

It is common on moulding machines to use the system which closes the mould to generate the clamping force. For example, a hydraulically driven toggle mechanism can close the mould at high speed and then, as it is a lever system, it can generate considerable clamping forces. In general, toggle systems are commonly used when the clamping force is below 500 tonnes and direct hydraulic lock machines are used for the larger sizes. Small, high speed machines (e.g. as used for packaging applications) are nearly always toggle machines; a good, fail-safe lubrication system should be incorporated on a toggle lock machine. Ease of setting and precise force adjustment are two of the major advantages of the direct, hydraulic lock system.

3.2 Clamping Force

The clamping force is the maximum force that the clamping system can exert on the mould. In Europe, the units of force are the tonne or the kilonewton (1 kN = 0·1 tonne). Force must be applied to the mould in order to stop the mould opening during material injection. The amount of force necessary is dependent upon the projected area of the moulding and the injection pressure is of the order of 2 tonnes in^{-2} (2 tsi). As the range of materials available grows broader and the complexity of the mouldings increases, there has been a tendency for clamping pressures to become higher; this has been off-set by more precise control of the mould opening forces resulting, for example, from the use of cavity pressure transducers and associated control instruments.

If the machine is fitted with a toggle locking mechanism, then occasionally the term ‘closing force’ is used and with such a toggle lock machine, the closing force is slightly higher than the clamping force. This is because once the toggles have been fully extended the links of the toggle system move past their fully extended position and it is the subsequent slight relaxation which causes the force reduction.

3.3 Mould Opening Force

This is often not quoted on machine specifications but this does not mean to say that it is unimportant. With some mouldings, e.g. deep-drawn components, the mould opening force required may be considerable and the exact amount available should be ascertained.

3.4 The Clamp Stroke

The clamp stroke is also known as the opening stroke and it is given in either inches or millimetres. It is the maximum distance over which the moving platen can be made to move and although a large opening stroke is desirable (e.g. for the moulding of deep-drawn components), it must now be forgotten that this stroke must be capable of being easily reduced if a shallow component is being moulded.

3.5 Clamp Speeds

The mould opening and closing speeds should be capable of independent adjustment so that ideally any speed at any point within the stroke can be obtained. Such an ideal flexibility of setting is not often obtained because of practical difficulties. However, the final mould movements should be slow or cushioned so that mould or component damage is avoided.

3.6 Mould Thickness

This is given in either inches or millimetres and is commonly quoted as a range, e.g. from 250 to 600 mm. The first figure is the minimum mould thickness that can be accommodated, whilst still maintaining clamp pressure, and the second figure is the maximum mould thickness that can be mounted on that particular machine. Any thickness in between can, of course, be accommodated.

3.7 Maximum Daylight

This again is quoted in inches or millimetres and it is the furthest distance that the machine platens can be separated from each other. It can be obtained by adding the maximum mould thickness to the maximum opening stroke.

3.8 Tie Bar Clearance

The clearance between the tie bars (or tie rods) determines whether or not a mould (of a given length or width) can be accommodated in that particular machine. On some machines a tie bar can be removed and this facility can be useful if large or awkwardly shaped moulds are being used.

3.9 Tie Bar Construction

This is not normally specified in the manufacturer's literature, as usually the size and strength of the tie bars are adequate. If, however, a fast cycling machine is being specified then the tie bars will need to be very strong and resilient (e.g. ion-nitrided). Platen parallelicity will also have to be checked very carefully so as to ensure that uneven tie bar loading does not result.

3.10 The Platens

The height and width of the platens are normally specified but of equal importance is the thickness. It is important because if it is too thin, or if it has been weakened by large holes, then platen deflection may occur during the moulding operation. This in turn could subsequently result in mould and moulding distortion. A pattern of drilled and tapped holes is available for mould mounting and the size of these holes, and the pattern in which they are laid out, is normally according to a standard (e.g. Euromap document No. 2). Other arrangements are, of course, possible but these are normally agreed between the buyer and the manufacturer and specified when the machine is ordered. The size and position of ejector bars or points should also be considered in the general platen layout at a very early stage.

3.11 Ejection

A number of ejection systems are available and what precisely is being offered in the basic machine price should be determined as, for example, some manufacturers offer hydraulic ejection as standard and others do not. The maximum ejection stroke should also be determined as if, for example, a deep-drawn component is to be moulded, then a long ejection stroke would be necessary.

When moulding complex components, it is sometimes found that a relatively simple way of freeing the components from the tool is to use multistroke ejection—in effect this shakes the mouldings out of the mould. If inserts are being moulded, then after the components have been ejected the mould should move forward, so that the ejector pins can move back, flush with the cavity wall, thus giving clear access for insert loading into the mould. On some machines, this is standard, whereas on others, it is not.

4 THE HYDRAULIC SYSTEM

Because of the ease of control and relative simplicity of hydraulically actuated mechanisms, most injection moulding machines are powered hydraulically. Despite the large amount of energy which is wasted as a result, each machine usually has its own, selfcontained hydraulic system which is powered by an electric motor.

4.1 Pump Drive

A direct displacement hydraulic pump is driven by an electric motor and this combination produces fluid flow within the system; when there is resistance to flow, then pressure, up to a preset maximum, will be developed. More than one electric motor and more than one pump may be installed on the same machine and the machine manufacturer's specification should indicate this together with the pump drive capacity. This is usually rated in kilowatts and on the 70 mm machine mentioned earlier, the drive capacity may be of the order of 70 kW. Some machines are offered with a range of electric motor speeds so that the most appropriate fluid flow rate can be chosen to suit the proposed cycle. As matching the machine to the job can save on operating costs, it is obviously worthwhile checking the cost and availability of such a system.

4.2 Accumulation and Intensifiers

In order to obtain the required rate of hydraulic fluid flow, many machines are fitted with accumulators instead of additional pumps. It is worthwhile checking that the capacity of the accumulator is large enough to complete the injection stroke without making a demand on the pump, otherwise erratic machine performance may result.

On some machines the high pressure necessary to achieve, for example, mould clamping is obtained by the use of an intensifier. Such a device may increase the line pressure by a factor of three so that in some parts of the hydraulic circuit the pressure may reach 35 $MN\,m^{-2}$ ($\simeq$5000 psi). Such devices present a neat solution to the problem of getting high pressures at relatively low cost but the likelihood of leaks is unfortunately increased.

4.3 Line Pressure

The maximum line pressure that should not be exceeded on the machine is usually quoted in the machine specifications but unfortunately it may be quoted in a variety of different units. At the present time, the units which most probably will be encountered are pounds per square inch (psi), bars, kilopons per square centimetre ($kp\,cm^{-2}$) or kilogramme force per square centimetre ($kgf\,cm^{-2}$). With the introduction of SI units, another system has been introduced and so for convenience, the equivalents of these various units are given below.

In SI units 1 $MN\,m^{-2}$ is the same as 1 MPa; if kilonewtons per square metre are the units employed then 1 $MN\,m^{-2}$ = 1000 $kN\,m^{-2}$.

1 $MN\,m^{-2}$ is roughly equivalent to 10 $kgf\,cm^{-2}$ or, 10 $kp\,cm^{-2}$ or, 10 bars or 145 psi. That is, 1000 psi $\cong$ 7 $MN\,m^{-2}$ $\cong$ 70 bars.

Most machines seem to have a maximum line pressure of 14 $MN\,m^{-2}$ (which is 2000 psi or 140 bars) but on many machines this may be reduced if there is no need for a higher pressure as reducing the pressure decreases the likelihood of leaks developing and saves on operating costs.

4.4 Hydraulic Components

Normally, hydraulic components are readily available but it is always worthwhile checking that the component parts of the hydraulic system are obtainable and can be serviced in the country in which the machine is to be used. Cases have arisen recently where users have experienced considerable difficulties in servicing closed-loop machines. The difficulties were experienced because the service engineers had little or no experience of that particular type of machine (which was a new design and the components came from another country). One should never buy a prototype machine, or allow prototype modifications to be performed on an existing machine, as otherwise the machine user may be supplying training facilities for the manufacturer's engineers.

4.5 Hydraulic Control

The pressures and speeds which are capable of being controlled on the hydraulic system should always be checked so as to ensure that the required degree of control for a particular operation and/or operating cost reductions are possible.

In recent years, it has been appreciated that oil temperature regulation can be of vital importance in ensuring satisfactory machine performance (Whelan and Craft, 1979). A readily visible means of checking the oil temperature should therefore be incorporated and a suitable system for controlling the oil temperature should be very seriously considered. Ideally, the oil should be preheated (e.g. before the machine is set or goes into production) and once production has started, the oil temperature should be precisely controlled, e.g. to $\pm 1\,^{\circ}C$.

4.6 Contamination

It is vitally important that the hydraulic system is protected against contamination of any sort. Most oil reservoirs are located in the base of the moulding machine and before purchase a typical machine should be inspected so as to ensure that obvious contamination cannot occur. On some machines, the oil is readily visible and will therefore become contaminated by dust and water. It must not be forgotten that water leaks or spillage will occur on injection moulding machines and it is very annoying to have to change the oil because of a stupid design fault. The oil system should therefore be enclosed and air and oil filtration should be employed; a magnet should be located in the base of the oil reservoir to pick up tramp ferrous metal.

4.7 Oil Cooling

During machine operation, considerable quantities of heat are generated and the vast majority of this heat finds its way into the hydraulic system. It is normally removed from the oil by means of an oil cooler and the amount of water which this cooler will require should be known so that adequate provision may be made before the machine is installed. Considerable interest is now being expressed in re-using the waste heat from injection moulding machines and systems are now available, for example, which use this heat to warm warehouse space (see also Section 4.5).

5 CONTROL AND SETTING

5.1 Type of Control System

For many years, injection moulding machines had their operating sequences regulated by mechanical or electromechanical systems, e.g.

timers, relays, etc. Many such machines are still in use throughout the industry and some machines of this type are still made, as, in the opinion of the manufacturers and users, they still have a place in the scheme of things. This is because they are relatively simple and thus can be easily understood and repaired. Relay-type machines were superseded by solid-state machines in which the control circuits contained no moving parts. Because of this, the systems are inherently reliable and because of their small size, a large number of control functions or options can be made available in a relatively small space. Such systems have an incredible number of joints and connections and as is usual with electrical systems, it is the connections which cause the problems. For example, they may take a long time to make and they may be a source of trouble in service.

Microprocessor-based systems are now available and as these require fewer connections, they are being widely adopted by the machine manufacturers, although many users are not so sure that this is a good thing. Such systems have a lot to offer the user in terms of ease of setting, reliability and control options but unfortunately, some manufacturers have introduced systems which appear very complex and difficult to understand. If such a machine is being contemplated, then a thorough investigation is needed in order to ensure that it will fit into a particular factory situation. Never, repeat never, accept a prototype machine; only accept a tried and tested system. Purchasing a microprocessor-based machine will not automatically result in better mouldings at a faster rate—more consistent mouldings and longer production runs should result because of the inherent reliability of such systems.

5.2 Control Location

The controls may be located either on the machine or in a separate control cabinet. Both systems are in widespread use throughout the plastics industry as each has its own advantages and disadvantages. For example, if the controls are mounted on the machine this gives a compact arrangement at a relatively low cost. On the other hand, if the controls are located in a separate cabinet then the elements of the control system are protected against shock and vibration, the controls are more accessible and therefore easier to service and, last but by no means least, they may be easier to see and to set.

This last point is very important as unless controls can be read and set easily (when required and not at the whim of the operator) it is unlikely that they will be used properly. All too often one sees injection moulding machines which require odd-shaped machine setters. More than one may be required because the controls are located on different sides of the machine and their shape may need to be odd because of the awkward places in which the valves or indicators are located. All controls should be grouped together and ease of reading and design should be a paramount consideration.

5.3 Fault Finding

All control systems should have an integral fault location and display system so that in the event of a machine stoppage the cause of that stoppage can be determined. Until the advent of microprocessor-based machines such fault-finding circuits were relatively simple—for example, with some machines it was possible to check if the heater bands had failed by pressing a button. Now with some machines the number of faults that can be detected is large as microelectronic-based systems can handle large amounts of information.

For example, on the Demag 'D' range of machines two full, diagnostic-fault analysis programs are built into the machine logic as standard so that both setting faults and machine electronic faults can be identified. A total of 20 setting faults are identifiable and clearly displayed in an address window on the central control panel. Common examples of identified setting faults which cause considerable downtime are guard doors out of position, hydraulic ejectors not returned and the hopper being empty.

Electronic faults are identified by using a plug-in test card which is supplied with each machine. It operates using luminous diode test lamps for checking inputs and outputs of the microelectronic plug-in cards. Seven of these replaceable cards are carried in a subframe which is located in the main control cabinet—compared to a solid-state or relay machine the number of control elements has therefore been drastically reduced. One card is utilised as a timer subassembly with microprocessor, two identical pairs are used for rectification and amplification and one each for core pulling and hydraulic ejector controls. The control processing unit (CPU) contains the storage module with the EPROM bit processor responsible for the preprogrammed machine sequences. (EPROM stands for erasable programmable read-only memory.)

The test card is plugged-in between a spring contact strip and the plug-in card concerned; flashing on the respective diodes indicates that the corresponding input or output is serviceable. By a relatively simple procedure it is possible to identify whether there is an electronic fault and to gain information on the fault location. Such ease of fault finding should drastically reduce the time for which a machine is out of production.

5.4 Repeatability

As the object of injection moulding is to produce identical mouldings, it is essential that the control system should be capable of repeating on each shot exactly what was done on the preceding shots. This is why there is considerable interest in machine data-logging systems (if necessary, associated with computers) as such systems can assemble and store vast amounts of information. Such information may be interpreted to yield production statistics or it can be used to reset a machine if ever that particular job is run again. With some computer directed machines, the information is stored on a card (which is approximately the size of a banker's card) and when the job is rerun the stored information on the setting card is used to set the machine. This may be done by inserting the card into a slot so that the stored information on the card may be 'read' and used to set temperatures, speeds, etc.

The control system which makes this possible is based on electronic components called large-scale integrated modules. Such modules are manufactured in large numbers and are therefore relatively cheap. They are used to build-up control systems for a variety of industries (e.g. automotive and home appliance) as they can be programmed to perform any series of events or control actions. As such modules contain many thousands of transistors the number of modules required is small and this saves on wiring and installation.

5.5 The No-load Cycle

The duration of the no-load cycle is measured in seconds and is sometimes referred to as the dry running time. During its measurement, the injection moulding machine is operated without any moulding compound and the time needed to complete a cycle of the following operation is determined. The no-load cycle comprises mould closing time, die or nozzle approach time, die or nozzle return time, mould opening time and switching time. The die approach and return times together make up the time necessary for carriage movements and the switching time includes all idle time which occurs during the no-load cycles.

6 OPTIONAL EXTRAS

The usefulness of a given type of injection moulding machine can be extended by equipping the machine with alternative or additional items. In the machine specification, the manufacturers should clearly state what is included in the price paid for the basic or standard machine and what optional items of equipment are available. If a particular feature or piece of equipment is not listed in the machine specification, then the manufacturer may be reluctant to supply it as the purchase and storage of the 'nonstandard' items complicates his spares and storage procedure.

Some of the items of equipment mentioned below may be supplied as standard by a particular machine manufacturer; the list is not exhaustive and is given for guidance only.

6.1 Oil Temperature Regulation and Control

The need for this has already been explained. With some systems, alarms may be actuated if the oil temperature falls outside prescribed limits and/or the machine may be shut down if the temperature varies by more than a preset amount.

6.2 Operating Hours Counter

Such a device may count the number of hours over which the electric motor has been running. It is useful for working out production statistics and also for maintenance purposes. A precise maintenance schedule, with details of what is to be done after specified times have lapsed, should be drawn up for each machine.

6.3 Alternative Ejection Systems

On most injection moulding machines, the parts are ejected from the mould by the action of knock-out pins which are actuated by the mould opening movement. Such a simple, cheap, reliable system may be adequate for most jobs but there are many cases where it is inadequate and alternative (or additional) ejection systems should be specified. Hydraulic ejectors are now available on most moulding machines and after checking whether they are included in the basic price or not, one should check what controls are supplied as standard. Ideally, it should be possible to regulate the speed on both the forward and the return stroke independently of the ejection force. Safety interlocks should be fitted, as with core-pulling devices, so that the machine cannot close and clamp with any item or part projecting.

With some materials and/or some moulds, an air ejection device is useful; this type of ejector introduces a blast of air for a preset period of time between the component and the mould. Such a device is extremely useful and so when injection moulding machines are installed, provision should always be made for a compressed air supply to each moulding machine.

6.4 Core-pulling Systems

Holes or slots are commonly introduced into injection mouldings by movable mould pieces or cores. Core-pulling facilities may be available as a standard feature on the machine but if they are not, it is worthwhile checking to see if they can be simply installed at a later date if required.

6.5 Additional Controls or Features

The number and type of additional machine controls that may be fitted to a standard machine should be indicated in the machine specification. Such additional controls could be:

1. Temperature controls. Additional temperature controllers may be needed if, for example, hot-runner tools are to be used.
2. Speed controls. Stepped injection speed or stepped screw rotational speed may be available.
3. Pressure controls. The injection pressure profile and/or the back pressure profile may be supplied in various alternative arrangements.
4. Timers. Additional timers may be fitted so that, for example, screw delay is possible.
5. Mains outlets. It is always useful to have a source of mains electrical power on any injection moulding machine and for this reason, the buyer may care to specify some mains outlet sockets, located at strategic points, around the machine. Such power-points are useful for vacuum cleaners, power-tools and ancillary equipment. Cleaning the machine hopper with a vacuum cleaner is far better than using compressed air as an air blast can cause extensive contamination.

7 SAFETY

Injection moulding machines can be extremely dangerous unless they are properly protected, operated and supervised. Machine guarding is just one way of making the machines safer and it should not be relied upon to the extent that inspection, maintenance and training are neglected.

7.1 Reasons for Concern

1. Hot equipment. Some injection moulding machines now operate with barrel temperatures of 400 °C and as any temperature above 60 °C is

unpleasant to the touch, then burning or burns, must be recognised as a common cause of accidents in the moulding industry.

2. Hot material. Polymers are organic materials which have high specific heats and relatively low thermal stabilities. At their processing temperatures, the materials may be easily degraded to give unpleasant, irritating odours from melts which contain a lot of heat. If some polymer melts come into contact with the skin, then they are difficult to remove as the material functions as a hot melt adhesive.

3. Moving equipment. Machine parts can be very heavy and may move with considerable speed and in a contest with a fast moving machine component, a human limb has little or no chance. Finger-trapping, and consequent bruising, is the simplest type of accident that can occur in such a circumstance but limb-loss or amputation is also possible. On some badly designed, or badly installed machines, the safety guards themselves function as finger-traps.

4. Electrically operated equipment. Most injection moulding machines are electrically operated and such machines draw their power from the mains supply; in the UK this may be 440 or 240 V. Such voltages can be very dangerous as they may not only cause, or give, electrical shocks, but they may be responsible for starting a fire in certain circumstances.

5. Hydraulically operated equipment. Most injection moulding machines use oil as the hydraulic medium and at certain times in the machine cycle this hot oil is under considerable pressure. In the event of a leak developing, accidents may be caused through people stepping on the slippery surfaces which result, or a fire may start if a spray or atomised oil is ignited by a hot surface or a spark. The fumes which are generated by the hydraulic medium should also not be forgotten as, for example, if excess fumes are generated through overheating, headaches and/or dizziness may result.

In common with other types of mechanical equipment, injection moulding machines are subject to rules or restrictions which seek to regulate their use. A machine purchaser must ensure that the equipment which he is buying conforms to local or national practice before he agrees to purchase the machine. Such legislation should be considered as the bare minimum necessary and a common sense responsible attitude towards safety should be encouraged. Apart from any personal considerations, any accident, no matter how small, is bad for business.

7.2 BPF Code of Practice (1978)

The initials BPF stand for the British Plastics Federation and in 1979 a code of practice issued by that body became operational in the UK. The code has been compiled to help designers, manufacturers, importers and suppliers conform with Section 6 of the UK Health and Safety at Work Act 1974 (HASAWA) and for employers to conform with Section 2 of that act. Also built into the aims of the code is compliance with Section 14 of the 1961 Factories Act as this act has not been fully superseded by the HASAWA.

The code was written with the aid and encouragement of the Factory Inspectorate but it is not an 'approved' code of practice as defined in HASAWA. However, the code will be used by HM Factory Inspectorate as a basis for that body's enforcement procedures. The UK moulding industry should therefore take this code of practice very seriously and as this code may form the basis for a European code of practice (because of the high standards that it stipulates) then it should be noted and acted upon by a much wider audience.

The code applies only to machines for moulding thermoplastics and thermosets and whose injection and clamping systems are in a horizontal plane; it relates only to safety devices and does not cover the area of industrial hygiene as this is the responsibility of injection moulding machine users. Although rubbers are excluded from the code, any moulder who handles rubbery materials on similar types of equipment would be well advised to take note of the recommendations laid down in this code.

Copies of the code may be purchased from the British Plastics Federation, 5 Belgrave Square, London SW1X 8PH; the publication number is 238/1. The code of practice is divided into three parts which are: Part I, general; Part II, machine use; and Part III, construction.

7.2.1 General

It is recommended that machines delivered before 1 July, 1979 should have, if necessary, their safety systems up-rated (e.g. by the provision of interlocking systems, interlocking guards, and fixed guards in high risk areas). For machines delivered during the period 1 July 1979 to 1 July 1982 all the measures specified in the code of practice should be applied apart from a concession on mechanical restraints in the mould closing area. New machines delivered after 1 July 1982 shall be provided with all the measures specified in the code of practice (see Section 7.2.3.3—this requirement may be dropped as it is very difficult to engineer).

7.2.1.1 Urgent action. For machines already in service and where the following safety arrangements have not been provided, urgent action shall be taken to install such safety measures. The first of these safety measures is a positively operated, hydraulic/pneumatic valve arrangement installed so that when the operator's front guard is part, or fully, open the supply of fluid to the platen closing cylinder is interrupted.

Second, the front guard should actuate two electrical limit switches, working in opposite modes, so that the platens are prevented from closing when the guard is other than fully closed.

Third, a positively operated, electrical limit switch must be installed so that it is actuated by the operator's rear guard—actuation should be such as to prevent the platens from closing when the guard is in any position other than fully closed.

Fourth, a guard must be provided and located beneath the platens so as to prevent access to the trapping region formed by the machine platens.

Fifth, the toggle area shall be guarded so as to prevent access to the platen closing mechanism.

7.2.1.2 Hazards. Examples of how a machine and/or its associated material may cause injury is illustrated in the code. Twelve types of hazard are listed and these include projecting parts, clamping mechanisms, core pull equipment, the screw, heater bands and traps or nips. Such traps or nips are created between fixed and moving parts, e.g. between the screw and the feed-throat, the nozzle and the sprue bush, the operator's front guard and the fixed platen, etc.

Hazards may be colour coded in accordance with British Standard 5378 (1980) and if labels are used they should be designed to conform to this standard.

7.2.1.3 Guards. Guards shall be provided to prevent access to the danger areas which exist on any injection moulding machine and such guards must be of sound construction and able to withstand the rigors of service. By definition, there are three types of guard and these are known as: (a) fixed guard; (b) interlocking guard (normal risk); and (c) interlocking guard (high risk).

A fixed guard is one which encloses trapping areas of the machine and such a guard should be secured to the machine during its assembly or installation by means of hexagonal headed bolts or socket-head cap screws, both of which should be captive. Such guards are used to cover projecting parts, traps or nips, the heated barrel, etc.

An interlocking guard is one which is sliding or hinged and which is interlocked electrically and/or pneumatically and/or hydraulically and/or power interlocked with the machine controls so that the machine will not operate unless the guard is fully closed. A normal risk interlocking guard is interlocked electrically with the machine controls so that the machine will not operate unless the guard is fully closed. This type of guard is one which is moved occasionally, e.g. in the nozzle area. An interlocking guard which covers a high risk area must have more than one type of interlock so that if, for example, the first or primary interlock fails, then protection is still given by the secondary or back-up interlock system. The primary interlock must be a hydraulic system; electrical interlocks can only be considered as secondary, back-up systems.

As far as is possible, machines should be designed and operated so that an inherently safe machine results—this is a far better approach than creating an unsafe situation and then relying on expensive, cumbersome guarding to solve the safety problem.

7.2.2 Machine use

All persons who set or use the moulding machine should appreciate what they are doing and should be protected by periodic safety checks.

7.2.2.1 Setting. The presence of the guard during loading, setting and/or try-out operations may cause inconvenience and result in a wastage of time. However, such considerations do not absolve the machine user from legal obligations which have been imposed by industrial legislation (e.g. the 1961 Factories Act in the UK).

Any person who performs a setting operation must understand what he is doing and he should appreciate machine functions and any particular hazards introduced by the material being moulded. During the setting operation, no safety devices should be overridden and the setter must be familiar with the manufacturer's instructions for setting and operating a particular machine.

7.2.2.2 Operational inspection procedures. In conjunction with the machine manufacturer's manual, the user must devise and conform with periodic safety inspections which shall be carried out by a person, or persons, who have been fully trained to the appropriate level. The code indicates what should be inspected and how often this should be done. Items such as guards, guard movements, emergency stops, etc. should be inspected daily, monthly and annually; due records of all inspections are to be kept.

If the machine is an immediate cause of danger, or if it may become a cause of danger, then the manager must be notified and the machine must not be used until the defect has been remedied. The action taken to correct the remedy should also be recorded.

7.2.3 Machine construction

This section is particularly important to the machine designer and to the machine manufacturer.

7.2.3.1 Principles of interlocking. The type of guarding necessary for a particular situation is decided after considering:

1. Accessibility. It must be decided if access is required during normal machine operation or if access is required for maintenance purposes only.
2. Frequency of access. It must also be decided how often access into the hazard area is required during normal operation.
3. The risk involved. The foreseeable risk and the severity of the injury which may result if the interlocking system should fail must also be considered.

An interlocking guard should prevent all dangerous movement of parts of the machine occurring within the guard area whilst the guard is open. All interlocks must operate before the guards can be opened and access gained to a dangerous area. The amount of interlocking required may be reduced if two or more guards are mechanically interlinked.

Five types of acceptable, hydraulic interlock are given in the code.

7.2.3.2 Actuation of valves and limit switches. The design of the valve mounting, the limit switch mountings, the cams and the method of actuation is as important as the interlocking system itself. The fittings used to secure a valve, limit switch or cam, must be designed to withstand arduous operating conditions. If this is not done, then the interlocking system may be unable to continue providing the protection required.

7.2.3.3. Mechanical restraints. To supplement the hydraulic and electrical interlocks on a moulding machine, a mechanical restraint must be provided on newer machines, i.e. those purchased after 1 July 1979. Such a mechanical restraint is also known as a scotch and two types are listed. These are:

1. An adjustable length scotch. The terminology is rather confusing as once the scotch has been adjusted to suit a particular machine, then it is sometimes referred to as a 'fixed length scotch'. Such a device should be interlocked with the operator's guard so that the guard cannot be opened until the moving platen has regained its rest position and the mechanical scotch or bar has dropped between the fixed and moving platen.
2. A progressive scotch. Such a device typically works on the ratchet principle so that it is capable of stopping the moving platen at any point in its stroke if the guard is moved from its closed position.

Machine manufacturers prefer the fixed length scotch because it is much simpler and does not have to resist large forces. The progressive scotch is much more difficult to engineer, particularly on large machines, as it may have to support the full closing power of the clamping system. According to the code new machines delivered after 1 July 1982 must be equipped with progressive mechanical restraint but because of engineering difficulties it is probable that this requirement will be amended or dropped.

7.2.3.4 Electrical apparatus. Good electrical practice must be followed during machine design and installation. For example, all installations shall be designed so as to ensure that components will fail safe. Power failure must not induce a dangerous condition in any part of the machine. The maximum power control circuit voltage shall be 110 V and shall be supplied by an isolated transformer suitably earthed. Emergency stop buttons must be provided in suitable places and such stop buttons should not restart the cycle when released or reset but they should only free the cycle-start button for operation.

7.3 Robots and Safety

Robots are becoming more widely used in injection moulding as they can, by taking over part or all of the jobs done by operatives, eliminate some of the hazards associated with injection moulding. They do this by, for example, handling hot mouldings and by removing large, heavy mouldings from the mould. This last point is very important as there is a growing feeling that operatives should never enter the space between the mould platens as a matter of course.

Robots (a pick-and-place unit is a comparatively simple robot) are particularly suited to mouldings which are difficult to handle because they are awkwardly shaped or because the surface can easily be spoiled by manual handling. Once programmed, such devices will repeat their schedules perfectly unless mechanical or electrical breakdown occurs. Because of the consistency of operation that such devices can introduce, shot to shot repeatability is often improved and as a result, productivity may increase. Robotic devices will become cheaper and more versatile due to the inevitable influence of the microprocessor. As a result of the operating flexibility which the microprocessor will introduce, the robot may, for example, be programmed to serve more than one machine or to sequence the moulding through a series of finishing operations.

7.4 Machine Installation and Safety

Factory lay-out and the provision of services, are very important considerations as they affect operating efficiency and safety. A common tendency is for the available space to become overcrowded either because the new machine is larger than the old one or because adequate space was not provided for ancillary equipment.

In siting machines of up to about 500 tonnes lock it is best to allow up to 2 m between them, but if space is limited it is possible to work in a clear space of 1·5 m. The best plan is to find from the machinery specification the actual machine dimensions and then to leave sufficient space between them to take in the gantry or fork-lift truck required for mould mounting (Bown, 1979).

8 MACHINE PURCHASE

Although the literature will provide the answer to basic questions, it must still be regarded as sales literature and should be treated accordingly.

8.1 Persons Responsible

Many factors influence the choice or selection of injection moulding machines and each of these factors will assume a different importance or value to each prospective buyer. It is therefore important to decide which person, or collection of persons, will make the final decision as to which machine should be bought, as personal preference will be of obvious importance.

8.2 Replacement of Existing Equipment

If the new machine is to be a direct substitute or replacement for existing equipment, and if the old machine has performed satisfactorily, then the choice of the replacement machine will be heavily influenced by standardisation considerations. If the machine is being purchased for a new type of job, then the decision is more difficult as the choice is then between a standard, or known machine, and a non-standard machine. If it is decided to look at alternative machines, then the whole process of machine selection will have to be performed. It should also be borne in mind that by introducing a different system, a number of other problems will be introduced, e.g. spares, maintenance, training, etc.

8.3 Specification Matching

Company policy may dictate that a certain specification has to be met and in such a case, that specification should be studied closely so that large amounts of time and effort are not wasted in choosing a system which would be unacceptable to the company. This does not mean to say that unacceptable specifications and machines should not be examined at all as useful ideas may be gleaned from such examinations.

In many cases, the equipment is being purchased for a new job and in such a case, an in-house specification should be drawn up so that it in turn can be compared with the machine manufacturer's literature. Such an in-house document should also make clear the reasons for purchase and the date from which the machine should be fully operational. A list of questions which the machine supplier will be expected to answer should also be drawn up at this stage.

8.4 Moulding Trials

Before purchase the proposed machine should be seen working; if it is a new machine then it is usually shown at a showroom and the customer should take full advantage of any offer to use his own moulds and materials of his choice.

8.4.1 Advice and consultation

Machine specifications are usually only consulted when it is proposed to purchase a machine and in this event, information from other sources should also be obtained.

For example, the opinion of people who you respect (these could be other moulders or representatives of material suppliers) should be obtained and, if after considering these and the machine specification, you are still impressed, then the machine should be tried out by you using, if possible, moulds of your choice. At such a moulding trial it is always advisable to have a colleague whose opinion you value with you as, in such a situation, two heads are often better than one. His presence will certainly help to spread the load so that the salesman cannot concentrate on one individual exclusively and, of course, he will want answers to his own questions. As far as possible, all questions and answers should be noted and a copy of the notes should be given to those present at the moulding trial so as to try and eliminate subsequent arguments. Safety provisions should also be checked at the moulding trials.

8.4.2 The moulding materials

The range of polymers which is processed by injection moulding is extremely wide and as new types of grades of material come on to the market, this range is gradually being broadened. Some materials are extremely easy to process as they require no blending or drying and can be handled on relatively simple machines using low temperatures and pressures. Other polymers may require, for example, extensive predrying before being moulded using high temperatures and pressures. The processing characteristics of the materials should therefore be known so that a machine appropriate to its task may be chosen (see also Section 2.12 and Chapter 7).

8.4.3 Moulds

The dimensions, service requirements and operational conditions should be readily available if existing moulds are to be run on the new machine. Such information may be used to decide if an existing mould can be bolted on to the platens of the machine and whether the minimum and maximum mould opening strokes available on the machine are suitable for a particular mould.

Machine platens are normally drilled and tapped with holes whose location has been specified by a national or international organisation. For example, in Europe, the Euromap document No. 2 (1969) is commonly used. If new tooling is being considered as well as a new machine, then the mould mounting points can be laid out according to the same standard and for this purpose the toolmaker should be supplied with a platen drawing. The machine manufacturer will supply this and can often arrange for the tooling to be made; some buyers prefer this as in the event of anything going wrong, then all complaints go to the same company.

If the same mould can be fitted on to different machines then it shall be possible to determine which machine is capable of moulding most accurately by using the procedure laid down in Chapter 6, Section 6.4.

8.5 Service and Maintenance

Wear and breakdown must be anticipated when a machine is purchased and for this reason, consideration of spares and service is of vital importance.

8.5.1 Spares availability

The price and normal delivery times of spares should be checked with the responsible organisation, i.e. the manufacturer or his agent. The cost of replacing or refurbishing items such as cylinders, screws, screw tips and back-flow valves should be checked and if the proposed machine is an imported model, then the 'normal' delivery time (for spares) from the other country should be ascertained. Although the range of spares carried by the agent may be large, Murphy's law will dictate that at some time he will not have a particular item in stock. Unforeseen delays when dealing with foreign equipment is a big disadvantage and one which should not be forgotten.

8.5.2 Service engineers

If a new machine is being purchased, then the cost of commissioning that machine should be determined as the rates charged for service engineers can be very high. The rates charged for routine or breakdown service should also be investigated together with the size and location of the service organisation. If the nearest engineer is located a long distance away, then the cost of service calls will be large because of the time and expense spent in travelling.

Such factors must be set against what is known of the skill and reliability of the service engineers: such knowledge is difficult to obtain, particularly if it is a foreign company with new or different agents whose engineers are unfamiliar with the product. Cases are known of service engineers who turned up to service a machine and who then spent large amounts of time, for which the customer was being charged, reading the machine manual initially supplied to the customer.

8.5.3 Training courses

Many companies prefer to use their own engineers for repair and/or maintenance and for this reason moulding company personnel may be sent on training courses which are offered by the machine supplier. The cost, availability and usefulness (this is sometimes difficult to determine) of such courses should be determined. Machine complexity and service requirements should be weighed against the quality of the moulding shop personnel available.

8.5.4 Maintenance

Inspection and maintenance of injection moulding machines must be carried out on a continuous basis and according to a preconceived plan. Every machine must be maintained if production targets are to be achieved and the advice of the machine manufacturer should be obtained before the machine is purchased.

Ideally, the experience of other moulders who have the same type of machine in production would be available but even if the moulders were prepared to divulge their 'secrets', it is unlikely that a great deal of value would be learnt. This is because few moulders keep accurate records of machine reliability and availability. In the UK it is thought that in many moulding shops, down-time averages approximately 40% of the total available production time. If machine availability could be increased from 60 to 70% of the total available production time, then this 10% extra machine time could double the pretax profits in certain cases (Walker, 1980). To do this, it is important to identify those machines which are not available for production and the reasons for their unavailability (Walker, 1980).

8.6 Delivery and Price

In Europe there are approximately 80 different firms making general purpose, injection moulding machines and details of what they produce may be obtained on request. Periodically, trade journals or newspapers carry surveys which give details, including addresses, of the number and types of machines available. For example, the trade newspaper *Plastics and Rubber Weekly* has published a European survey of injection moulding machines (Anon., 1979, 1981). What this means is that the choice of machine available to a prospective purchaser is large and as machine purchases are normally made by 'engineering-types' large piles of technical specifications are obtained and happily waded through. A short-list of machines is then drawn up and inspection visits are made to appropriate factories or exhibitions. The decision as to which machine to buy is then made on technical merit.

However, there are two more factors which are of overriding importance and these two factors are delivery and price.

For various reasons, the many purchasers of injection moulding machines do not anticipate their needs and find themselves in the unenviable position of virtually taking any machine that can be delivered immediately. A moulder who is in such a position will be fortunate indeed if he gets the right machine at the right price. Adequate production can only be maintained if machine purchase and installation is meticulously planned.

It must not be forgotten that the object of injection moulding is to produce acceptable mouldings at the right price. The cost of the total machine package (which includes machine price, commissioning cost, delivery charges, etc.) together with the terms offered by the supplier are very important considerations and when comparing machines the advice of a cost accountant should be sought. Unless the cost of owning different types of machine have been quantified over the expected life of that machine, it is difficult to imagine how a sound decision can be reached. To make a success of injection moulding, it is necessary to install the right equipment at the right time but it is also just as important to know when equipment should be replaced.

REFERENCES

Anon. Injection moulding machines, *Plastics and Rubber Weekly*, p. 17, 9 March 1979.

Anon. Injection moulding, *Plastics and Rubber Weekly*, p. 7, 7 March 1981.

Bown, J. *Injection Moulding of Plastics Components*, McGraw-Hill Book Co., New York, 1979.

BPF Code of Practice, The British Plastics Federation, London, Publication No. 238/1, June 1978.

British Standard 5378. Safety signs and colours, 1980.

Euromap document No. 1. *Description of Injection Moulding Machines*, Available from The British Plastics Federation, London, Jan 1969.

Euromap document No. 2. *Mould Fixing Dimension for Injection Moulding Machines*, Available from The British Plastics Federation, London, July 1969.

Euromap document No. 5. *Procedures for the Determination of Important Production Data for an Injection Moulding Machine*, Available from The British Plastics Federation, London, 1972.

Ireland, R. A. and Smith, J. B. *Plastics and Rubber International*, p. 225, 2(5); 1977.

SPI. New standard for plasticating performance of injection moulding machines, *Modern Plastics*, March 1968.

Walker, A. Life cycle costing—complex but for your own good, *British Plastics and Rubber*, p. 18, May 1980.

Whelan, A. and Craft, J. L. Improving machine performance. Paper presented at a PRI conference in Athlone, Eire, Nov 1979.

CHAPTER 4

Component Finishing and Decoration

Many mouldings are ejected from the mould ready to use; others may require extensive finishing and/or decoration before they can be put into service.

1 FLASH REMOVAL AND BUFFING

On many injection mouldings the mouldings are automatically separated from the feed system. There are, however, a very large number of mouldings produced which must be cleaned of flash and/or separated from their feed system in a postmoulding operation. Rubber mouldings commonly require a great deal of attention simply because of the nature of the material—it is soft and flexible and so automatic ejection is difficult. The ease with which a rubber compound can penetrate very small gaps makes flashless moulding difficult in many cases. Some rubber compounds have been known to penetrate gaps which are less than 0·025 mm. After flash removal the appearance of the mouldings may be improved by a buffing operation.

1.1 Hand Trimming

The removal of excess material by a hand trimming operation is still common practice in cases where, for example, the mouldings have a complex configuration or an awkward shape. A rubber facemask intended for sub-aqua use could be a case in point. Excess material could be removed by an operator using scissors and a sharp knife or scalpel.

Flash removal may also be necessary on thermoplastics mouldings and in such a case a sharp knife or scalpel is usually sufficient. Some companies have used a flame for this purpose as it is capable of giving smooth, shiny edges. However, the safety problems are considerable because many thermoplastics burn easily producing considerable heat, fumes and smoke.

Deflashing with a hot, inert gas is better from a safety point of view. Because of the brittle nature of thermosetting materials, flash may be most easily removed using a file or linisher; sprues or runners could be removed by a variety of techniques, e.g. side-cutters, a junior hack-saw or a small electric saw. The scar may be cleaned by sanding, buffing or polishing.

Similar tools and equipment are used to remove the feed system from many thermoplastics mouldings. Where, for example, the operator is required to work in conjunction with a moulding machine (i.e. the machine is working semi-automatically) then the operator may be employed as a finisher. The finisher or operator may have at his disposal power tools which enable flash removal to be performed more quickly or efficiently. For example, the flash may be removed from rubber mouldings (such as mats) by the action of two rollers which are driven together so that the rubber is sheared between them. When rubber mouldings have been produced by compression–injection moulding the resultant mouldings are connected together by a thin film of material which in the case of items such as O-rings may need to be removed from both the inside and the outside of the moulding. In such a case this may be done by a punching machine as the punch and die-sets used are capable of giving very accurate trimming. In such a case the operator would load and locate the sheet of components into a fixture or jig and then pack the assembly into the machine.

When inserts have been incorporated during the moulding stage then trimming or cleaning of the moulded component may be required because, for example, thread fouling has occurred. In such a case the excess material can be removed from the metal by a wire brush which may be electrically driven. Buffing with wire brushes is used on mouldings which have to be bonded to other components as the surface roughening so produced aids adhesion. Textile-wheel buffing is also used to remove thin flash and to improve the surface finish of polymer mouldings. For example, if mouldings have been sanded (as part of a machining operation (see Section 1.3)) then the surface may be brought to a high polish by the use of buffing equipment. The recommendations in Section 1.3 are intended for use with nylon mouldings (e.g. Du Pont's Zytel®) but they provide a useful guide for buffing plastics in general.

1.2 Deflashing by Barrelling

If flashed plastics mouldings (made from brittle materials) are placed inside an octagonal barrel, together with some wooden pegs, then it will be found that as the barrel is rotated the thin, brittle flash will be broken or abraded away. If the sides of the barrel are perforated then the flash will drop through so that only mouldings will be removed at the end of the deflashing operations. After the flash has been removed from, say, aminoplastics mouldings, the semifinished mouldings may be loaded into an enclosed chamber where they are slowly rotated whilst being bombarded by a stream of crushed nut shells; this polishes the rough edges. Such a technique cannot be used to deflash mouldings made from soft, flexible mouldings unless the temperature is reduced.

This can be done by loading the mouldings into a barrel which has solid, insulated walls, together with dry ice (which is solid carbon dioxide). Before use the drum or barrel should be precooled by allowing it to rotate when loaded with dry ice alone. After this precooling period the drum should be filled to 60 % of its total capacity with rubber mouldings and dry ice in the ratio of two parts mouldings to one part dry ice by weight. The drum is then rotated slowly (e.g. at 30 rpm) and after a specified time (which may range from 15 to 30 min) the components are removed. Flash thicknesses of up to 0·33 mm (0·013 in) may be removed in this way.

For many years the most popular refrigerant for sub-zero deflashing was dry ice but in recent years producers of liquid nitrogen have entered the market offering a package deal which includes tumbling barrel, control equipment and the hire of a tank in which to store the liquid nitrogen (Anon., 1972). Such liquid nitrogen systems can reduce the temperature of the barrel (and its contents) to very low temperatures very quickly and so a liquid nitrogen system can deflash most rubber components including those which have been designed to be flexible at very low temperatures. However, weight for weight the cooling capacity of dry ice is 67 % greater than that of liquid nitrogen (see Table 1) and so in the majority of cases, a dry ice system would probably be cheaper.

TABLE 1

A COMPARISON OF THE COOLING CAPACITIES OF DRY ICE AND LIQUID NITROGEN SYSTEMS (after Anon., 1972)

Item	*Dry ice*	*Liquid nitrogen*
Temperature at atmospheric pressure (°F)	−110	−320
Latent heat required to vaporise 1 lb of product (Btu/lb)	247	86
Heat required to warm gas from vaporising temperature to −40 °F (Btu/lb)	14	70
Total cooling capacity (Btu/lb)	261	156
Ratio	1·67	1·00

1.3 Buffing and Polishing

The buffing operation is normally considered as three steps: ashing, polishing, and wiping (Du Pont, 1979). Ashing is done with a ventilated wheel of open construction, made up of alternating layers of 450 and 200 mm (18 and 8 in) diameter muslin discs. In this way, an ashing wheel of some 100 to 125 mm (4 to 5 in) in width may be built up. The ashing wheel is kept dressed with a slurry of pumice and water during the buffing operation. The nylon part is held lightly against the wheel and kept in constant motion to prevent burning or uneven ashing. Speed of the wheel should be approximately 1000–1200 rpm for wheels of 300–400 mm (12–16 in) diameter. It is essential that the wheel be operated slowly enough to retain a charge of the slurry.

The polishing operation is performed in a similar manner and on a similarly constructed buffing wheel, the difference being that the wheel is operated dry and a polishing compound is applied to one-half of the surface of the wheel, the other half remaining untreated. The part is first held against the treated half of the wheel for polishing and then moved to the untreated side to wipe off the polishing compound. Optimum speeds for the polishing wheel range from 1000 to 1500 rpm for a 400 mm (16 in) diameter wheel.

2 MACHINING OF PLASTICS

It is sometimes advantageous to eliminate some of the more difficult details on a mould, so as to save tooling costs and production time, and to remove material by machining, i.e. a postmoulding operation may be more convenient or cheaper. Mouldings are machined for other reasons; for example it may be cheaper to produce a range of components by moulding a common shape and then finishing into the shape required by machining. Such a procedure would enable the tooling costs to be shared or distributed over the range of components and may result in injection moulding becoming economically attractive. In many cases the attraction of being able to use existing moulds is a tremendous incentive to use machining as the job may be finished more quickly and at a low cost.

2.1 Range of Material Properties

Within the family of materials called polymers, one can encounter a very wide range of properties as not only do the materials themselves differ but they can also be combined with other additives. The hardness or stiffness of polymer mouldings can therefore vary from very soft to hard but even a hard, stiff plastics moulding will be 'soft' when compared against a traditional, engineering metal. In general the modulus of elasticity of plastics (which is a measure of their rigidity) is very much less than those of metals. The thermal properties of plastics materials are also very different to those of metals; in general plastics materials, compared to metals, have very low thermal conductivities and high coefficients of thermal expansion. These differences in properties must be taken into consideration when machining plastics as compared to machining metals.

2.2 Deformation Recovery

Because of their low modulus of elasticity polymer mouldings can be easily deformed during clamping and machining. When clamping plastics components ready for machining, it is important to ensure that the clamping forces are not excessive and that the work piece is well supported. Excessive force also should not be applied during the cutting or machining operation (e.g. caused by the use of dull or blunt tools) as this again could cause a change of shape or deformation. Furthermore, when resilient materials are being machined elastic memory may become a problem, particularly if dull or blunt tools are being used.

As the cutting tool is pushed into the moulding, the polymer may find it easier to move out of the way. Because the elastic limit has not been reached, the material will attempt to recover and this means that a combination of the following may be observed:

1. Excessive heat build-up. Rapid heat build-up may occur as a result of the increased friction that elastic memory can cause between the tool and the work piece.
2. The finish hole is under-sized. This is commonly found when plastics materials are drilled or reamed and it is therefore common practice when materials such as nylon 66 are being machined to use tools (e.g. reamers and taps) which are slightly over-sized. For example, when threads are being tapped in nylon 66 it may be necessary to use a 0·1 mm (0·004 in) over-sized tap so as to compensate for plastic memory.
3. There is a minimum quantity which can be removed. If the hole needs opening out, by drilling or reaming, then it is commonly found with many thermoplastics materials and rubbers that it is difficult to remove small quantities of material. For example, as the reamer is passed through the hole, the material distorts and then returns to its original dimensions when the reamer is removed. For nylon 66 it is necessary to remove at least

0·1 mm (0·004 in) by final reaming if a hole of the correct size is to be produced (Du Pont, 1979).

4. Cutting tool wear can be excessive. Because of the binding action produced by plastics memory, the wear on cutting tools can be very much more severe than that experienced for metals if rubbing rather than cutting occurs during machining.

2.3 The Importance of Cutting

If rubbing rather than cutting occurs, then the rubbing action can cause problems as the heat which is generated cannot be conducted away through the mouldings as these have very low coefficients of thermal conductivity. Even if the polymer is not degraded by the heat generated then the hole will be of a different size to that required because of the high thermal expansion of polymers, unless compensation is applied. Because of the high coefficient of thermal expansion, it is important to check hole size at the appropriate operating temperature of that particular component and, if necessary, after the component has been conditioned or annealed. Some plastics materials absorb water and as this absorption occurs the moulded components change in size—again this should be checked.

Therefore, when machining polymers it is necessary that the tools suit the material being machined—recommendations on machining specific polymers can be obtained from the raw materials manufacturer or by consulting specialist references (e.g. Metcut Research Associates, Ltd, 1978–79).

As it is important to cut rather than to rub during machining, it is vital that very sharp cutting tools with plenty of clearance are used. Clearance is important, as once the material has been cut then it is necessary that it should be got out of the way as quickly as possible so that it cannot melt or clog at the cutting interface. For this reason it may prove desirable to employ cutting tools which give a continuous strip of waste material, rather than chips, as such a strip will tend to move out of the way provided that a clearance path is available. For example, in the case of drills very wide polished flutes will help the cut material to escape from the hole. The surfaces on all cutting tools should be smooth, well polished and if necessary coated with a nonstick surface so that binding or clogging is minimised.

The rate at which material is fed to the cutting tool should be adjusted so as to minimise heat generation and to obtain the surface finish required. In milling, for example, too slow a feed rate can produce excessive temperature rises through a rubbing action. During many machining operations a

TABLE 2

RECOMMENDED CUTTING CONDITIONS AND CUTTING TOOL GEOMETRY FOR THE MACHINING OF THERMOPLASTICS MATERIALS (after Hoescht, 1977)

Machining method	*Cutting speed m/min*	*Rate of feed mm/rev*	*Rake angle γ deg.*	*Clearance angle α deg.*	*Tool angle, notes*
Planing and turning	200–500	0·1–0·5	0–15	5–15	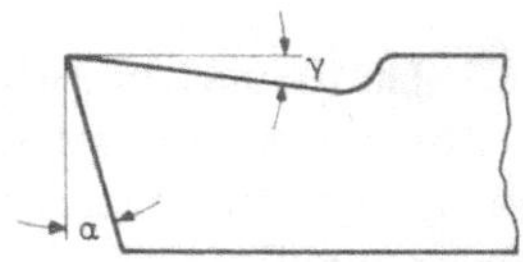
Drilling	50–100	0·1–0·3	25–30	about 16	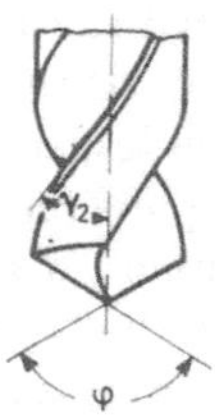 Spiral angle γ_2 at the chisel edges Clearance angle measured at the chisel edges Drill point angle should be 110°
Sawing					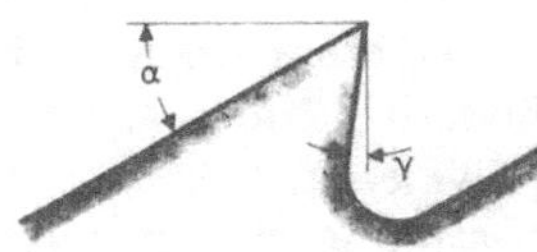 Travel per tooth (Tooth pitch $t = 3$ mm)
Circular saw	up to 3 000	0·1–0·3	0–15	10–15	
Band saw	up to 3 000	0·1–0·3	0–5	30–40	
Milling	up to 1 000	0·3	0–15	5–15	

constant stream of coolant will minimise many of the problems mentioned. Tests should be performed so as to ensure that the cutting lubricant does not affect the properties of the moulding.

Lasers have been used to cut and to drill plastics but at the present time the practice is not widespread. Some recommended cutting conditions and cutting tool geometries, for the machining of thermoplastics materials, are given in Table 2 (Hoescht, 1977).

2.4 Thread Cutting

Threads are sometimes cut directly in a plastics moulding as opposed to using a metal insert or a moulded-in thread. Some recommended values for the dimensions of drill holes which are to be threaded are shown in Table 3 (Hoescht, 1977).

TABLE 3

RECOMMENDED VALUES FOR THE DIMENSIONS OF DRILL HOLES TO BE SCREW THREADED (after Hoescht, 1977)

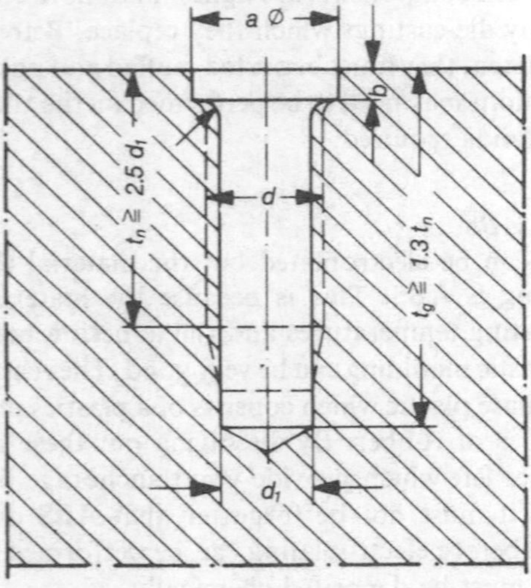

d mm	d_1 mm	t_g mm	t_n mm	a mm	b mm	r mm
2·6	2·16	7·58	5·83	2·9	0·2	0·2
3·0	2·55	8·96	6·89	3·4	0·3	0·2
3·5	2·95	10·36	7·97	4·0	0·3	0·3
4·0	3·30	11·58	8·91	4·5	0·4	0·3
5·0	4·20	14·74	11·34	5·5	0·4	0·4
6·0	4·90	17·20	13·23	6·5	0·5	0·5
8·0	6·60	23·17	17·82	8·6	0·6	0·6
10·0	8·30	29·13	22·41	10·7	0·8	0·7

Sharp V-threads, with a nonrounded root should not be used as the 'notches' thus produced will introduce stress concentration which will cause early failure.

3 METALLISATION

Many plastics mouldings are metallised in order to improve their appearance, usefulness or service life.

3.1 Electroplating of Plastics

Electroplated thermoplastics parts are now commonly used as metal replacements as such components are lighter and more corrosion resistant than the zinc alloy die-castings which they replace. Before zinc alloy die-castings can be plated, they must be sorted, buffed and polished; secondary machining operations may need to be performed on the alloy castings if, for example, fine detail is required.

3.1.1 The use of ABS

Most plastics can be electroplated but the material that is associated with electroplating is ABS. This is because the material is tough, can withstand the plating temperatures and the adhesion between the plated metals and the plastic moulding can be very good. The reason for this is that ABS is a heterophase plastic which consists of a plastics phase in which are embedded particles of rubber. By dissolving out these rubber particles, undercut holes are left which provide good anchorage for the deposited metal. However, it must not be forgotten that ABS does not conduct electricity and so before electroplating can be performed (as for metals) a conductive layer must be deposited chemically.

3.1.2 The metals used

Most electroplated plastics parts have a chromium plated finish and such plated mouldings find widespread use in automobile and bathroom applications. This finish can be very attractive but, unfortunately, chromium is a rather brittle metal and can be easily chipped or flaked off as the ABS moulding expands or contracts as the temperature changes. (Like most plastics, ABS has a very high coefficient of thermal expansion.)

In order to minimise this thermopeeling, it is usual to put a softer, more resilient metal layer in between the chromium and the plastics substrate; the two most commonly used metals for this purpose are copper and nickel. Copper is always used in conjunction with nickel whereas it is now common practice, in the plastics industry to use nickel without using copper. The metals on the surfaces of the plastics component could therefore be copper (0·02 mm thick), nickel (0·02 mm thick) and chromium (0·0003 mm thick) (metal thicknesses are often expressed in microns: 1 μm = $3{\cdot}94 \times 10^{-5}$ in which equals 0·001 mm); alternatively, to avoid problems caused through copper corrosion, just nickel and chromium could be used.

The chromium thickness is kept thin, not only to save on money, but also to avoid stress in the system. Improved corrosion resistance can be obtained if the surface of the chromium is not continuous on a molecular scale (i.e. it is micro-discontinuous) so that some 'give' is possible.

3.1.3 Production of mouldings

It must be emphasised that the plating of plastics components will not hide or disguise any surface imperfections; if any surface imperfections are present they will be accentuated by the deposited metal. The object of the moulding process is therefore to produce mouldings with a perfect finish and which contain spherical rubber particles at the surface of the moulding. This means that the mouldings must be produced under conditions which result in long cycle times, i.e. high melt temperatures, slow fill rates and high mould temperatures. Large gates should also be used and a comparatively low packing pressure should be employed for as short a period as possible. For ABS this means that melt temperatures are in the region of 270 °C and mould temperatures are of the order of 75 °C.

The use of such moulding conditions will also result in the production of mouldings which have lower moulded-in stress and this, in turn, will mean that the mouldings will either be more stable during the plating process and/or more stable in service.

The temperatures in the plating solutions can reach 70 °C so that the part must be designed with sufficient rigidity to withstand such temperatures. Sharp corners and blind holes, which will result in nonuniform plating, should be avoided as far as possible, as should large flat areas as they will readily show surface defects. For large mouldings suitable places (in unimportant areas) for mounting the moulding on the plating rack should be provided. Before moulding, the ABS polymer should be dried, for approximately 3 h at 85 °C, so as to minimise the formation of blisters which could otherwise appear after plating has been performed.

As the presence of silicone mould release can interfere with the plating process, it is best to avoid the use of such release systems and the moulds must therefore be designed so that ejection is easily accomplished. Upon removal from the machine, the mouldings should be handled as little as possible and where handling is necessary, this should not be done with the bare hand. With some components it is best if they are individually wrapped immediately after moulding.

3.1.4 Tests performed on mouldings

If the ABS material has not been sufficiently predried, then after plating small blisters may be observed on the surface. Therefore, it makes sense to test components periodically so as to ensure that the predrying operation is

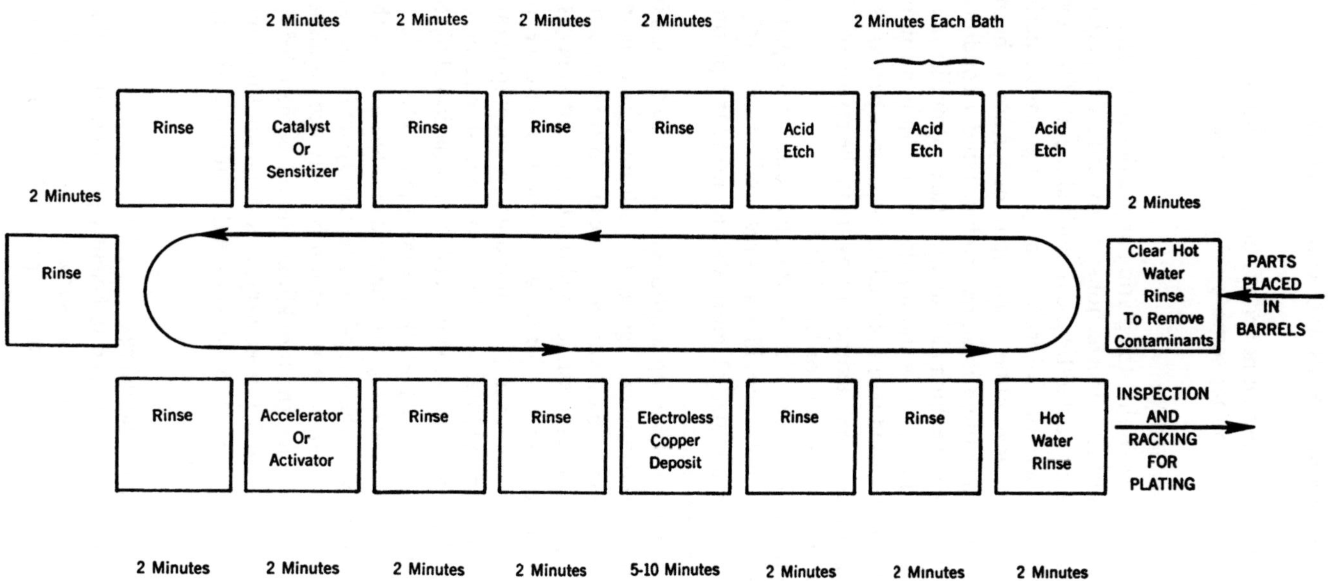

Fig. 1. An electroless copper line (after Anon., 1966a).

being performed satisfactorily, An easy way of doing this is to immerse a component in boiling water for 1 h and then inspect the surface with a magnifying glass. If small blisters can be detected, then the material has not been predried sufficiently.

In order to detect whether or not the components contain high levels of stress, the ABS mouldings may be dipped in glacial acetic acid for a specified period, e.g. 3 min. After rinsing thoroughly in cold water, the moulding should then be allowed to air dry before being examined. Where there is a high level of internal stress, then a series of white, deep cracks may be seen. However, this test must be treated with caution as rubber-rich areas (which are desirable) can also yield white patches. Care, and therefore experience, is needed in interpreting the results of this test.

3.1.5 The plating process

Before plastics can be plated, it is necessary to apply, or to deposit, a conductive coating of metal on to the surface of the plastics moulding. Once this has been done, then electroplating proceeds in the usual way except that the electroplating current is low initially so as not to burn off the thin, chemically-deposited layer of metal. However, in order to get the best results, i.e. satisfactory adhesion, the plastics components must first be etched and then chemically treated.

The total electroplating process consists of a large number of steps or stages (as can be seen from Figs 1, 2 and 3 (Anon., 1966a)) which can be divided into a smaller number of main steps or stages which are:

1. Cleaning
2. Surface conditioning
3. Sensitisation and activation
4. Electroless metal deposition
5. Electroplating

The parts are conveyed from one step or stage to another by means of a conveyor belt system and the time spent at each step is rigidly controlled; in between each step, thorough rinsing or neutralisation of the components may be required.

An electroless copper system will now be described; such a process is illustrated in Fig. 1.

3.1.5.1 Cleaning. The object of this part of the process is to remove substances which would interfere, for example, with the chemical etching stage. It is best not to rely on this stage, rather the object should be one of trying to make it unnecessary by handling the mouldings as little as possible, handling them with gloves when absolutely necessary and keeping them clean and free from contamination, e.g. from airborne dust, moulding powder, etc.

Cleaners which are alkaline in nature are capable of removing such things as fingerprints but they are incapable of removing silicone mould releases completely and, as such release agents interfere with plating, it is best if their use is avoided.

Many plastics mouldings are loaded into barrels (made from PVC coated wire) which are shaken or moved automatically during their immersion time so as to provide necessary agitation. The same barrels are not used throughout the entire electroplating operation as, for example, once they have been dipped into the activating solution, then copper would be deposited on the barrel.

Larger plastics mouldings may be placed on racks so as to ease handling problems. Again, the mouldings may be reracked after they have been dipped into the activating solution.

3.1.5.2 Surface conditioning. This stage is also called etching or deglazing as it produces a surface which is dull without being rough. This is done by dipping the mouldings into a strongly oxidising acid bath which consists of a mixture of chromic, sulphuric and phosphoric acids. This treatment renders the surface of the plastics mouldings hydrophilic and also causes surface etching. In the case of ABS, this means that the surface rubber particles are removed so as to create pockets or undercuts in the surface of the moulding into which the metal is deposited. This may take 10 min at a temperature of 60 °C but once etching has been completed it is very important that the mouldings are rinsed very thoroughly, otherwise the chromic acid will destroy subsequent plating solutions.

3.1.5.3 Sensitisation and activation. During this part of the process, a film of precious metal (e.g. palladium) is deposited on the surface of the plastics mouldings; the precious metal is necessary to cause copper deposition. This seeding or activation stage is done in two steps. The mouldings are dipped into a solution of stannous chloride (which is acidic) and then the mouldings are dipped into an acidic solution of palladium chloride. The precious metal ions are reduced to basic metal on the surface of the mouldings.

3.1.5.4 Electroless metal deposition. In order to obtain mouldings with the greatest resistance to thermal cycling, it is usual to build-up a thick layer of ductile copper between the plastics moulding and the chromium surface finish. The first layer is built-up by a chemical deposition technique whereby copper is deposited on to the surface from a solution of a copper salt, e.g. copper sulphate. The solution is carefully formulated and the operating conditions are chosen so that random reduction does not occur in the bulk of the solution but only on the specially prepared (i.e. catalysed with precious metal) surfaces. Once the deposition is started, the reaction is self-sustaining as the deposited metal is a catalyst for the reducing system employed. An adequate thickness of copper, e.g. 0·0003 mm can be deposited in approximately 5–10 min; because of the autocatalytic nature of the reduction reaction, it is important to ensure that this stage is timed accurately.

3.1.5.5 Electroplating. After the mouldings have been coated with a thin layer of metal they are removed from the barrels or racks, inspected and then placed on the racks which will convey them through the electroplating stage (Fig. 2). Contacts must be attached to the surfaces of the mouldings, at points that do not matter, so that electricity may be conducted through the thin surface layer of copper—such contacts should be not more than 20 mm (0·8 in) apart.

Because this surface layer of copper is very thin, it is important not to apply high current densities initially as otherwise this thin surface layer may be burnt off. Once the metal thickness has increased to approximately 0·03 mm plating can proceed in the usual way (i.e. as for metals) with high current densities being employed.

To assist in a rapid build-up of metal, the plating solution must have good throwing power—acid copper baths are commonly employed (cyanide baths cannot be used as such solutions would burn off the electroless copper layer).

Subsequent metal depositions are built-up using techniques developed for the electroplating of metal components. The total electroplating process (plating of copper, nickel and chromium) may take approximately 30 min.

3.1.6 Alternative systems

The system described is not the only one possible as plastics other than ABS are employed and other plating metals are commonly specified. Because of its cheapness, rigidity and heat resistance, polypropylene is now a major plating material. Keying is again achieved by immersion in strong oxidising acids; these attack the amorphous regions.

Electroless nickel is commonly specified in place of electroless copper as the plated mouldings have a better resistance to corrosion. The plating process is also shortened and this has obvious economic attractions. As the electroless nickel solutions are also more stable than the electroless copper

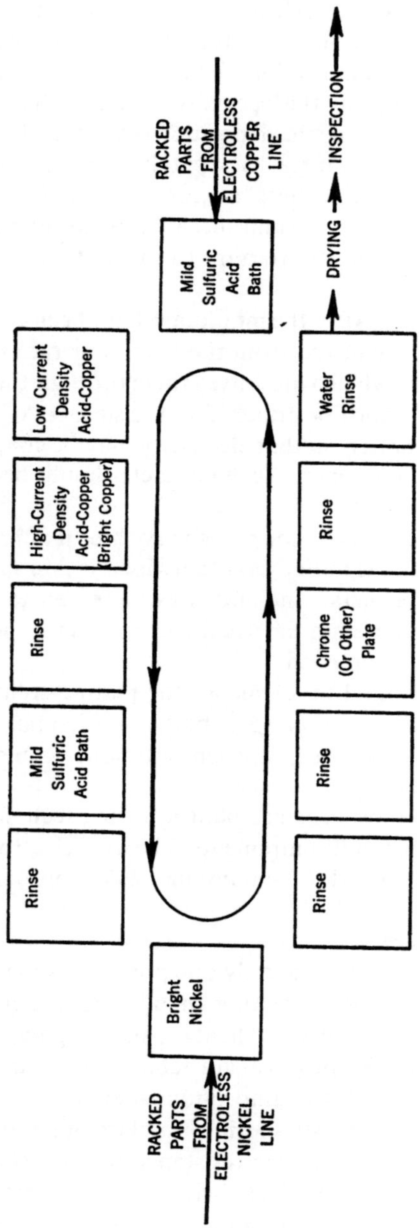

Fig. 2. An electroplating line (after Anon., 1966a).

solutions, the electroless nickel process is probably now the most widely used plating process (Fig. 3).

3.1.7 Applications of plated plastics mouldings

Plated plastics mouldings are widely used in the automobile industry for parts such as mirrors, light housings, grilles, decorative badges, etc. Such components are strong, light and corrosion resistant.

Plated plastics mouldings are widely used in other sectors such as plumbing and bathroom fixtures, e.g. handles, levers, tap components, etc.

3.2 Vacuum Metallising

Within the plastics industry, the term vacuum metallising is applied to components which have had a layer of metal applied to them by evaporating a metal (usually aluminium) under a high vacuum. Adhesion of the metal to the moulding is improved by surface treatments and/or application of a lacquer; the metal may be protected, against oxidation, abrasion, etc. by another coating or lacquer.

The metallising step may therefore only be one of several steps or stages and for convenience these have been divided into the following:

1. Moulding production
2. Base coat applications
3. Vacuum deposition
4. Top coat application

3.2.1 Moulding production

In general, the recommendations given for the moulding of parts which are to be electroplated should be followed. As far as possible, the components should be free from stress as the presence of such stresses may cause cracks to appear during the plating process. Sharp corners or edges should be avoided as the presence of such features will result in nonuniform coatings—due to surface tension effects, the coating will be thinner at these sharp corners or edges.

The moulds should be very highly polished as this will ensure that the deposited metal has a high gloss or reflectance. Good surface finish and careful part design are essential as mould releases cannot be employed to ease removal of the mouldings during the injection moulding process. If mould releases are used, then they may prevent metal adhesion. The plastics material should also be selected carefully as many such plastics materials contain lubricants or flow promoters and such additives can again cause adhesion problems.

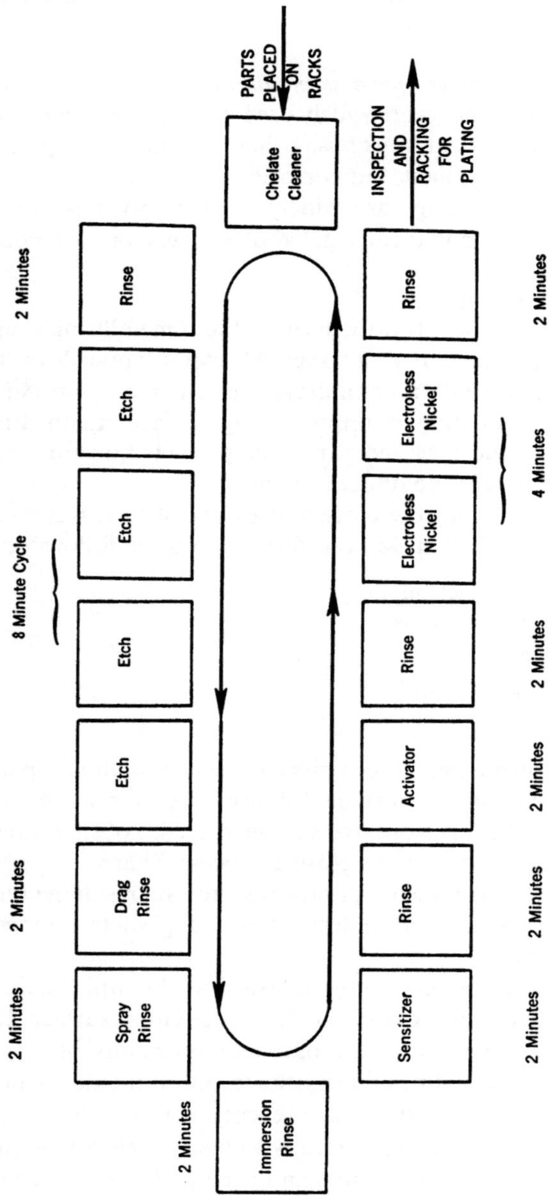

Fig. 3. An electroless nickel line (after Anon., 1966a).

It is important that the material is also clean and dry as otherwise gas pockets, or bubbles, may be formed in the resultant moulding and this volatile matter may 'out-gas' during the vacuum metallisation stage. To eliminate the need for extensive predrying, some moulders prefer to use machines equipped with vented units. Mould filling is usually done comparatively slowly as such slow fillings gives components with a high, uniform, surface gloss and uniform surface density. Such surface properties are important if uniform metallic coatings are to be obtained.

Once produced an injection moulding should be touched as little as possible and when this is necessary they should be handled only with gloved hands. Sprues and runners may be left attached in many cases as they may be used for mounting or jigging purposes.

3.2.2 Moulding preparation

Many plastics mouldings are produced in one location, stored and then transported to the vacuum metallisation plant. Even if stringent precautions are taken the mouldings may still become dusty or dirty and this surface contamination must be removed before metallisation can proceed; the first stage in the process is usually one of cleaning. In the simplest case the mouldings are cleaned by immersing them in a suitable detergent and then the mouldings are rinsed before being dried.

With some materials, where good adhesion between the metal and the substrate is required, it may prove necessary to employ more elaborate techniques. For example, with acetal mouldings it may be advantageous to clean the mouldings with solvents such as trichloroethylene so as to ensure that the mouldings are free from grease or oil. Annealing may then be employed in order to relieve some of the moulded-in stresses which could otherwise cause problems. Annealing may be done by heating the moulding at 150 °C for a specified time (e.g. 15 min for every 3 mm (1/8 in) of wall thickness.) After annealing, the mouldings should be slowly cooled and when cold may then be acid etched by immersing the moulding in an oxidising acid mixture (e.g. which contains sulphuric acid and potassium dichromate). The mouldings need only be immersed for a comparatively short time (e.g. 15 s) before being removed and rinsed thoroughly in cold, clean water. Drying is then performed.

3.2.3 Base coat applications

Traditionally plastics materials were coated with a base coat (which is also known as a lacquer or varnish) as this base coat stopped any volatile matter from vaporising during the metallisation process (out-gassing) and also improved the surface finish of the mouldings. Nowadays for some applications, e.g. sealed light units, such base coats may not be necessary as both the materials and the processing machinery have been improved.

Base coats are still, however, widely used and a major reason for their use is that they improve the adhesion of the deposited metal to the component. They also stop out-gassing, improve surface finish and stop top coat solvent attack against the plastics moulding.

Many materials have been used as base coat lacquers but generally polymer solutions which are capable of forming a clear, high-gloss coating after drying are used. If thermosetting plastics are used (e.g. based on alkyd resins) as the base coat then stoving at 70 °C for 30 min may be necessary.

In order to reduce handling costs, particularly for small components, the mouldings are mounted on a carrier (called a spider) during base coating, metallisation and top coating. The base and top coats may be applied by dipping, spray coating (probably the most common system) and flow coating. In this last system the components are exposed in an enclosed area to streams or jets of the lacquer whilst being rotated on the carrier or spider; this process is claimed to give very thin, uniform coatings.

3.2.4 The metallising process

After the lacquer has been dried and then stoved at the appropriate temperature the coated mouldings are mounted on to racks (if this has not already been done) which are then in turn placed in the metallising chamber so that either the first surface or the second surface may be covered with metal.

3.2.4.1 First and second surface metallisation. In first surface metallising, that part of the moulding which is closest to the eye in service is coated with metal, i.e. the outer or upper surface. Instrument panels and control knobs are common examples of parts covered in this way.

In second surface metallising the reverse side of a clear plastics moulding is coated so that the metal is viewed through the body of the moulding itself. Items such as badges and emblems, made from PMMA, are commonly coated in this way as during the service life of the component the plastics material protects the coating against scratching or abrasion.

3.2.4.2 Equipment considerations. The vacuum metallising machine consists of the following major components:

1. A chamber into which the plastics components, which are mounted on spiders racks, may be loaded. Such chambers usually consist of horizontal cylindrical vessels which are fitted with a front opening door. Chambers

with diameters of up to 2 m are known. The chambers are commonly lined with removable shields so that deposited metal can be readily removed with the shield.

2. A pumping system is connected to the chamber so that nearly all the air in the chamber may be removed within a reasonable time, e.g. 10 min. The required vacuum level is of the order of 0·2–14 μm or 2–4 × 10^{-4} mmHg (Storms, 1978–79). To achieve this level of vacuum a three-stage pumping system may be required with the third stage or pump being a fractionating type, oil vapour diffusion pump.

3. An evaporation or heating source is fitted inside the chamber and this vaporises or boils the aluminium. This is most commonly done by using tungsten wire filaments on to which aluminium wire or tape is crimped.

4. A substrate or carrier rotation system is also fitted so as to ensure that all mouldings are exposed to the coating metal. When power is passed through the tungsten wire the aluminium boils and then evaporates. Under the high vacuum conditions the aluminium atoms fly away from the tungsten elements and coat anything within an optical line-of-sight. To ensure uniform coating of the components it is usually necessary to rotate them during the evaporation process. Some small mouldings are not mounted on racks but are bulk vacuum metallised in baskets or barrels. Items such as curtain rings, golf tees and pen components are sometimes coated in this way.

3.2.5 Top lacquer application

The layer of metal which is applied by vacuum metallising is extremely thin and porous. Because of this, the aluminium would easily tarnish (in less than 1 day) if it was left unprotected. This thin coating (0·05 μm or 0·000 002 in) of aluminium could also be easily removed by rubbing.

A coating of a protective lacquer is therefore applied over the top of the deposited metal and for many purposes an air-drying lacquer (based on cellulose) is often good enough. The best protection is given, however, by stoving type lacquers or varnishes.

If the top coat is left clear then the moulding will have a bright, silvery finish; by using coloured, transparent lacquers the colour of the metal may be changed so that it resembles gold, brass, etc. By flattening the top coat a range of brushed metal finishes can be created, e.g. brushed aluminium. Such brushed finishes do not reflect light very well and because of this they are commonly specified in components which form part of a car's interior.

If a thin, pigmented, opaque colour (based on another lacquer) is applied to the top-coated moulding then this additional coating is commonly referred to as an overlay and its presence masks the metallic colour so that a wide range of effects is possible, e.g. a bright metallic spot in the centre of a dark overlay.

3.2.6 Areas of use

Vacuum metallising is used to decorate a very wide range of mouldings which are employed in many different industries. The automotive industry utilises vacuum metallising a great deal for both decorative and functional reasons. The process is commonly used to decorate trim and is also used to create light-reflecting surfaces such as those found in lamp housings. Closures, containers and decorative emblems for the cosmetic industry is another major applicational area.

3.3 Hot Foil Marking

Hot foil marking is a low-cost, simple, attractive, dry process which is much used for the marking or decorating of plastics materials. It is known by several alternative names, e.g. hot stamping, gold stamping, gold blocking, product marking, hot-press marking and dry printing (Jolley, 1978). All these names refer to the process whereby a legend or mark is made on a substrate by pressing a heated die against a film which carries the desired colour effect. Where the heated die touches the film the colour is transferred. As there are no solvents involved, the product can be handled immediately.

3.3.1 The carrier film or foil

A wide variety of pigmented, metallic, brushed metallic and wood-grain finishes are now available on film or foil. Silver and gold marking is extremely popular and both of these metallic effects are based on aluminium which has been deposited on to the film by a vacuum metallisation technique—the gold colour is obtained by using a transparent amber lacquer. Many special effects are produced, a brushed metallic appearance is obtained by scratching the film surface and then vacuum metallising (Robb, 1978–79). Almost any surface that can be photographed can be reproduced with gravure cylinders and made into a stamping foil pattern. In this way a complete range of woodgrain and marble patterns is reproduced.

3.3.2 Foil composition

The hot stamping foil composition depends upon a number of factors, for example the plastics material to which it is to be bonded, the colour and gloss required, etc. The composition for a metallised foil will typically be as follows:

1. A carrier film. This is usually based on a polyester as such plastics materials resist the solvents that are employed to build-up the subsequent layers. Because of the high gloss possible with this material it makes an excellent substrate for metallising and its high strength and heat resistance are also necessary properties. If the film surface is 'scratched' then brushed or satin finished metallics are possible; typical film thicknesses are of the order of 0·025 mm ($\simeq$0·001 in).
2. A release coat. This may be based on a wax coat or layer and as the name implies its function is to permit separation of subsequent layers after the hot stamping process has been performed. Release coat thickness may be of the order of 0·0012 mm ($\simeq$0·000 05 in).
3. A protective lacquer. This is based on a plastics material and its function is to protect the metallised layer against abrasion and/or oxidation; lacquer thickness may be of the order of 0·0012 mm ($\simeq$0·000 05 in).
4. A metallised layer. This is usually aluminium which has been deposited in a vacuum chamber and which is extremely thin, e.g. 0·000 25 mm ($\simeq$0·000 01 in).
5. A hot-melt adhesive. This may be a blend of waxes and polymers and its function is to bond the surface finish to the plastics moulding; typical thicknesses are 0·0025 mm ($\simeq$0·0001 in).

3.3.3 Application methods

A die which is of brass, steel or silicone rubber, which has the image to be printed, is mounted on a heated platen at a temperature compatible with the release characteristics of the foil, e.g. 100–200 °C (Jolley, 1978). This die is then pressed against the foil which in turn is pressed against the plastics moulding so that under the influence of heat and pressure the image is transferred to the moulding. After a preselected time (e.g. 0·2 s) the pressure is released so that after a cooling time (e.g. 2 s) the part may be removed.

It is important that the components are well supported (so that distortion does not occur during the hot stamping process) and that accurate control over time, temperature and pressure is possible. The foil is supplied in rolls so that after each stamping operation it is possible to automatically advance the foil whilst another moulding is being inserted into the stamping area.

By using heated, silicone rubber rollers it is possible to apply foil finishes around corners, e.g. as on injection moulded TV cabinets; this is especially useful if the cabinets are moulded from structural foam as such foamed mouldings traditionally have a 'swirled' finish.

3.3.4 *Applicational areas*

Many plastics mouldings are decorated by vacuum metallising as, being a 'dry' process, the mouldings can be handled virtually immediately they are printed. Accurate registration of additional colours is possible with this process. As there are no solvents involved in the stamping process the technique is very attractive as solvent handling is eliminated, toxicity problems are very much reduced, stoving and drying ovens are not necessary, etc.

The technique is widely used to decorate components intended for the automotive, cosmetic, furniture, radio and TV industries. Such mouldings may be decorated in part (e.g. with a legend) or they may be covered completely (e.g. with a wood grain).

3.3.5 *Moulded-in foils*

Postmoulding decoration may be eliminated if a preprinted foil is incorporated during the moulding stage. The technique is thought to have originated in 1913, when a moulded-in overlay was used for decorating a celluloid shaving brush (Anon., 1966b). Its real development, however, dates from 1953 when a Swiss inventor, Gustav Hessel, introduced a range of foils compatible with melamine, to be incorporated in the increasingly popular melamine tableware. Such moulded-in decoration, based on rayon papers, is still most closely associated with compression moulded tableware.

When printed foils are used in injection moulding the problem of locating and holding the foil in position, during the mould filling stage, must be overcome. The common way of foil location involves the use of static electricity, e.g. the foil is given a strong electrostatic charge just before it is placed in the mould. The nonprinted side of the foil (which may be based on PP, PE, PS, etc.) is placed against the mould so that the incoming polymer melt flows over the printed surface.

In order to obtain adequate adhesion it is necessary to have nonprinted areas so that the incoming plastics melt can key to the foil overlay.

The technique gives mouldings which have a durable finish but the process is limited to relatively simple shapes, e.g. flat surfaces or cylindrical components. Spot area decoration (e.g. a small arrow) is also possible with this technique. As the foils are relatively thick, e.g. 0·1 mm (0·004 in), and foil insertion lengthens the cycle time, it follows that this is a relatively expensive process but one that is capable of giving very durable finishes.

3.4 Vacuum Sputtering

Vacuum or metal sputtering is now established as a commercially available finish for plastics mouldings as reasonable coatings can be obtained at a comparatively low price. Metal coatings can be applied more cheaply by this process than by electroplating and the coatings are more durable than those produced by vacuum metallisation. Moulding complexity, shape, size and base material also often restrict the application of metals by electroplating.

3.4.1 Process description

If an electrical discharge (e.g. 450 V DC 100 amps) is struck between two electrodes at low gas pressures (3×10^4 mbars) then erosion of the cathode occurs. The cathode is often referred to as the target (or as the target electrode) and it is eroded by the bombarding stream of ionised gas molecules. These strip metal from the target and it is this metal which covers the mouldings (which are mounted on racks) as the metal condenses on the surrounding surfaces.

To reduce the electrical input, and therefore heat generation, magnetic fields are used to concentrate the ionised gas around the water-cooled cathode. This improves the efficiency of the process and allows those plastics which have low heat distortion temperatures to be covered.

If argon is used as the ionisable gas then the vaporised metal is unaffected during the covering process; if, for example, nitrogen is introduced into the chamber then nitrides (e.g. titanium nitride) will be deposited on the mouldings. This process is referred to as 'reactive sputtering' and the coatings produced have different properties—the colour or hardness can be markedly different to that of the unmodified metal.

3.4.2 Metals used

The sputtering technique can be used with a wide range of metals and metal alloys and it is this versatility which makes the process so useful. Aluminium, aluminium–bronze, phosphor bronze, copper, chromium, nickel and stainless steel have all been successfully deposited. Because of their inherent corrosion resistance chromium and stainless steel are probably the most widely used metals, for example, in the automotive industry.

A 'silver' finish based on chromium applied by sputtering would be more durable than one based on aluminium and applied by vacuum metallisation. Water can penetrate vacuum metallised coatings and cause darkening on ageing.

3.4.3 Coating thickness

The metal thickness applied is usually twice as thick as that applied by vacuum metallisation, i.e. the sputtered coating is of the order of 600 Å thick where the product is to be used outdoors. Such a deposit would take approximately 2 min to deposit and the coating would need to be protected by a lacquer (for some coatings, e.g. indoor applications, lacquering is not necessary). Much thicker coating (e.g. up to 10 000 Å) can be applied if required by this technique. Heat build-up can be a problem if thick coatings are applied to some materials.

3.4.4 Pretreatment and posttreatment

To get the best appearance it is necessary to use mouldings of the highest quality, i.e. free from any surface imperfections. Such mouldings may be covered directly and the coating has good adhesion to the substrate, presumably because of the speed of impact. Adhesion is much better than that obtained with vacuum metallisation.

For many mouldings it is necessary to apply a base lacquer to the moulding before sputtering commences (see Section 3.2). This seals and smooths the surface of the moulding. The mouldings are then placed on jigs or racks and placed inside the sputtering chamber where the pressure is reduced and metal coating is performed. Good coverage of recesses and internal surfaces is obtained with this technique. After sputtering a protective lacquer is applied, e.g. by spraying, flow coating or dipping; the top-coat lacquer imparts abrasion and UV light resistance.

3.4.5 Areas of use

It is the automotive industry which has made use of vacuum sputtering, e.g. for the covering of grilles, headlamp surrounds, badges and emblems. Such mouldings (e.g. made from ABS) may be covered with chrome or chrome-based alloys as these materials are inherently corrosion resistant. Plastics such as polystyrene, nylon, polycarbonate and polypropylene have all been coated. Flexible mouldings made from thermoplastic polyurethane have also been metal covered in this way, e.g. car grilles. The grilles are 'snapped' into position and it is claimed that a metal coating can be applied which will withstand this treatment.

The process is also expected to be used in the aerospace, communications, packaging and home-appliance industries, e.g. for handles, knobs, etc.

4 PRINTING ON MOULDINGS

Eight distinctly different methods as well as a great number of hybrid methods are utilised to print plastics (Coningsby, 1978–79). The eight methods are: letterpress, letterflex, flexography, dry offset, offset lithography, gravure, stencil and silk screen.

Of all these 'wet' printing methods silk-screen printing is probably the most important printing process used for injection mouldings. The other printing methods which can follow the contours of a three-dimensional moulding are letterflex and dry-offset printing. Such techniques are commonly used to print legends on to the plastics mouldings so that (a) the appearance of the moulding is enhanced, (b) operating instructions are incorporated (for an appliance) or (c) the component is identified.

4.1 Silk-screen Printing

This process is capable of giving very dense, opaque printing on to mouldings and either rotary or flatbed printing systems may be used. Basically, the idea is extremely simple and involves the use of a screen which was formerly made from silk but which now can also be made from synthetic or metal fibre. The pores of the screen are blocked so that the ink can only strike through on to the moulding, which is placed beneath it, in a predetermined pattern. A high viscosity ink is placed on the screen in front of a flexible blade or squeegee which is then drawn across the screen; this action causes the ink to strike through the open area of the screen thus transferring a part, or all of the image to the moulding. Only part of the total image may be transferred as only one colour can be transferred through one screen at a time.

Careful formulation of the printing ink is important and factors such as toxicity, cost, the solvent resistance of the moulded component and the stress-crack resistance of the component should be taken into account.

PMMA is liable, for example, to stress-cracking and so before moulded components are printed tests should be performed so as to assess the ink's suitability. Screen printing inks for this material would be based on polyacrylic resins in combination with alkyd resins; suitable solvents could include medium and high boiling point aliphatic solvents in conjunction with high boiling point alcohols. These materials provide printability without causing stress-cracking (Kohlepp, 1976).

With many thermoplastics it is possible to select a combination of organic solvents which results in the ink being well-bonded to the moulding because of some slight solvent attack. However, care should be taken as, for example, polycarbonate is sensitive to stress-cracking caused by solvent.

With polyolefines adequate adhesion may only be obtained if the component is pretreated. Flame pretreatment may be used to render the polyolefine surface more receptive to ink and momentary exposure to a flame with a temperature of approximately 1000 °C may be necessary. For such plastics materials it may be necessary to use a printing ink which dries by chemical reaction, i.e. a two-component system is employed. Once the ink has been applied it is dried and then cured or set; a lacquer may then be applied over the printing so as to protect the print against damage by abrasion. Such inks have the best resistance to UV light, solvents, etc.

4.2 Letterflex Printing

In this printing process a high viscosity ink is applied to a flexible (i.e. based on rubber) printing plate. This plate carries a mirror image of the required legend and it is this legend which is pasted by the ink. When the plate is brought into contact with the plastics component, the ink is transferred and correct-reading copy is obtained.

4.3 Dry-offset Printing

Once again a printing plate, which is usually rigid, has ink applied to a raised image. The ink plate is then brought into contact with a smooth rubber pad or blanket so that the legend is transferred to this blanket. By bringing the flexible blanket into contact with the moulding, the image is transferred. Surprisingly enough it is possible to get very accurate registration of patterns and colours with this process.

4.4 Wood Graining

Wood graining or wood printing is now applied to plastics mouldings and the wood effect may be obtained by painting or printing techniques for example (Casadei, 1980):

1. Apply a first coat of an undercoat by means of a roller or spray.
2. This undercoat is then dried in air for 10–20 min or dried in an oven for 3–5 min at 50–60 °C.
3. Ink is then applied to this dried undercoat by means of one or two rollers on to which the veining has been etched. With two cylinders

use is made of two types of ink, one of a soft shade and one of a more intense shade, thus giving more realistic graining or veining.

4. The ink is then air dried before a top coat of lacquer is applied by curtain coating or spraying.

4.5 Labelling

When printed labels are used in conjunction with plastics mouldings, it is important to ensure that the required degree of permanence is obtained. Many plastics mouldings have a self-adhesive label attached, either for decorative or information purposes, and such labels may be intended to be temporary or permanent features of the mouldings. If they are temporary labels, then it is annoying for the consumer to find that on removal a layer of adhesive is left which is difficult to remove without marring the surface of the moulding. Such marring may occur because the adhesive deteriorates, for example, during storage. Adhesives based on natural rubber may suffer from this defect.

Further testing of the moulding plus label is usually necessary so as to ensure that unforeseen changes do not occur. Traces of solvent left in the adhesive may cause wrinkling or stress-cracking of the moulding and metallic labels may peel due to the difference in thermal expansion between the label and the substrate.

Labels are sometimes used to secure a moulded lid to a container base, e.g. such as those used to package cigarettes.

5 PAINTING

In recent years there has been growing interest in the painting of plastics mouldings as if the components are produced from one basic colour, moulding is obviously simplified. Normally the paint is applied to the moulded component but in some processes, e.g. reaction injection moulding, it is possible to apply the paint to the mould and the coating is then transferred to the component during the moulding operation, thus giving good bonding.

5.1 Reasons for Painting Plastics

Some of the more important reasons for painting plastics have been summarised (Casadei, 1980):

1. To produce a practically limitless range of colours, even for small-scale production, thus giving stylists a freer hand

2. To create a less 'plastics' image.
3. To reduce or increase gloss according to requirements
4. To produce polychromatic effects (such as wood)
5. To create light finishes on plastics of a dark nature (e.g. phenolics)
6. To hide moulding defects
7. To improve the resistance of certain materials to weathering, wear, solvents and other aggressive agents

However, the painting of plastics is not easy, as with all painting jobs it is not the application of paint which is the difficult part of the operation it is the surface preparation which takes the time and effort. When polymers are being painted one has the problem of getting satisfactory adhesion without causing, for example, excessive solvent attack or stress-cracking of the substrate. As plastics materials are good electrical insulators, dust collection (through electrostatic charges) is always a problem. Because of such considerations the painting of polymers is a multistep operation.

5.2 The Painting Process

The use of silicone mould releases is not normally recommended because it is thought that the use of such finishes will result in poor paint adhesion. The moulding process must be well regulated or controlled so that parts of the desired surface finish with a constant amount of frozen-in strain are produced. This is because mouldings are commonly painted using solvent-based systems and the solvents may cause wrinkling of the surface through stress relaxation. In order to eliminate this the mouldings may be annealed by being heated to a temperature which is just below the glass transition temperature for that particular material. For example, acrylic mouldings may be heated to 70 °C and then slowly cooled in order to relieve some of the moulded-in stresses.

If the mouldings have become contaminated then surface cleaning (e.g. using solvents) may be necessary. With some materials etching of the surface (e.g. with acid solutions) may also be necessary if adequate adhesion is to be obtained (see Section 3.1).

In order to eliminate problems caused through static electricity (i.e. dust attraction) any electrostatic charges on the surfaces of the mouldings may be removed by either washing with an appropriate solution or by 'washing' with ionised air. By blowing ionised air over and into the component, the dust is removed and re-attraction of the dust is minimised. As these two treatments have only limited effectiveness, they should only be performed shortly before painting.

The paint may be applied to the moulding in various ways. For example, dip, flow, roller and spray coating have all been used. In general, spray coating is the most widely used technique as the equipment is readily available and can be adapted to suit a wide range of components. Hand, airless and electrostatic spraying are all used, the latter technique is being developed as it offers a way of coating large numbers of parts very uniformly and efficiently. To make the mouldings electrically conductive, they may be coated with a suitable solution or undercoat. With some mouldings, the same effect can be obtained by placing them over a large mass of electrically earthed metal which attracts the charged paint particles.

5.3 Paint Curing or Setting

As solvent-based systems are still the most common paint type used for painting injection mouldings the solvent must be removed, e.g. by air drying and/or oven drying. Because of problems associated with the use of solvents (e.g. toxicity, fire risks, solvent recovery, etc.) other types of paint system (e.g. water-based systems) are being actively investigated.

Some paints cure by reaction with air. If the paint contains drying oils, then setting is accomplished by oxidation; the moisture in the air may be used to cure or set certain types of polyurethane. The use of two-part polyurethane paints is growing in popularity as such systems are very attractive and durable, e.g. for automotive and motor cycle applications.

An alternative to painting is to dye or shade the plastics parts after they have been moulded. This may be used, for example, to recolour plastics components that fail to meet a specification, or it may be used to avoid frequent colour changes on the injection moulding machine.

A wide range of thermoplastics materials (e.g. PP and PA) have been dyed by heating the mouldings in aqueous solutions of the dye at approximately 75 °C. Due allowance must be made for any change in dimension that this heating stage can produce. It would be expected that materials such as nylons would be very receptive to surface dyeing because of the structure of the nylon molecule.

6 WELDING

It is often simpler, and in many cases cheaper, to produce a component by joining together several smaller components, as by doing so mould costs and delivery times can be substantially reduced. The joints so made are either intended to be permanent (nonrecoverable) or temporary (recoverable).

Welding gives nonrecoverable joints and can be a very fast and efficient way of joining thermoplastics. Because it is fast and clean it is often the first joining system to be considered.

6.1 Ultrasonic Welding and Assembly

Ultrasonic welding may be used to assemble thermoplastics components together in one of three basic ways:

1. Thermoplastics welding
2. Metal inserting
3. Staking or riveting

Plastics welding and staking provide permanent joints whereas metal inserting can be used to provide joints which may be dismantled. Metal inserting and staking may be used to join plastics to other materials, e.g. metal to plastics mouldings.

6.1.1 Equipment requirements

Ultrasonic assembly of plastics is achieved by transmitting ultrasonic vibrations through, and into, the components which are to be joined. This causes rapid motion or vibration of one component against another so that frictional heat is generated very quickly. If two thermoplastics mouldings are being welded together then this can easily be achieved in less than 1 s for most components. A schematic layout of an ultrasonic press is shown, together with the major components, in Fig. 4 (Intertherm, 1979).

6.1.2 Plastics welding

Ultrasonic plastics welding may be conveniently divided into two main classes which are contact welding and remote welding. Contact welding is sometimes referred to as near-field welding and traditionally the term means that the ultrasonic vibrations need only travel 6 mm (0·25 in) before reaching the surfaces which are to be welded together, i.e. contact welding is usually applied to film and sheet. If the vibrations travel more than 6 mm from the point of probe contact then the term remote or far-field welding is employed. (The probe is sometimes called a 'horn' or 'sonotrode'.)

Remote welding is usually the technique used to joint thermoplastics components together and welds have been achieved with the interface at 250 mm ($\simeq$ 10 in) from the point of ultrasonic contact. Remote welding will seal irregularly shaped thermoplastics components and does not require that the ultrasonic energy be applied over the total area of the mouldings. The welding characteristics of various thermoplastics are shown in Table 4.

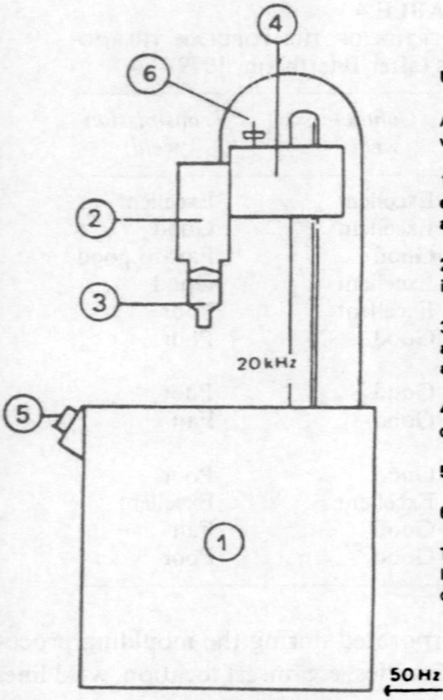

ULTRASONIC WELDING

A typical ultrasonic equipment used for the welding of thermoplastics consists basically of

1. An electronic generator which changes the mains electricity of 50/60 Hz to 20 kHz and boosts it to a given power.

2. A transducer to convert the electronic power into mechanical vibrations.

3. A probe that, when attached to the transducer, amplifies the original mechanical vibrations and allows them to be applied to the work.

4. A pneumatic press for forcing the contact area of the moulding together during the welding cycle.

5. Timer and controls.

6. Depth stop.

The system should operate as close to the upper limit of human audibility as is consistent with the operator's comfort, which is 20 kHz nominal.

Fig. 4. An ultrasonic welding machine or press (after Intertherm, 1979).

As far as possible the components should be designed to suit the ultrasonic welding process. For example, since welding is caused by the friction generated at the interface it is advantageous to have a large proportion of the mating surface area parallel to the direction of pressure application. Such parallel, mating surfaces should be made to be sliding fits as these will weld quicker than tight assemblies since tight assemblies can transmit the ultrasonic vibrations directly to the second component without very much friction being generated. Many good welds have been produced using tongued and grooved assemblies, as such assemblies assist in part location during welding. Some examples of joint design for ultrasonic welding are shown in Fig. 5 (Intertherm, 1979).

6.1.3 Metal inserting

Metal inserts are commonly incorporated into plastics mouldings if, for example, a nonpermanent joint is required. In this case the metal insert would carry a threaded hole into which the retaining screw could be

TABLE 4
ULTRASONIC WELDING PROPERTIES OF THE POPULAR THERMOPLASTICS MATERIALS (after Intertherm, 1979)

Plastics material	*Contact weld*	*Transmission weld*
Polystyrene	Excellent	Excellent
ABS	Excellent	Good
Nylon	Good	Fair to good
Acrylic	Excellent	Good
Polypropylene	Excellent	Poor
low density	Good	Poor
Polythene		
high density	Good	Poor
rigid	Good	Fair
PVC		
plasticised	Good	Poor
Polycarbonate	Excellent	Excellent
Acetal	Good	Fair
EVA Copolymer	Good	Poor

tightened. Such inserts can be incorporated during the moulding process but such a technique can lead to difficulties, e.g. insert location, weld lines, cracking around the insert, etc.

Nowadays many such inserts are incorporated into the moulding in a postmoulding operation. An easy way of doing this is to use ultrasonic equipment.

A hole is moulded in the plastics component and this hole is slightly smaller than the metal insert. The metal insert is then placed in position over the hole and an ultrasonic probe is brought into contact with the insert. The frictional heat developed at the interface causes the plastics material to melt and the pressure applied to the probe causes the insert to be driven into place. Incorporation takes less than 1 s and if the insert is suitably designed (e.g. with a knurled or fluted surface) good keying is obtained between the insert and the substrate.

This technique has been used to incorporate metal hinges into polymers, e.g. spectacle frame hinges, terminals, pivots, hubs, etc.

6.1.4 Staking

Ultrasonic staking or riveting is a technique for the fastening of metal or plastics components to other plastics components. Most staking applications involve attaching a metal component to a plastics moulding. A

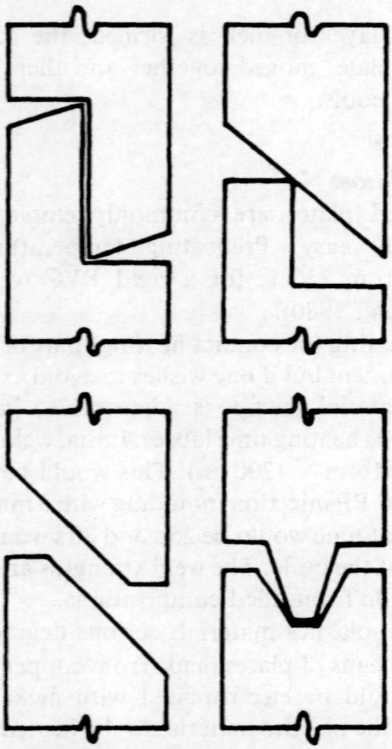

Fig. 5. Typical joints designed for ultrasonic welding (after Intertherm, 1979).

hole in the metal component receives a plastic stud or boss; the ultrasonic probe is brought into contact with this stud so that the tip of the stud is melted and shaped by the probe into an overlapping bead or head. Good, tight, consistent joints have been obtained using this technique in, for example, the automotive industry.

One application of this technique (Walklin, 1970) consisted of securing four coloured acrylic lenses to an ABS panel. The panel had been designed so that a pair of lugs could be swaged over two slots, one in each end of the lenses, by means of a heated tool. Although the heated tool method was reasonably successful, the results were inclined to be erratic and the process was slow. Using ultrasonic staking good results were obtained on a 0·5 s cycle.

6.2 Hot-plate Welding

In this process the thermoplastics components are held against a heated metal plate until a layer of melt is formed; the components are then removed from the plate, moved together and then held under pressure whilst the weld line cools.

6.2.1 Welding conditions

Electrically heated plates are commonly employed as temperature control is relatively easy. Preheating temperatures for thin-walled components vary from 230 °C for a rigid PVC to up to 400 °C for a polycarbonate (Anon., 1980).

During this preheating (or contact heating) part of the cycle pressure is applied to the component but if one wishes to avoid excessive displacement of heat-softened material then it is advisable to gradually reduce this contact pressure as the heating time lapses. Initial welding pressures may be of the order of 0·2 $MN\,m^{-2}$ (200 psi). This would be the sort of pressure required if an unfilled PP injection moulding with 3 mm (0·125 in) walls was being welded; preheat time would be 25 s and 20 s would be allowed for the weld and cool part of the cycle. The weld strengths are often reduced if the components are made from filled compositions.

With some thermoplastics materials serious degradation may occur if they are heated by means of plates made from copper or brass and in such cases the plates should be electroplated with nickel or chromium. To prevent adhesion of the plastics material to the heated plate, it is advisable to make the surface nonstick and this is most commonly done by coating the plate with polytetrafluoroethylene (PTFE).

6.2.2 Applications of hot-plate welding

It is in the car industry that one sees many hot plate welded assemblies. This is because in that particular industry components are required in very large numbers and it is worth while tooling-up with semi-automatic or fully automatic assembly systems for hot-plate welding. The components may be too large to be welded by other techniques and/or they may be moulded from materials which do not weld satisfactorily by, for example, the ultrasonic method. The relative simplicity, and therefore the low cost, of hot-plate welding should also not be forgotten.

Typical automotive components made by hot-plate welding include petrol tanks and filters, brake fluid reservoirs, radiator expansion tanks, interior trim, bumper protectors, water pumps, vacuum reservoirs, plenum valves, horn bodies, fascia panels, carburettor floats, starter batteries and rear lamp clusters (Anon., 1980).

6.3 Friction Welding

As the name of this assembly technique implies, the heat necessary to cause fusion of two thermoplastics surfaces is generated by rubbing the two surfaces together. Successful joints have also been made in certain cases where only one of the components was a thermoplastic and in such a case the thermoplastics material functions as a hot-melt adhesive.

6.3.1 Spin welding

This is probably the most common friction welding technique as it is possible to get pressure-tight joints quickly and reliably. If identical plastics components are being joined, then the weld strength can approach the strength of the parent material. Such strong, consistent welds may be produced in less than 0·5 s using comparatively cheap equipment. Some idea of the speed of assembly possible can be obtained by considering bottle production—one way is to produce bottle halves (top and bottom) which are then spun together. One comparatively small machine can produce 18 000 bottles per hour.

The components must be rigid enough to withstand the welding pressures and this normally means that each of the two halves must be supported during the welding operation. Whilst one component is held stationary the other is rapidly brought up to welding speed whilst being held in a specially designed fixture. The two components are brought together and a predetermined force is applied for the set time, whilst rotation ceases.

Melt production is confined to very thin layers and so cycle times can be extremely rapid.

6.3.2 Welded assemblies

Components which have been produced in this way include bottles, gas lighters, aerosol containers and floats. Components have been based on acrylics, acetals, nylons and polyolefines and toughened polystyrene. Noncircular components may be joined by oscillating, or reciprocating, one component against another at high speed.

6.4 Induction Welding

Induction welding is also known as electromagnetic or magnetic welding as the welding temperature is reached by using a magnetic field.

6.4.1 The basic process

If a magnetic material is surrounded by a coil which carries a high-frequency alternating current, then the temperature of the magnetic material will increase because of the resultant energy changes within that material. If the magnetic material is placed at the interface of two plastics mouldings then the heat generated may be used to weld those two components into one assembly. Originally the magnetic material was in the form of a wire or strip but weld strength was limited by the small area over which the actual weld was made. Consistently good welds may be obtained by using the magnetic material in the form of a fine powder (e.g. iron oxide powder) dispersed in a thermoplastic, hot-melt adhesive. Mouldings may therefore be joined to other plastics mouldings (which may or may not be the same material) or they may be joined to other materials, e.g. wood.

6.4.2. Equipment considerations

The induction generator may be capable of giving up to 5 kW of output power at a frequency of up to 30 mHz (30 mc s^{-1}); the actual value of the power and frequency depends on the job in hand. The high-frequency, alternating current is fed to copper coils which are usually imbedded in a nonconductive material whose external shape suits the components which are to be welded. In this way the induction coils form part of the holding jigs or fixtures so that when pressure is applied to the assembly the moulded components are well supported.

6.4.3 Applications

Induction welding can be used to join the same or dissimilar materials rapidly and consistently. Components with irregular surfaces or edges may be joined in this way using relatively low contact pressures, e.g. the pressure is usually below 0·7 MN m^{-2} (100 psi). Structural foam components have been assembled in this way, e.g. boxes and containers. Pipe fittings have also been welded on to hoses and drums using this technique.

7 MECHANICAL JOINING

Joining by mechanical means often provides a simple and economical way of joining, for example, a plastics moulding to a metal component.

7.1 Snap Fits

This technique of joining is widely used on thermoplastics materials, particularly those which can be classed as high strength, resilient materials, e.g. nylons and acetals. Permanent or recoverable joints can be produced by mating a moulded undercut on one part with a lip on another component so as to provide a strong mechanical joint.

7.1.1 Material properties

During assembly the thermoplastic component is stressed and snap fitted; snap fitted joints therefore demand plastics materials which have good elastic properties and high, permissible elongations. If the joint is to be recoverable, i.e. frequent coupling and uncoupling is necessary, then the fatigue resistance of the material should also be adequate.

Plastics materials have high coefficients of thermal expansion and so when snap fitted joints are made with metals it is important to realise that severe temperature changes can cause problems. For example, a severe temperature change may cause the strength of the plastics material to be exceeded and joint failure may result. The relatively poor creep resistance of many thermoplastics materials should also not be forgotten when snap fit assemblies are being considered.

7.1.2 Joint design

The limitations of this technique lie only in the ability of the tooling to produce the undercut in each given material (Madan, 1969). Naturally, the undercut angle cannot be more than 45° or shearing will occur on ejection, and there must be provision in the tooling for the plastics material to stretch.

Joints of circular cross-section should always be designed in such a way that the elongation caused by coupling or uncoupling should be equally distributed around the whole circumference (Scherer, 1975). This is of course only possible if the wall thickness of the plastics part is uniform.

7.1.3 Application of snap fits

Snap fitted joints permit simple design solutions for diaphragm-controlled devices such as those used in hydraulic and pneumatic-controlled systems. In place of crimped-metal rings, or screws, the parts may be joined by segments of undercuts and corresponding shoulders, which ensure leak-free joints.

Acetal gear wheels have been joined to metal shafts using press or interference fits. All-plastics chains and conveyors have been assembled from acetals using snap fits; bicycle pedals are another example of a combination of plastics and metals.

Snap fits are commonly used to joint a container lid to its base and in such a case the base may be made of polypropylene and the lid may be made of low-density polyethylene.

7.2 Riveting

The use of the traditional metal rivet is well established as a jointing system for plastics materials as it is quick and cheap. Nowadays the rivets are made from both metal (e.g. aluminium) and from plastics materials themselves; when the rivets are made from thermoplastics then the 'clinch' (the head which locks the rivet in place) may need to be formed whilst the material is being heated.

7.2.1 Solid riveting and staking

The simplest type of rivet is the solid type and this is shown in Fig. 6. Very high strength joints are possible in certain circumstances with this type

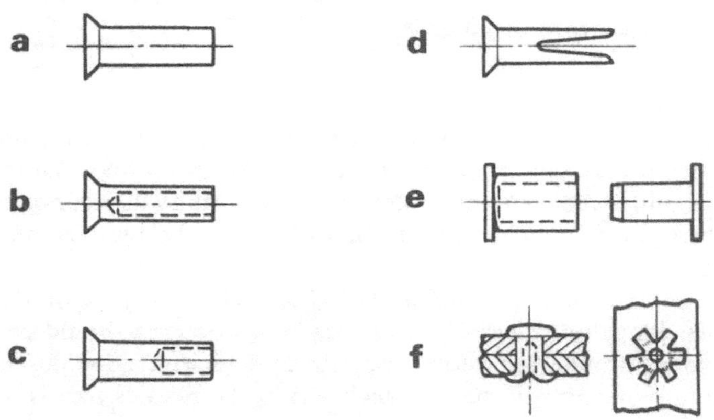

Fig. 6. Basic rivet types: (a) solid rivet, (b) tubular rivet, (c) semi-tubular rivet, (d) bifurcated rivet, (e) compression rivet, (f) rolled clinch, (g) star clinch (after Booth, 1969).

of rivet but care must be taken when the rivet is clinched (i.e. headed-over) as this action can cause fracture in many plastics materials. When used with plastics materials the solid rivet is often stronger than necessary and assembly can be slow because of the strength of the rivet.

Traditionally the rivet is a separate piece but it is sometimes convenient, and certainly economical, to mould rivet lugs on one component and to provide location holes on the other component. With ductile materials (e.g. acetals) then cold heading may be possible; for many materials it is necessary however to head the rivet using an ultrasonic probe or a heated tool in order to clinch the rivet. This process is often referred to as staking; ultrasonic staking is capable of giving very tight joints in which there is no play after assembly (see Section 6.1).

7.2.2 Tubular and bifurcated rivets

Tubular rivets, developed originally as self-piercing fasteners for soft materials such as leather, eliminate many of the failings of the solid rivet without sacrificing too much strength (Booth, 1969). The purpose of the hole, which runs the length of the shank, was to provide a recess which accepted the material being punched out. If the hole only runs part way up the shank then the rivet is known as a semitubular rivet and this type of rivet is capable of being easily set or clinched to give strong, shear-resistant joints.

As with all riveting operations it is important to ensure that correct hole sizing, as specified by the manufacturer, is observed. Too small a hole will result in component fracture whereas too large a hole will result in a loose or sloppy joint. Correct rivet length is also important as if the rivet is too short then there is an inadequate amount of material available to make the clinch whereas if the rivet is too long then clinching is difficult and slow. After assembly, and before clinching, the shank of the rivet should protrude from the far side of the assembly by approximately 0·6 D (where D is the shank diameter).

Tubular and semitubular rivets can be set or clinched in either one of two basic ways: roll set and star set. These are shown in Fig. 6. Where a tight joint is required then roll set clinches are used. Star set clinches are generally speaking not so strong but are useful in certain circumstances, for example, the clinch shape helps to resist twisting or turning and can accommodate a wider range of component thickness variations.

By combining a tubular rivet with a solid rivet a compression rivet assembly is obtained. The deeply drilled tubular rivet is pressed into the hole in the component and then the solid rivet is driven in from the other side. As the tubular rivet is a close fit in the moulding, and the solid rivet is a close fit inside the tubular rivet, then on assembly a strong, neat joint results.

The bifurcated rivet is a self-piercing rivet which can be driven through soft materials and then clinched in one operation thus saving assembly time. Bifurcated rivets are cheaper than semitubular and tubular rivets but more expensive than solid rivets. Their assembly costs are, however, fairly low.

7.2.3 Blind rivets

Blind rivets were developed for situations where access was limited to one side of the workpiece only but because of their ease of application they are now used where access is not limited. Basically the rivet consists of a hollow body through which runs a metal stem or pin. The shank of the rivet is inserted through the hole until the head of the rivet lies flush with the top surface of the moulding. Clinching or setting is done by pulling the stem out and this causes the protruding shank of the rivet to be deformed into a clinch or head.

7.3 Screwed Assemblies

In many cases the use of a threaded system results in a very strong joint which can be taken apart as the occasion demands. Because of this feature, such systems are often referred to as 'recoverable assemblies'.

7.3.1 Nuts and bolts

The humble nut and bolt can give a very good joint in plastics materials provided that due allowance is made for the properties of the base plastics material. In this context it should be remembered that plastics have a higher coefficient of thermal expansion and a lower modulus (or stiffness) than most metals. It is very easy therefore to over-tighten the threaded assembly. Because of this a thread with too fine a pitch should not be used as excessive stress can be produced during tightening and this will either cause breakage or distortion. Washers should be used as these distribute the load. If more than one bolt is involved then for thermal expansion reasons it is better to use individual washers on each bolt rather than one large plate with several appropriately located holes.

It has been pointed out that an unsatisfactory assembly (Booth, 1969) may result during service because of temperature changes. Because of their high coefficient of thermal expansion, plastics materials can generate considerable thermal stresses on the assembly and as a result the final stress may be far above that which was envisaged or recommended. An idea of the magnitude of the thermal stress can be obtained by considering the following simple calculation of thermal stress (Booth, 1969):

$$(A_p - A_s) \times \Delta T \times E_p$$

where A_p is the thermal coefficient of expansion for plastics, A_s is the thermal coefficient of expansion for steel, ΔT is the temperature change and E_p is the modulus of elasticity for plastics. If $A_p = 3{\cdot}75 \times 10^{-5}$, $A_s = 0{\cdot}65 \times 10^{-5}$, $\Delta T = 100\,°F$ and $E_p = 3{\cdot}22 \times 10^1$, then thermal stress would equal

$3{\cdot}1 \times 100 \times 3{\cdot}22 = 1000$ psi (7 MN m^{-2}). If the initial clamping stress was 2000 psi (14 MN m^{-2}) then the total stress would be 3000 psi (21 MN m^{-2}) and this could be far above that recommended.

The problems associated with the use of nuts and bolts can be minimised by using plastic ones as these have similar coefficients of thermal expansion to the base plastics components and are inherently corrosion resistant. Such nuts and bolts are commonly made from nylon.

In order to spread the load more evenly, it is recommended that washers (plain or spring loaded) be used and that excessive force should not be employed during tightening. On many mouldings (e.g. electrical connectors) a recess is moulded into one of the components to accommodate the nut as this facilitates assembly and gives a neater finish.

7.3.2 Speed nuts or clips

Speed nuts or clips are made from spring steel and are usually concave with a through-hole machined in them. The speed nut is used with a conventional bolt as the hole in the speed nut is contoured to suit the screw thread. When the bolt is tightened, then the speed nut is deformed (i.e. it is straightened) and a vibration-proof assembly therefore results.

Speed or spring clips are similar in construction to speed nuts and are usually used when one of the components has a moulded, protruding stud or boss. This stud or boss passes through a clearance hole in the other component and assembly is completed by driving the spring clip over the stud so that the two components are locked together.

7.3.3 Self-tapping screws

This method of assembly is very quick and cheap as, for example, no additional items such as inserts must be purchased and incorporated. Where frequent dis-assembly is envisaged, then this system of jointing should not be employed as thread stripping or cross-threading will eventually occur.

Self-tapping screws are commonly used to joint plastics components using the following technique: the screw passes through a clearance hole in one component and is then driven into the hole in the second component where the thread is either cut or formed. The hole is commonly located in a protruding stud or boss and in such a case (see Fig. 7; Madan, 1969) the strength of the joints are optimum for thermoplastics when the hole size is approximately 0·8 times the screw's diameter (the engaged length and boss diameter being 2·5 times the screw's diameter).

Being an integral part of one component makes assembly, storage and

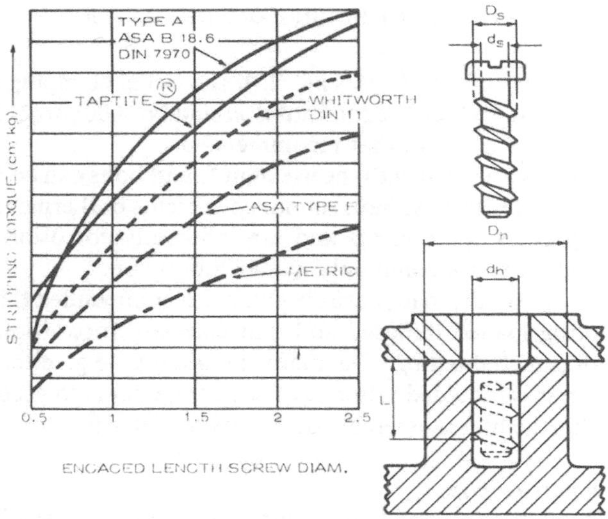

	Screw No.	6	7	8	10	12	14	16
Screw	D_s mm	3.6	4	4.3	4.9	5.6	6.5	7.1
	in.	0.141	0.158	0.168	0.194	0.221	0.254	0.280
	d_s mm	2.6	2.9	3.1	3.4	4.1	4.7	5
	in.	0.096	0.108	0.116	0.126	0.155	0.178	0.189
Hub	D_h mm	8.9	10	10.8	12.2	14	16.2	17.7
	in.	0.348	0.394	0.425	0.480	0.550	0.640	0.698
	d_h mm	2.9	3.3	3.5	4.1	4.7	5.5	6.0
	in.	0.114	0.130	0.138	0.162	0.185	0.217	0.236
Pull-out load kg	Screw type A or B	300	400	500	650	800	1000	1200
	Screw BF	250	300	400	500	600	800	1000
Stripping torque cm kg	Screw type A or B	20	30	40	50	70	100	130
	Screw BF	16	25	30	40	55	80	100
Driving torque cm kg	Screw type A or B	6	10	12	16	20	30	40
	Screw BF	4	6	8	10	14	20	25

Fig. 7. The use of self-tapping screws with acetal resins. Types A and B are thread forming and are recommended for maximum holding power and repeated assembly. Type BF is thread cutting and may be used for minimum driving torque and reduced stress in the boss. The graph shown above compares the strip length of self-tapping screws with the other types of screws. Note that type A screw has over twice the holding power of the standard metric machined screw in the hub of Delrin. The hub dimensions shown in the table are designed to give excellent retention using the given screw sizes. The table also includes the approximate values of pull-out load, stripping and driving torques which can be obtained from the recommended screws. For maximum strength, the engaged length *L* of the screw should be 2·5 times the screw diameter (after Farago, 1969).

handling, relatively easy but it must not be forgotten that the presence of the stud or boss can complicate mould design and will most probably lengthen the cycle time.

Self-tapping screws can be divided into two basic types: thread cutting and thread forming. Thread-cutting screws are normally used with hard brittle materials (for example, thermosetting materials) and such screws contain features, e.g. cutting edges, which cut or machine the thread as the screw is being tightened. As material is being cut or removed from the base plastics component, it is important that the hole be longer than the screw so that this waste material can be accommodated. A chamfer or lead-in should be provided at the entrance to the hole and the hole taper should match that of the screw.

With plastics materials which are tough and resilient (e.g. acetals) then thread-forming screws may be employed as this type of screw gives very high joint strengths.

When a thread-forming screw is driven into a plastics material the driving action displaces the material away from the hole. Provided that the stress is not excessive, then this displaced material will attempt to return to its original position and this restoring action grips the screw, thus giving a tight and secure joint.

7.3.4 Moulded-in inserts

The use of threaded metal inserts is widespread in the polymer industry as it enables good, strong joints to be obtained with a soft material and/or frequent assembly and dis-assembly. If the thread was carried in the plastics material directly then after several coupling operations wear would become a serious problem in most cases.

Moulded-in metal inserts are used in both the plastics and rubber industry as they are capable of giving very strong joints when used properly. In the case of thermoplastics, this is because the insert can be designed to give maximum holding power, e.g. by the use of heavy knurls, ridges, flats, etc. Such changes in shape on the outside surface of the insert are necessary as with thermoplastics the insert is held in place by a mechanical-keying action which is generated when the plastics material cools down from the moulding temperature. When using moulded-in inserts it makes sense to avoid a postmoulding, thread-cleaning operation by designing the mould and the insert to eliminate this time consuming and expensive step. Some suggestions for ways in which this step can be minimised are shown in Fig. 8 (Booth, 1969).

The insert should be kept as small as is practical so that it does not cause a

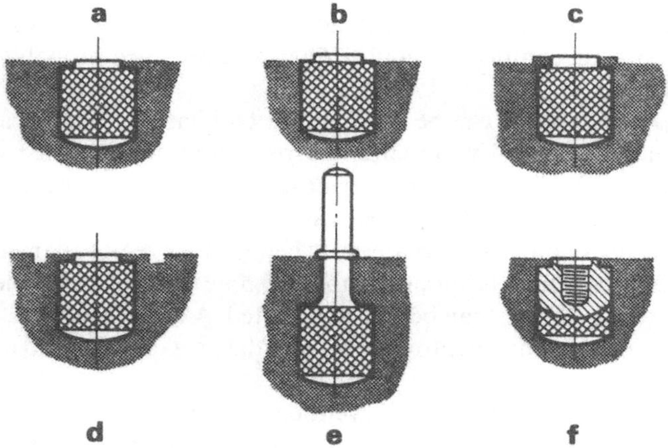

Fig. 8. Good practice when using moulded-in inserts. It is wise to avoid a postmoulding thread-cleaning operation by making provision in mould and component design. (a) Is an example of a situation in which the thread will almost certainly be fouled. With (b) the insert has a shoulder which enters the mould, whilst (c) and (d) use thin sections which harden quickly to prevent material flow. (e) Is a stud with a squeeze ring, whilst (f) shows an insert with a counterbore, using a close fitting mould pin to prevent material ingress (after Booth, 1969).

large cooling effect; the plastic wall around the metal insert should be fairly thick, e.g. half the greatest diameter of the insert.

When moulding rubber around metal inserts the problems of thread fouling become even more severe as rubber can penetrate extremely fine gaps, such as those formed between the insert and the mould-locating pin. Both the insert and the mould must, therefore, be made extremely accurately and the moulding operation must be tightly controlled if thread fouling is not to become a very serious problem.

Although direct bonding of rubber to metal (e.g. steel) is possible, it is found more convenient and/or consistent to employ a bonding medium or system which is applied to the metal surface after all the surface contamination has been removed. Removal of surface contamination may be combined with a step which also improves bonding. For example, several proprietory treatments are available in which removal of oil and oxide films is combined with the deposition of a thin adherent film of iron or zinc phosphate. Thick, powdery, phosphate films must be avoided, and it is also important to include a final chromic passivating rinse (Morton and Quinton, 1971). Once the metal inserts are chemically cleaned a bonding cement is applied in an even coat, or coats, immediately after cleaning so that recontamination of the substrate metal surface does not occur, e.g. by oxidation.

Most bonding cements are proprietory systems and the manufacturers offer a number of grades, the choice of which depends upon factors such as the type of polymer, the substrate employed, the function of the cement (primer or secondary coat), etc. Such bonding systems have largely replaced electrodeposited brass which at one time was considered essential for rubber to metal bonding applications. Rubber and brass can have a tremendous affinity for each other and this is used, for example, to make tyre valves which consist of a threaded brass stem surrounded by rubber. This strong bond may be improved, or made more reliable, by the use of bonding cements.

Injection moulding of bonded components presents no problems provided that there is no trace of precure on the surface of the stock (the melt) as it reaches the bonding agent and provided the bonding agent coated metal remains in contact with the hot mould for the minimum period (Cox, 1969). As the bonding cements are based on rubbery materials (e.g. chlorinated rubber) then precure can occur if the coated metal is prebaked as internal crosslinking of the bonding agent film occurs. This results in the loss of some, or all of the 'chemical bridges' which are necessary if high bond strengths are to be obtained. It has been shown (Cox, 1969) that loss of peel strength occurs if coated metal is prebaked for more than 1 min at 180 °C or for more than 30 s at 190 °C and 200 °C.

As the bonding cement is basically a rubber solution it may be applied in a variety of ways, e.g. spraying, brushing, dipping, flow-coating, etc. Which method is chosen would depend upon factors such as the number of inserts involved, the area to be covered and the shape of the substrate. After application, the solvent is removed by drying; the dried adhesive coating is normally nontacky and treated components may therefore be stored collectively in air-tight tins if required. It is important to stop contamination of the treated surface, e.g. by air-borne dust, handling, etc.

Injection moulding of rubber to metal bonded components is now widely employed (e.g. for items such as engine mountings and tyre valves) as such components are of heavy cross-section. Injection moulding gives more uniform and rapid cure than compression moulding and during the mould filling operation (in injection moulding) a clean rubber surface, free from dusting agents, flows over the metal components thus helping to ensure good bonding.

Because of their corrosion resistance brass inserts are also commonly used with plastics materials. Care should be taken when brass is used in conjunction with materials such as PP as if the temperature exceeds 50 °C, degradation of the PP can occur. In such a case the brass insert should be plated with tin, nickel or cadmium and/or a copper-stabilised grade of PP should be specified; it is the copper in the brass which causes the degradation.

7.3.5 Postmoulded inserts

Because of the difficulties involved in moulding-in metal components, many inserts are assembled into holes which have been made during the injection moulding process. Moulding a blind hole is cheaper, simpler and faster than moulding around a metal insert.

The postmoulded insert may be made from a metal or a nonmetal and can be incorporated by pressing, screwing, ultrasonic assembly or spin welding.

With thermoplastics mouldings the holding efficiency of the insert is improved by knurling the outside surface of the insert and/or by expanding the insert once it has been positioned. For example, the action of a screw being driven into the insert may push the sides of the insert into close contact with the hole walls, or a spreader may be pushed down and this action causes the insert to be opened out. The big advantage of this type of insert is that equipment requirements are minimal. A very simple type of insert consists of a coil of wire which is pressed into the moulded hole and which acts as a thread for the driven screw.

8 ADHESIVE BONDING

Like welding and riveting, adhesive bonding (or cementing) is classed as a permanent method of joining plastics components. With some materials there are many obstacles to be overcome before good, strong, permanent joints can be obtained, for example, the thermoplastic may be stress-cracked by a particular solvent system. Expensive testing and surface preparation may therefore be required.

8.1 Wetting and Hardening

Most of the adhesives systems now in use are based on synthetic polymers and examples can be found from all the major divisions of polymeric materials, i.e. from thermoplastics, thermosetting materials, rubbers and thermoplastic elastomers. The adhesives may be solids or liquids. After application the adhesive must go through a fluid stage so that it can wet the surface completely; ideally it should also cover the surface completely, even if the surface is rough, as otherwise any voids will be a source of weakness.

After the adhesive has been spread its viscosity increases and it may go through a tacky stage. With pressure-sensitive tapes the adhesive does not proceed beyond this stage, i.e. it remains as a permanently sticky or tacky material. With other types of adhesive it is necessary after application that the adhesive be set and this may be done by causing the change from liquid to solid to occur by cooling, liquid removal or polymerisation.

8.2 Pressure-sensitive Adhesives

Pressure-sensitive adhesives are commonly used in the form of tapes and as such are widely used for packaging and repair. The adhesive system consists of high molecular weight polymers (e.g. natural rubber, reclaimed rubber, etc.) lower molecular weight tackifying resins and stabilisers such as antioxidants. After mixing the compound is dissolved in a solvent blend and then spread on to a carrier film. When both sides of the carrier are coated with the adhesive it is referred to as a double-sided tape. The carrier system need not be a solid plastics tape; it could, for example, be a plastics material in expanded or cellular form. Expanded PVC and expanded PE are both available coated with a pressure-sensitive acrylic adhesive and such tapes may be supplied as rolls or pads (interleaved with release paper).

Such adhesive systems are useful as they adhere to any surface instantly and, because of the cellular form of the substrate, they have gap filling properties. Apart from bonding plastics mouldings to themselves, such pads may be used to bond mouldings to other surfaces, e.g. metals, paper, etc. Despite their usefulness this type of adhesive has a relatively low strength.

8.3 Solvent Systems

Solvent-based systems can only work with thermoplastics materials as crosslinked polymers cannot be softened sufficiently by solvent systems. The solvent is applied sparingly to the edges of the thermoplastics mouldings and then the two components are brought together. Pressure is applied until sufficient bond strength has been developed by solvent loss or evaporation from the joint area. Great care must be taken to ensure that the solvent does not come into contact with any other visible surface as marring or marking on that surface will result.

Interdiffusion of polymer chains leads to the formation of a bond in a solvent promoted joint and if the bond is to be a strong one then a good

TABLE 5

POLYMERS SUITABLE FOR SOLVENT WELDING (after Titow, 1978)

Polymer		*Solvents employed*	
Name	*Normal morphology*	*Type*	*Examples*
Acrylic (polymers and copolymers)	Glassy	Chlorinated hydrocarbons	Dichloromethane, dichloroethane
PVC and some vinyl chloride copolymers	Glassy	Homopolymer: selected ketones	Tetrahydrofuran
		Copolymers: some ketones and esters	Tetrahydrofuran, dioxane, butyl acetate
Polystyrene and styrene copolymers	Glassy	Aromatic hydrocarbons (especially for the homopolymers), chlorinated hydrocarbons, ketones (especially for the copolymers)	Toluene, dichloroethane, methylethyl ketone
Modified polyphenylene oxide	Glassy	Chlorinated hydrocarbons	Dichloromethane, dichloroethane
Polysulphone	Glassy	Chlorinated hydrocarbons	Dichloromethane, dichloroethane
Polycarbonate	Glassy (can be partially crystallised by heating or solvent treatment)	Chlorinated hydrocarbons	Dichloromethane, dichloroethane
Cellulose esters (especially nitrate, acetate and acetate/butyrate)	Some may be partly crystalline	Selected ketones, esters, and chlorinated hydrocarbons	Methylethyl ketone, acetone, butyl acetate, dichloromethane
Polyamides	Partly crystalline	Phenolic compounds, some bivalent metal salt solutions	Phenol, resorcinol, cresols (aqueous or alcoholic solutions), calcium chloride (alcoholic solution)

solvent is required and each component should be made from the same polymer. By using a good solvent rapid sorption into the polymer, and strong plasticising action in the joint area, will occur. As a result chain mobility will be increased to such an extent that interdiffusion of polymer chains, forming bridges from one component to another, can happen relatively quickly.

Once diffusion has occurred then the solvent should be quickly lost from the joint region as otherwise a region of weakness will be present in the assembled component. A good solvent shall therefore be capable of rapid sorption, rapid diffusion away from the load region and quick evaporation. Examples of solvent systems are shown in Table 5 (Titow, 1978).

In order to get gap-filling properties some of the parent polymer may be dissolved in the solvent and such systems may be referred to as cements or bodied adhesives.

Solvent cementing systems are commonly associated with polystyrene or toughened polystyrene mouldings which are glued together so as to make models of ships, planes, etc. Thermoplastic pipes are often joined using pipe couplings of the same material and where permanent joints are required a solvent-based cement may be employed, e.g. for ABS and UPVC pipe systems. The polymers mentioned are all amorphous thermoplastics and such materials give the best results with solvent systems as there are no crystal structures to restrict interchain diffusion.

8.4 Hot-melt Adhesives

Many thermoplastics materials are very sticky when they are in the melt form, as the melt is sufficiently fluid to achieve good wetting on to the surface of a substrate. Hot-melt adhesives are commonly based on low viscosity melt materials (such as polyethylene and polyamide materials) and such systems are used widely in the packaging industry to join and seal a wide range of packages, e.g. for milk, bread, etc. Electromagnetic bonding (as described in Section 6.4) is one way of using hot-melt adhesives. Metal plates or labels have been attached to thermosetting mouldings using hot-melt adhesives but in general such adhesive systems do not appear to be widely used for joining mouldings to themselves or to other substrates.

8.5 Reactive Adhesives

This category of adhesives is set by a chemical reaction so that polymerisation and/or crosslinking occurs after the adhesive, in liquid form, has been applied. The main advantage of these materials over solvent-based adhesives is that faster production, lower cost, and stronger bonds may result from the elimination of solvents.

For the joining of the polymer mouldings to themselves, or to each other, those materials which set by addition polymerisation are the most useful as contact pressure only needs to be applied during the setting process. This is because there are no volatile byproducts eliminated during polymerisation which necessitate the use of large, opposing forces.

8.5.1 Epoxides

Epoxides, or epoxies, are the adhesives which are associated with very high bond strengths with a very wide range of materials. They can be used to join or anchor mouldings based on thermoplastics, thermosetting materials and rubbers to substrates such as metals, glass, concrete or to each other.

Epoxides have sufficient flexibility to allow materials with different coefficients of expansion to be bonded together. Because they shrink very little on curing, surface contact is maintained during setting and the glue line contains very little strain. The materials contain no solvents which would attack, or stress-crack, a thermoplastics moulding and the set resin has good resistance to heat, corrosion and moisture.

When epoxides are supplied as adhesives they are liquids and these are mixed with liquid hardeners in definite proportions to give a mixture which has a specific pot life. By using different hardeners it is possible to get set resins whose properties differ markedly and the range can be extended by blending low molecular weight rubbers into the epoxide resin so as to give toughened epoxide systems.

8.5.2 Cyanoacrylate adhesives

When very rapid bonding is required, for example in less than 1 min, then cyanoacrylates are used. These are acrylic monomers that are chemically related to methyl methacrylate and as such are sometimes referred to as acrylic esters. The monomer does not polymerise until it is exposed to the atmosphere when the water present causes polymerisation. Bond strength may be improved by curing at slightly elevated temperatures (although this is not essential).

As cyanoacrylate adhesives have limited gap-filling abilities it is necessary to ensure that the substrate surfaces mate well before the adhesive is applied.

If the plastics mouldings will not bond well enough with this type of adhesive then in some cases a simple solvent wipe (e.g. with acetone) may be sufficient to improve adhesion as the solvent wiping action removes surface contamination such as mould release agents. In other cases a light sanding of the edges to be joined may be necessary but again a solvent wipe should follow the sanding operation. With polyolefine mouldings more expensive surface treatment may be necessary. For example, the surface of the moulding may require to be treated with chromic acid or with an oxidising flame so as to introduce polar groups on to the otherwise inert surface.

Such adhesives are not outstandingly durable, although they will withstand a wide range of common solvents. They have been used, for example, to bond rubber gaskets and sealing rings to electrical plugs and washing machines. To achieve the best results it is sometimes advised that the working area be maintained at a selected humidity level, for example, 50 % relative humidity.

8.6 Design Considerations

If adhesives are being considered as a joining method for plastics components then it is advisable to make the decision at an early stage in the design operation so that the required strength of joint may be obtained at the lowest possible cost. Despite the large number of adhesives that are now available, it does not follow that adhesion joining is an automatic replacement for moulding or mechanical systems of joining as many plastics materials are 'low energy surfaces'. This means that adhesives will not wet the surface of the moulding and so a poor join will result. Extensive surface modification may therefore be necessary before any sort of bond is possible.

As far as possible the components should be designed so that when the joint is made the adhesive layer is subjected (if it must be subjected to anything) to shearing forces in end-use. Adhesives do not function well when subjected to tension, peel, or cleavage forces and when such forces are applied early joint failure is normally the result. It is commonly found that the highest joint strengths are obtained with a large area and thin glue line as this helps to spread the load. A thin glue line can help minimise in-built stresses which result when the adhesive sets or hardens. In order to stop stress being concentrated in the glue line, when the assembly is flexed, it is best to choose an adhesive which has a lower modulus than the components. A flexible adhesive may also withstand stresses which are induced as the result of differing coefficients of thermal expansion. For example, this may be found when a plastics component is joined to a metal substrate.

Before any adhesive is used, it is important to ensure that there is no ingredient contained in the adhesive system which will cause either of the components to fail. For example, many thermoplastics materials will crack in the presence of specific organic liquids, i.e. they will stress-crack. Such liquids may be solvents or nonsolvents for that particular plastics material. Therefore, when solvent sytems are being used it is best to allow the solvent to evaporate almost completely before the joint is made.

Where the completed assembly must have a certain degree of chemical resistance it is important to ensure that the adhesive layer has the same degree of chemical resistance. The heat resistance of the adhesive is also important as the adhesive must be capable of maintaining the assembly over the anticipated temperature range of use.

REFERENCES

Anon. *Plastics Design and Processing*, Marbon Chemical, West Virginia, March, 1966a.

Anon. Making a good impression, *Plastics and Rubber Weekly*, p. 30, 9 Dec 1966b.

Anon. Self-fastening cuts costs, *Plastics and Rubber Weekly*, p. 18, 28 Nov 1969.

Anon. Sub-zero deflashing of moulded rubber, *Plastics and Rubber Weekly*, p. 17, 10 Nov 1972.

Anon. Hot plate gives a capital solution, *Plastics and Rubber Weekly*, p. 21, 21 June 1980.

Booth, M. Mechanical at fore, *Plastics and Rubber Weekly*, p. 12, 28 Nov 1969.

Casadei, S. Parts finishing, *Plastics and Rubber Weekly*, p. 11, 14 June 1980.

Coningsby, A. R. Printing, *Modern Plastics Encyclopaedia*, Modern Plastics Int., USA, p. 446, 1978–79.

Cox, D. R. Some aspects of rubber-to-metal bonding. *Rubb. J.*, p. 49, **151**(4), 1969; Influence of polymer type and compound ingredients, *Rubb. J.*, p. 73, **151**(5), 1969; Bonding agent chemistry and related aspects, *Rubb. J.*, p. 36, **151**(6), 1969.

Du Pont, *Machining Operations for Zytel*, Du Pont (UK) Ltd, 1979.

Farago, P. J. (Ed). *Design Engineering* (*Plastics*) *Handbook*, Morgan Grampian Publishers Ltd, Kent, 1969.

Giradet, C. *Assembly Techniques for Delrin and Zytel*, A Du pont (UK) publication TR/Z21, November 1967.

Hoescht. *Fabrication of Semi-finished Products.* Trade literature by Hoescht Plastics, W. Germany, January 1977.

Intertherm. *A First Guide to Plastics Welding*. Trade literature by Intertherm, London, 1979.

Jolley, R. Hot foil marking, *Plastics and Rubber Weekly*, p. 23, 6 Oct 1978.

Kohlepp, G. Printing on plastics—problems can be solved. *Plastics and Rubber Weekly*, p. 12, 7 May 1976.

Madan, M. S. Assembly methods, *Design Engineering Handbook*, Chap 5, Ed Wilson, C. Morgan Grampian Publishers Ltd, Kent, 1969.

Metcut Research Associates Ltd. Machining. Article written by staff of Machinability Data Centre (operated by Metcut Research Associates Ltd), *Modern Plastics Encyclopaedia*, Modern Plastics Int., USA, p. 453, 1978–79.

Morton, G. F. and Quinton, G. B. Manufacturing techniques, *Rubber Tech. Manufacture*, Chap 10, published for the IRI (now the PIRI) by Butterworths, London, 1971.

Robb, E. G. Hot stamping, *Modern Plastic Encyclopaedia*, Modern Plastics Int., USA, p. 440, 1978–79.

Scherer, R. Rights and wrongs of snap fits, *Plastics and Rubber Weekly*, p. 10, 12 Dec 1975.

Storms, C. D. Vacuum metallising, *Modern Plastics Encyclopaedia*, Modern Plastics Int., USA, 1978–79.

Titow, W. V. *Adhesives 2*, Chap 12, Ed Allen, D. W. Applied Science Publishers, London, 1978.

Walklin, P. *Ultrasonic Plastics Welding in Car Production*, British Plastics, Oct 1970.

CHAPTER 5

Effect of Processing on Properties

1 INTRODUCTION

Given an injection moulding machine, a mould and a material it is possible to produce a range of mouldings which look similar but which have markedly different properties. If flashed mouldings at one end and short mouldings at the other end of the scale are excluded then it is possible to adjust the machine controls so as to produce a series of mouldings with an incredibly wide range of properties.

Consider the results shown in Table 1 (Allen and Van Putte, 1974).

The results quoted show that the mechanical properties of the mouldings varied quite considerably when changes were made in the moulding conditions or in the operating procedure of the moulding machine. However, account must also be taken of the dimensional stability of the moulding and its appearance if acceptable mouldings are to be produced.

TABLE 1

TOTAL RANGE OF MEASURED PHYSICAL PROPERTY VALUES (after Allen and Van Putte, 1974)

Property	*Polystyrene*	*Polypropylene*
Tensile load at failure, pounds	157–252	130–202
Flexural strength, psi	5080–8227	2340–4092
Izod impact strength, foot-pounds per inch	1·4–4·0	0·9–8·3
Ball drop impact strength, inch-pounds	1·59	46–120
Shrinkage across flow, inches per inch	0–0·006	0·007–0·016
Shrinkage with flow, inches per inch	0–0·006	0·010–0·014
Stiffness modulus, psi $\times 10^{-5}$	2·2–3·1	0·9–1·8

It must be emphasised that the object of injection moulding is to produce consistent mouldings of the required quality at the lowest possible cost. This means that the quality required must be capable of being defined and that it is known how polymer properties are influenced by changes in processing conditions.

2 GENERAL EFFECT OF ORIENTATION AND CRYSTALLINITY ON PROPERTIES

The changes in properties mentioned in Section 1 arise because polymer morphology can be altered if the processing conditions are altered.

'Morphology' means the way in which the polymer molecules are assembled. The polymer mass may be amorphous (with the assembly of polymer chains having no real order) or they may be crystalline, in which case segments of the polymer chain aggregate into regular structures. Amorphous materials are generally hard, clear and brittle whereas crystalline plastics are usually opaque, tough materials.

However, even in amorphous materials chain orientation can occur and this can have a dramatic effect on properties.

2.1 Effect of Molecular Orientation

Polymer molecules in their normal or unstressed state exist as coiled, intertwined chains; as pressure, and therefore stress, is applied to the polymer melt (i.e. the material is at its processing temperature) the long-chain macromolecules are stretched in the flow direction. This stretching or distortion is referred to as orientation and it is important for polymers simply because polymer molecules are so incredibly long.

On cessation of shear the random wriggling of the molecules causes the molecule to return to a random coiled up state providing the material remains molten for a sufficient time. In processes such as injection moulding the molecules are not given sufficient time and so complete relaxation cannot occur. The residual orientation is sometimes called a 'frozen-in-strain' or a 'frozen-in-stress'.

As the polymer molecules are aligned more in one direction than they are in another it would seem reasonable to assume that the strength properties are different in different directions. If, for example, tensile strength samples were cut in two directions at right angles to each other from an anisotropic plastics mouldings, the strength parallel to the flow direction would be much higher than the strength measured perpendicular to the flow direction. In the parallel direction the applied load would be sustained by main-chain bonds and a high tensile strength would result.

When the samples cut from the other direction are tested then the applied load would be supported, or resisted, by much weaker bonds (e.g. chain interactions) and the strength measured will be correspondingly lower.

If Izod or Charpy impact strength samples were cut from the same anisotropic plastics moulding then the impact strength measured would also have been different. Parallel to the flow direction the impact strength would be very high and this result has sometimes been taken as meaning that orientation in a moulding is good as higher strengths result. However, something for nothing is not obtained as the strength in the 'across' direction would be weaker; in practice when mouldings are struck or dropped the mouldings always fail in the weakest direction.

Many research papers have appeared in the past in which Izod impact tests have been used to study the effect of processing on impact strength and applied to cases where a 'falling weight' situation was more appropriate. This has meant that conclusions have been drawn which are the complete reverse of what they should have been (Brydson, 1977).

In processes such as extrusion and calendering it is possible to control the direction and/or the amount of orientation; by stretching the product in two directions (at right angles to each other) biaxial orientation is possible. Such biaxial orientation can give products with very high strengths and stiffness. The introduction of multi-axial orientation into injection moulded products is difficult. One way in which it has been done used a rotating core (Cleerman, 1976) but such an approach is only suitable for simple components, e.g. beakers. By quickly quenching the moulding after multidirectional orientation had been introduced, components with very high strengths were produced.

2.2 Effect of Crystallinity on Properties

Polymer crystallisation is a highly complex subject and for those who wish to learn more about this interesting aspect of polymer science work by Sharples, 1966 and Treloar, 1970 should be consulted. For a given material the effects on properties of an increase in the crystallinity level is generally as follows:

1. As the crystallinity increases then the polymer molecules pack closer together and properties such as density, hardness and shrinkage increase.

Because of crystallisation the shrinkage of crystalline thermoplastics is always much higher than that of amorphous thermoplastics. The shrinkage of polypropylene may be five times higher than that of, for example, polymethacrylate.

2. Because the molecules are packed closer together when there is increased crystallinity it follows that greater forces are required to prise them apart so that the effect of an increased crystallinity level is to increase rigidity, tensile strength and softening point. The effect of increasing crystallinity on rigidity as measured by tensile modulus is staggering. On going from 70 to 95 % crystallisation the tensile modulus may be more than doubled, e.g. 448 to 1034 MN m^{-2} (65 000 to 150 000 psi) (Shearer *et al.*, 1961).

3. Surface gloss and transparency may be increased if the crystallinity level is also increased. This last comment on increased transparency is surprising but it can be true if the size of the crystal structures is kept small, e.g. by the use of nucleation agents.

Increased crystallinity usually results if a nucleating agent is added to the polymer and as the nucleating agent causes the development of a larger number of smaller crystal structures, the moulding may be more transparent than the non-nucleated grade. Nucleation may also reduce the tendency to warpage as a more uniform crystal structure is developed. In many mouldings made from crystalline materials, the level and type of crystallinity changes from one part of the moulding to another. This can cause differential shrinkage to occur within the moulding and the resultant stresses so developed can easily cause the moulding to warp.

4. Because the molecules are packed closer together in mouldings which have high levels of crystallinity the permeability to gases and liquids is lower. This means that a low molecular weight molecule finds it difficult to diffuse through the polymeric material.

5. Generally speaking, increasing the crystallinity level for a given material decreases its resistance to impact and also decreases its environmental stress-cracking resistance (ESC). This reduced resistance is presumably due to the fact that the presence of large uniform crystal structures do nothing to stop a growing crack propagating through the moulding. If the individual crystals grow to a large size then there will be a plane of weakness at the interface between crystals. Low molecular weight material (which, therefore, has inferior properties) may be squeezed out by the growing crystals (as it is different from the material from which the crystal is made) and this is deposited at the interface between crystals.

3 MOULD FILLING, PACKING AND STRESS RELIEVING

3.1 Mould Filling

The flow of polymer melts into a cold mould cavity is a typical problem of unsteady-state nonisothermal flow of viscoelastic fluids. Rigorous analysis of the problem is very difficult, if not impossible, to carry out for the following two primary reasons. One is the complicated form of the rheological equations of state and the other is the complex flow geometry encountered in many practical situations (Han, 1976).

3.1.1 A simple system

It is necessary to understand how an injection mould fills as this understanding helps when one tries to appreciate the properties of injection mouldings. Most of the studies performed in an effort to understand mould filling, have been done using thermoplastics materials and relatively simple moulds in which the mould is filled by a single small gate, e.g. a tensile specimen mould. Consider what happens when this long, thin cavity is filled through an end gate. The shot size has been set to a low value and the injection speed has been set low enough to stop jetting. Under these conditions the mould will not fill and a short moulding will result. It is the shape of the short moulding which is of interest at the present time; on a side view the ends of the specimen will be rounded and on a plan view the shape of the specimen will be as shown in Fig. 1. If the shot size is gradually increased then the mouldings will get longer and this increase in size will occur by the specimen 'growing' from the gate. A series of mouldings produced under such conditions is shown in Fig. 1.

During the course of this simple experiment it is also instructive to write on the surface of the mould with a soft wax crayon as it is found that this wax marking is transferred virtually intact to the surface of the moulding. As the mark appears to be in the same place, and as it is intact, this seems to show that the layers of plastics material at the surface are moving very slowly if they are moving at all. This fact taken in conjunction with the first half of this simple experiment seems to indicate that the mould is filled by material which has been pushed over other layers of the same material.

When the polymer melt is forced into the mould the hot material wets the wall of the mould and is rapidly cooled until it solidifies. There is now a skin or layer of material adjacent to the cold wall of the mould which cannot move. As we have two mould halves, we therefore have two layers of frozen material which obstruct the flow of incoming polymer. This incoming polymer can only reach a far point in the moulding by flowing through, and over, these frozen layers. It is the action of pushing, or shearing, one polymer layer over another cooling layer which causes orientation.

The centre of the flowing melt stream will be travelling much faster than the material at the edges. In the body of the moulding, the flow direction is parallel to the mould wall but as the faster moving layers reach the melt

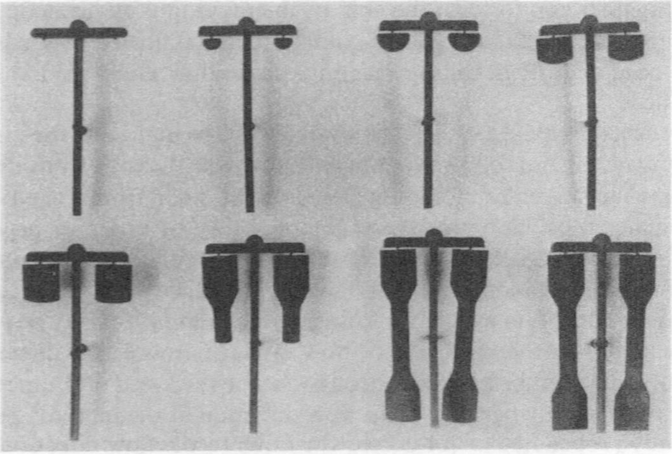

Fig. 1. Range of mouldings produced from a tensile mould.

front, the flow direction alters; flow starts towards the wall and laminar flow changes over to radial flow. This has been called the 'fountain effect' (White and Dee, 1974) and is responsible for the surface of mouldings being at low levels of orientation. The surfaces are formed from hot material from the centre of the melt stream and this hot material has a low shear history. Orientation is not therefore uniform throughout the thickness of a moulding.

Ballman and Toor (1960) proposed a mechanism for nonisothermal plastic flow based on observations from birefringence measurements. They assume that the incoming melt which contacts the mould walls first, is unoriented because the flow pressure gradient will be zero at the wave front. Since this material quickly solidifies into a skin on contact with the mould wall, it will not be subject to large shearing forces when it is in a sensitive state (i.e. above its glass transition temperature, T_g); this results in this layer of material remaining unoriented. Adjacent layers cool with increasing slowness, as they are insulated from the mould by previously deposited layers, and are subject to shearing forces which result in an increase of orientation.

3.1.2 More complex systems

When the simple mould (described in Section 3.1.1) was filled, then the 'fountain effect' was introduced. This implied that the polymer flow was two-dimensional, i.e. with material curling out from the centre of the flowing melt stream to form the skin of the moulding by being deposited against the walls of the mould. In-coming material flowed over what had already been deposited, thus generating orientation whilst the mould was being filled.

In such a moulding, orientation will always be highest in the direction of flow but it does not follow that this will always be the case. Consider what happens when a circular moulding (e.g. a disc) is filled from a central gate. The polymer must flow in a very complex way in order to produce a complete moulding. The polymer must flow both radially and circumferentially if the mould is to be filled. This can produce some very complex orientation patterns in which, for example, the orientation may be greatest at right angles to the direction of flow. What is more, the direction of orientation may not be the same throughout the thickness of the moulding, i.e. it may be mainly parallel to the flow direction at the moulding surface but mainly perpendicular (or across the flow) to the flow direction in the body of the moulding.

The direction of orientation has been assessed, in many cases, by studying mouldings made from polymers which contain fibrous fillers. The polymers used in these cases have been transparent or translucent and in such instances a relatively simple visual examination can reveal the fibre orientation. (When the polymer is opaque, then examination using, for example, x-rays is usually necessary.) What complicates such studies may be the need to prepare microtomed samples; interpretation of photomicrographs is also not easy.

In one study (Wheelans, 1979) a cavity was filled from a centrally located gate with a transparent rubber mix. This mix contained coarse nylon fibres and their size was such that they could be seen with the naked eye. The final orientation of the fibres could therefore be determined relatively simply, i.e. either by a visual examination of the mouldings or by studying photographs of the mouldings.

The fibres in the sprue were found to be oriented parallel to the flow direction and this direction of orientation was also found in the area immediately adjacent to the sprue—in this area the fibres possessed a radial orientation. However, very soon after leaving the sprue area the fibres were found to be almost fully oriented in the circumferential direction—they had turned through 90° so that most were lying across the flow direction. Some random orientation was also observed.

In a later study performed at the National College of Rubber Technology, London, the direction of fibre orientation was observed by using black cotton fibres (of 5 mm length) dispersed in a transparent rubber compound (Craft *et al.*, 1980). Originally glass fibres were used but these did not survive passage through the screw and barrel assembly. The direction of fibre orientation could be easily discerned with the naked-eye and was found to be markedly dependent on processing conditions. For example, if a slow injection speed was used (the total shot weight of 40 g was injected in 20 s) then transverse fibre orientation, as previously described by Wheelans (1974), was observed. If a fast injection speed (1 s) was used then a more random fibre distribution pattern was observed.

If it is assumed that the direction of fibre orientation is similar to the polymer orientation then certain conclusions may be drawn, i.e. during the filling of the feed system (sprue and the runner), the orientation will be in the flow direction. However, in this centre gated moulding the flowing polymer stream is required to expand outwards in order to fill the cavity and such 'expansion flow' generates orientation transverse to the flow direction. Once the cavity has been filled and the gate has sealed, Brownian motion can cause relaxation of the oriented polymer molecules. The relaxation forces may disturb the fibre orientation introduced during the mould filling thus producing a more random orientation pattern. (It must be emphasised that fibre orientation does not necessarily correlate with polymer orientation.) A more random orientation pattern should result in a more isotropic moulding.

If glass fibres are added to some plastics materials then the affect of the addition can result in a material which appears to be relatively insensitive to changes in processing conditions. Orientation will still be produced but changes in processing conditions may not produce a large change in a particular property. For example, a change in melt temperature of 40 °C may not cause a change in the value of the tensile strength of glass fibre reinforced polypropylene.

3.2 Packing and Relaxation

A considerable amount of pressure must be applied to polymer melts in order to get them to flow and to fill an injection mould. The pressure which must be applied is not constant but increases as the mould filling operation continues. As the pressure is increased, stress starts to build-up within the moulding and when the mould fills there is a sharp increase in the resistance to flow. This can cause an increase in stress and this increase occurs during the mould packing part of the cycle. If the material is cooled to below its glass transition temperature with a stress acting upon it, then the molecular chains will be set or frozen in an unnatural state.

At the high pressure encountered in injection moulding, polymer melts may be compressed by significant amounts. The more melt that is forced or compressed into the mould then the lower will be the moulding shrinkage. However, in many cases overpacking can result in mouldings which are difficult to eject and/or which fail in service. They are difficult to eject because of the reduced shrinkage which overpacking causes but the reason why they should fail more easily in service is more difficult to explain. Such failure usually occurs in the gate area as a result of an impact and it could be that the packing pressure causes orientation (through shearing 'cold' material) or that it prevents relaxation of the orientation. This would apply if large gates, long dwell times and high melt temperatures are being used.

The pressure applied during the dwell part of the cycle (and the time for which that pressure is applied) is therefore very important as it has a large affect on the properties of a moulding. Apart from the magnitude of the pressure itself, the gate cross-sectional area and the length of the gate must also be considered as such dimensions affect the pressure drop between the runner and the mould. The gate dimensions also affect the rate at which the gate cools and this is important as once the gate has sealed, then no further pressure can be applied to the material, e.g. in an effort to combat shrinkage.

Provided that the polymer is above its glass transition temperature (T_g) then stress relaxation can occur once the pressure has been removed (e.g. by the pressure being dropped or by the gate sealing). However, as a result of the high elasticity of the material, the stress built up does not relax instantaneously, as would be the case for inelastic materials (Han, 1976). The polymer chains cannot therefore get rid of the molecular orientation and thus get back to a more relaxed or natural position. Time is against them and so injection mouldings nearly always contain residual orientation. This orientation is not uniform throughout the moulding.

Orientation increases as one moves from each of the two outer surfaces of the moulding towards the centre of the moulding. The maximum amount or degree of orientation occurs somewhere between the surface and the centre as orientation is relaxed by Brownian movement within the polymer melt. The orientation is not uniform along the length of the moulding because of the interaction between shear (which causes orientation) and Brownian motion (which relaxes orientation). Orientation is usually highest in the gate area as melt flow is longest here. The level of orientation is therefore not the same throughout the moulding but differs from point to point. Such orientation affects both amorphous and crystalline materials. For example, PP exhibits clearly defined differences in crystallinity between the skin and the core of a moulding. Near the mould surface the PP molecules are highly oriented in the flow direction and the structures are nonspherulitic, whereas in the core of the moulding the crystallinity is spherulitic.

3.3 Differential Shrinkage

When a moulding cools then not all parts of the moulding cool at the same rate as, for example, they may not have been at the same temperature initially. Such uneven cooling can introduce stresses into the moulding or, in the case of crystalline materials, it can cause different types of crystal structure to be developed throughout the mass of the moulding.

Poor impact resistance of some nylon 66 mouldings has been attributed to this pattern of behaviour; the uneven temperature distribution may be caused by exceeding the plasticising capacity of the machine for that material.

The stress, introduced by differential shrinkage, can be introduced in other ways, e.g. through using sections of different thicknesses, improper gate location, poor mould cooling, etc.

Removal of the stresses caused by such differential shrinkage may be responsible for the improved resistance of annealed polystyrene mouldings to materials such as white spirit.

3.4 Annealing

Within the context of injection moulding the term annealing usually implies that the sample has been heated to a temperature just below its softening point and then the moulding has been cooled slowly; heating may be done in an oven or a fluid heating bath. If thermoplastics are annealed then it is found that the properties of mouldings have been changed by the annealing process. For example, if polystyrene is annealed then it is found that the resistance of the mouldings to white spirit, or heptane, has been altered. Before annealing the samples may have cracked or crazed after being immersed in white spirit; similar mouldings which have been annealed are more resistant.

It is unlikely that the annealing step removes the orientation (as the sample is below its T_g (see Section 2.2)) and it could be that annealing reduces

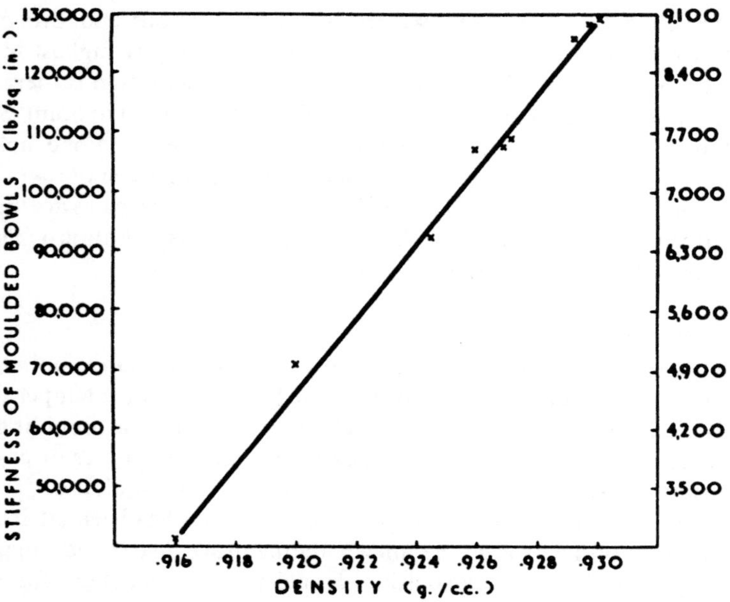

Fig. 2. Variation of moulded-part stiffness with density (after ICI, 1960).

surface stress through rearrangements of chain segments. In practice it is probably preferable to slow down the cooling rate of the moulding, rather than reheat it subsequently, as this minimises the danger of distortion. However, cycle times are increased if this procedure is adopted.

Annealing of mouldings will increase the crystallinity level of a particular moulding, e.g. for LDPE this may be done by heating at 100 °C. (The effect on density will not appear to be very much; the effect of small changes in density on properties can be enormous.)

TABLE 2
EFFECT OF COOLING RATE ON DENSITY OF POLYETHYLENE
(Density, $g\,cm^{-3}$) (after ICI, 1960)

Cooling rate	*'Alkathene' WRM 19*	*'Alkathene' XRM 21*	*'Alkathene' XRM 32*
Annealed at 140 °C cooled at 5 °C h^{-1}	0·918	0·923	0·930
Annealed at 100 °C cooled at 30 °C h^{-1}	0·916	0·921	0·927
Injection moulded fast cooling	0·913	0·919	0·924

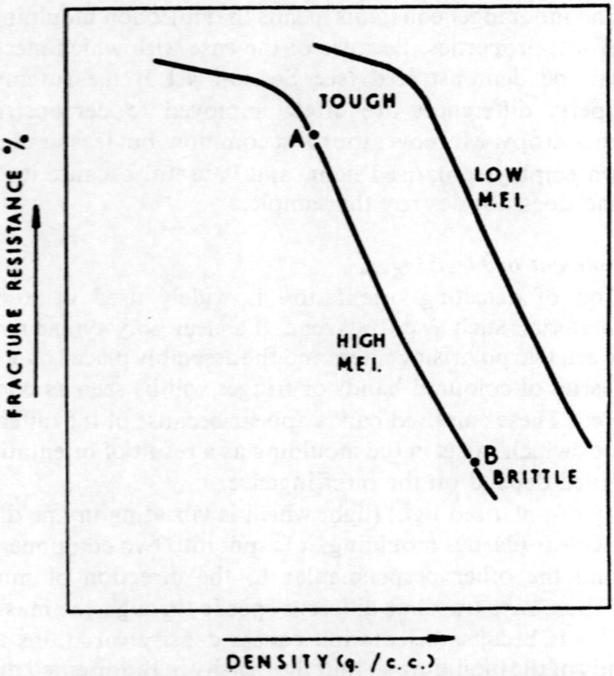

Fig. 3. Variation of fracture resistance with density (after ICI, 1960).

For example, the following results were obtained by ICI workers using three grades of polyethylene. These were injection moulded with a rapid cooling cycle and in some cases the mouldings were annealed to allow crystallisation to develop further. This caused density variations of only 0·005 g cm^{-3} (see Table 2) but the effect of this annealing step on stiffness and toughness is shown in Figs 2 and 3 (ICI, 1960).

4 DETECTION OF ORIENTATION AND CRYSTALLINITY

It is difficult to overestimate the affect of changes in orientation level and/or of crystallinity on the properties of injection mouldings. Very large changes in some properties can occur because of, for example, quite small changes in the operating procedure or moulding method.

4.1 Detecting Orientation

The products of injection moulding have a structure which is both anisotropic and inhomogeneous; this means that injection mouldings have very non-uniform properties. Because of the ease with which mechanical properties can be demonstrated (see Section 4.1.3) measurements of physical property differences are often employed to demonstrate the presence of anisotropy. Moreover the most common, but least understood, demonstration employs polarised light, and is useful because it is easily performed and does not destroy the sample.

4.1.1 Measurement of birefringence

This method of detecting orientation is widely used in studies on transparent materials such as polystyrene. If a clear polystyrene moulding is placed between two polarising filters and the assembly placed over a white light then a series of coloured bands or fringes will be seen as one of the filters is rotated. These coloured bands appear because of the difference in refractive index which exists in the moulding as a result of orientation; the colours observed depend on the birefringence.

When a ray of polarised light (light which is vibrating in one direction only) strikes a clear plastics moulding, it is split into two components, one parallel to and the other perpendicular to the direction of molecular orientation. These rays travel at different speeds through the mass of the moulding. This is because orientation causes density variations to arise within the body of the moulding so that the density in the oriented direction is different to that found across the direction of orientation. On emerging from the moulding the two rays which had travelled at different speeds throughout the moulding, are out of phase and elliptically polarised light results.

The polaroid which is closest to the eye (i.e. the analyser) transmits the component of this light which lies along its own axis of polarisation.

White light is polychromatic and on dispersion into its component wavelengths the retardation at any point will be just sufficient to cancel out a particular wavelength from the spectrum causing the eye to see the combination of the remaining colours or the complement of the cancelled colour (Ballman and Toor, 1960). A coloured band or fringe therefore results. Each fringe or band connects together points which have the same level of birefringence or orientation and the closer together the fringes are, the more rapidly the orientation is changing.

If a short, or incomplete, moulding is examined in this way (i.e. under plane polarised white light) then it will be found that the short edge of the specimen is black and this is interpreted as meaning that there is no orientation at that position. If the eye is then moved along the centre line of the moulding towards the gate a yellow coloured fringe will be seen; next to it will be a red one and next to that there will be a green one. This group of three fringes is referred to as 'first order'; the group next to that as second order and so on.

The position of maximum orientation is usually indicated by a roughly circular fringe which is normally located between the centre of the moulding and the gate. After the position of maximum orientation has been passed, the sequence of colours reverses as the orientation decreases.

The birefringence may be calculated so that the level of orientation within the moulding may be expressed numerically. Usually the numerical value of the birefringence is not calculated for a given moulding but the total number of orders is counted and this number is used to indicate the overall level of orientation. Looked at in this way the fringes, or bands, may be likened to the contour lines seen on a map; the more there are the greater is the orientation.

Birefringence may be calculated by using the following formula

$$\eta = \frac{\lambda R}{t}$$

where λ = the wavelength of light; R = the retardation in phase; and t = the sample thickness.

The wavelength of white light may be taken as 5500 Å and the retardation may be determined by counting the number of orders, up to the position of maximum orientation, and noting the last colour seen. To each colour is assigned the following fractional order values: yellow 0·3; red 0·7; and green 1·0. This means that, for example, a twelfth order yellow would have a retardation of 11·3, a sixth order green would be 6·0, etc. (Ballman and Toor, 1960).

Let us imagine that the highest order seen in a particular moulding was a tenth order red and the sample thickness was 2 mm (0·08 in), then the birefringence would be:

$$\eta = \frac{\lambda R}{t} = \frac{5{\cdot}5 \times 10^{-7}}{2 \times 10^{-3}} \times 9{\cdot}7 = 26{\cdot}6 \times 10^{-4}$$

This method of measuring orientation is restricted to transparent mouldings but it does have a number of advantages. If required the orientation may be expressed numerically and the measurement may be performed quickly and without destroying the sample. The test is commonly used on general purpose polystyrene mouldings. This material has a highly polarisable side group and it is the orientation of this side group that is responsible for the observed birefringence. It does not necessarily follow that side group orientation and main chain orientation are one and the same.

4.1.2 Orientation stress relaxation

When making these measurements advantage is taken of the fact that when a thermoplastics material is heated it shrinks as a result of orientation relaxation. If a moulded sample is therefore clamped whilst being heated then a force will be exerted on the clamps as the locked-in stresses are released. The magnitude of the relaxation force has been used as a quantitative measure of built-in stress.

Injection mouldings may be clamped in the jaws of a tensile testing machine and the heat applied by means of radiant heaters. Using this equipment the force may be easily measured although it is rather an expensive way of doing so.

Equipment has been designed specifically for measuring relaxation forces which consist in effect of two arms between which the sample is clamped (Jensen and Whisson, 1971). The mouldings were heated by submerging the assembly in a heated fluidised bed of glass ballotini beads. When the specimen shrank between the grips the deflection was registered on a dial micrometer.

The magnitude of the retraction force and the reproducibility of the test were found to be dependent upon test temperature and the temperature differed from one material to another. An appreciation of the magnitude of the forces which result on heating, and of the temperatures required for different materials, can be obtained from Table 3 (Jensen and Whisson, 1971).

This method appears to be suitable for use with all materials but it does require that the sample is destroyed; a certain amount of experience also appears to be necessary in order to get the best results.

A test which is similar to the above in some respects is known as the heat reversion test and in this test the sample is heated and allowed to shrink. The amount of reversion or shrinkage that occurs is measured and is used to indicate the presence of orientation. More local heating can be achieved by the use of a hot wire adjacent to the surface of the moulding. If a series of lines has been previously marked on the moulding surface then line movement may again be used to detect orientation.

4.1.3 Variations in physical properties

Orientation implies that the moulding has different properties in

TABLE 3

EFFECT OF INJECTION PRESSURE ON ORIENTATION STRESS PARALLEL TO FLOW IN SAMPLES 3 in × 1 in (after Jensen and Whisson, 1971)

Material	*Injection pressure (psi)*	*Test temperature (°C)*	*Orientation Stress (kNm^{-2}) Centre of plaque (3 in × 1 in)*	*Side of plaque (3 in × 1 in)*
Polystyrene	5 800	120	593	440
	7 800		404	277
Polypropylene	5 000	170	591	513
	6 000		560	588
	7 000		419	432
Polysulphone	13 500	215	278	178
	15 000		286	246
	16 500		214	218
ABS	8 000	140	550	560
	9 000		575	580
	10 000		575	565

different directions, i.e. the moulding is anisotropic. The presence of orientation can therefore be shown by measuring a property in two directions at right angles, e.g. the flexural strength may be measured in the flow direction and across the flow direction. The difference between the along and the across direction should be noted together. It was found that the sample with the highest molecular orientation (as measured by birefringence) gave results which differed most in the two directions.

4.1.4 Ultrasonic measurements

A measure of average anisotropy at a particular point within an injection moulding can be obtained by determining the velocity of an ultrasonic wave, in different directions, within the sample (Thomas and Meyer, 1978).

The velocity of an ultrasonic wave (5 kHz) propagated through an elastic material varies with the elastic moduli of the material. The propagated wave may be of longitudinal or transverse (shear) type, when its velocity will be largely determined by longitudinal or shear modulus respectively in the direction of propagation. In isotropic material, modulus = density × (wave velocity)^2; in nonisotropic material, this relationship is approximately true, and the value of (longitudinal wave velocity)^2 in three principal orthogonal directions gives a convenient measure of anisotropy.

Details of how to conduct ultrasonic measurements have been given (Thomas and Meyer, 1974, 1976). It should be noted that the ultrasonic elastic modulus is not the same as the elastic modulus determined by more conventional means—this is due to the viscoelastic behaviour of polymers.

Rapid measurements and result calculations can be made using this technique but because of the equipment requirements (holding jigs, an accurately controlled water bath, ultrasonic transducers, etc.) it is unlikely that this technique will become generally used in the plastics moulding industry.

4.1.5 X-ray microradiographs

Many thermoplastics materials are now modified by the incorporation of glass fibres; during mould filling, alignment of the glass fibres occurs. The arrangement of the fibres can be revealed by taking x-ray microradiographs of selected portions or sections of the moulding.

The differential absorption of x-rays by glass fibres permits a shadow graph of the positions of the fibres to be obtained. A fine, high intensity source of x-rays is required, and this is obtained by using the electron beam in a scanning electron microscope (e.g. a Cambridge S.600) to irradiate a 4 μm-thick foil of cobalt. The specimens are placed in contact with a photographic emulsion at a distance of 20 mm from the foil. The thickness of the sample is determined by the resolution which is needed, and it has been found satisfactory when a general impression of the structure is required to use sections up to 1·6 mm thick. Detailed measurements of fibre distribution can be made on microradiographs of thinner sections, or on optical photographs obtained from polished sections, but such measurements are extremely laborious (Thomas and Meyer, 1978). Such measurements do reveal, however, detailed structural features.

4.2 Detecting Crystallinity

Materials such as polyethylene and polypropylene consist of both amorphous and crystalline regions. The properties of such materials are affected by the amount of amorphous material, the amount and nature of the crystalline phase and the orientation (Rubin, 1972).

Crystallinity has been measured using the following properties:

1. Specific volume
2. Specific heat
3. Specific enthalpy
4. Specific enthalpy of fusion
5. Infra-red extinction coefficient
6. Specific line widths in NMR spectroscopy
7. X-ray diffraction intensities

The most common method of detecting crystallinity is density determinations as these are relatively simple to perform, e.g. at 70% crystallinity a sample of PP had a density of 0·896 and at 95% crystallinity the same material had a density of 0·903 g cm^{-3}. Therefore the density determination must be done extremely accurately as the changes are relatively small.

It is not only the amount of crystallinity that is important but also its texture, e.g. fast cooling gives small, fine crystal structures whereas slow cooling gives much larger, coarser structures.

5 EFFECT OF MOULDING CONDITIONS ON PROPERTIES

5.1 General Observations

It is difficult to establish precisely the relationship between injection moulding variables and molecular orientation in a given moulding as the position is complicated by the fact that the orientation observed is a result of (i) the initial orientation induced during the initial mould filling operation (sprue, runner and gate filling), (ii) the orientation introduced by cavity filling, (iii) the orientation induced during packing, and (iv) relaxation of the orientation after the shear stress has dropped to zero.

For a given size of mould and gate the net amount of molecular orientation decreases as the mould and melt temperatures are increased. Orientation increases as dwell pressure and dwell time are increased. Increasing injection speed should decrease orientation as this increased speed may result in an increase in melt temperature and this, together with the shorter cavity fill time should result in low residual orientation levels. In a slowly filled part, cooling occurs during flow and this introduces shear, and therefore orientation, which is 'frozen-in'—the low melt temperature means less relaxation. When thick-walled components are being moulded slow filling can be used as the cooling rate is very slow; slow filling improves surface finish.

In general, crystallinity levels will be increased if the dwell pressure is increased as the effect of the extra pressure will be to cause orientation and force the chain segments together thus creating regions of order. High mould and melt temperatures also increase the probability of higher crystallinity contents, as the slow cooling which results increases the probability of the chains coming into the right configuration for crystallinity to occur and develop. Conversely, quick cooling from a high injection temperature reduces crystallinity and also reduces the size of the crystal structures—this affects transparency (see Section 2.2).

The rate of crystallisation is influenced by the temperature; there is a temperature at which the rate of crystallisation is a maximum. Therefore, the longer the polymer is kept at this 'optimum' temperature (which is approximately midway between the glass transition temperature (T_g) and the melting temperature (T_m) the more crystallisation there will be within a given sample.

5.2 Experimental Results for PS and PP

The previous section has indicated very briefly how compression, orientation and crystallisation occur and what effect they have on properties of mouldings. Through this we are generally able to explain the results (Brydson, 1977) obtained by Allen and Van Putte (1974) when they studied the effects of melt temperature, injection time and injection pressure on several properties.

In their work they varied these three operating variables, measured the change in properties and then calculated the correlation coefficients. (This is a statistical property indicating the degree of correlation between two variables. It has a maximum value of +1 to indicate perfect positive correlation, a minimum value of −1 to indicate perfect negative correlation and a value of 0 to indicate zero correlation.) Their results are summarised in Tables 4 and 5.

Let us first consider polystyrene. This is an amorphous material and the main variable will be the amount of residual orientation. An increase in melt temperature (keeping everything else equal) will normally mean a decrease in residual orientation. Hence we would expect for end-gated tensile,

TABLE 4
CORRELATION COEFFICIENTS FOR MODIFIED POLYSTYRENE

Property	*Melt temperature*	*Injection time*	*Injection pressure*
Tensile load at failure	−0·83	0·50	0·02
Flexural strength	−0·90	0·47	−0·01
Izod impact strength	−0·90	0·28	0·13
Ball-drop impact strength	0·90	−0·11	−0·11
Shrinkage across flow	0·27	0·04	0·58
Shrinkage with flow	0·25	0·19	−0·55
Stiffness modulus	−0·75	0·45	−0·13

TABLE 5

CORRELATION COEFFICIENTS FOR MODIFIED POLYPROPYLENE

Property	*Melt temperature*	*Injection time*	*Injection pressure*
Tensile load at failure	−0·72	−0·39	0·63
Flexural strength	−0·72	−0·45	0·63
Izod impact strength	−0·58	−0·17	0·11
Ball-drop impact strength	0·49	0·01	0·57
Shrinkage across flow	0·31	−0·31	0·94
Shrinkage with flow	0·12	−0·15	−0·95
Stiffness modulus	−0·77	−0·45	0·61

flexural and Izod mouldings a decrease in these three types of strength with increased melt temperature. On the other hand ball-drop impact strength would be expected to increase. These results are borne out by the strong correlation coefficient which is negative or positive as appropriate. (When the correlation coefficient is negative this means that, for example, an increase in melt temperature will cause a decrease in flexural strength.) No such strong correlation occurs with injection time (see Section 5.3.2).

Being amorphous the greatest effect of increasing injection pressure is to increase the compression packing of the polymer so that the decreased moulding shrinkage is expected.

Turning to polypropylene we have a crystalline material where the crystallisation is somewhat more dominant than orientation of the molecules. For this reason melt temperature has a less definite influence on impact strength (both Izod and falling ball types). Higher injection pressures are clearly having a major influence on the degree of crystallinity as reflected in higher strengths and stiffness. The effect of reducing shrinkage is all the more marked.

A summary of experimental results is shown in Table 6 (Allen and Van Putte, 1974). These results clearly show that the physical properties associated with polystyrene and polypropylene are, to a major degree, a function of the choice of moulding conditions. If, for example, the cavity pressure is varied, or varies, then the mouldings produced will be of a different size as shrinkage is sensitive to changes in cavity pressure.

If the hydraulic line pressure is set at the same value then the cavity pressure can still vary enormously as line pressure alone does not decide what the cavity pressure shall be. Allen and Van Putte (1974) produced two sets of samples from the same material in the same mould using the same

TABLE 6

SUMMARY OF EXPERIMENTAL RESULTS (after Allen and Van Putte, 1974)

Property	Range	Standard error of estimate	Controlling variable	
			Primary	Secondary
Tensile load at failure, pounds				
Polystyrene	95	7	Melt temperature (−0·83)	Injection time (0·50)
Polypropylene	72	5	Melt temperature (−0·72)	Injection pressure (0·63)
Flexural strength, psi				
Polystyrene	3147	210	Melt temperature (−0·90)	Injection time (0·47)
Polypropylene	1752	205	Melt temperature (−0·72)	Injection pressure (0·63)
Izod impact strength, foot-pounds per inch				
Polystyrene	2·6	0·2	Melt temperature (−0·90)	None
Polypropylene	7·4	1·0	Melt temperature (−0·58)	None
Ball-drop impact strength, inch-pounds				
Polystyrene	58	6	Melt temperature (0·90)	None
Polypropylene	74	6	Injection pressure (0·57)	Melt temperature (0·49)
Shrinkage across flow, inches per inch				
Polystyrene	0·006	0·0008	Injection pressure (−0·58)	None
Polypropylene	0·009	0·0003	Injection pressure (−0·94)	None
Shrinkage with flow, inches per inch				
Polystyrene	0·006	0·0009	Injection pressure (−0·55)	None
Polypropylene	0·004	0·0003	Injection pressure (−0·94)	None
Stiffness modulus, psi × 10^{-5}				
Polystyrene	0·9	0·12	Melt temperature (−0·75)	Injection time (0·45)
Polypropylene	0·9	0·08	Melt temperature (−0·77)	Injection pressure (0·01)

line pressure. In one case the cavity pressure was 21 MN m^{-2} (3000 psi) and in the other case it was 70 MN m^{-2} (10 000 psi). Differences in pressure drops across the cavity were also found. Due to the effect of pressure on shrinkage, a large pressure drop in a cavity will cause non-uniform shrinkage and therefore warping.

Variations in cavity pressure, at the same nominal line pressure, can occur for many reasons; for example:

1. Variations in the operating characteristics of the hydraulic components cause line pressure variations (from shot to shot) which pass unnoticed because the gauge used may be inaccurate, slow to respond or, located in the wrong place.
2. Melt temperature variations, caused through using an inadequate controller (e.g. on–off) or nonuniform plasticisation.
3. Material changes may cause the screw cushion size to vary and this will affect the pressure drop from the mould to the cavity.
4. The amount of melt lost from in front of the screw may vary, e.g. due to valve nonsealing.

5.3 Weld Strength

In Section 5.2 it was pointed out that a large correlation coefficient could not be obtained between various tests and injection time. It does not automatically mean that injection time is unimportant as, for example, until fairly recently it was difficult to 'compartmentalise' the moulding process. A change in injection temperature could cause a change in mould filling speed and therefore a change in initial mould filling time and final mould filling time or packing, both events occurring within 'injection time' (see Chapter 6). Alternatively, a change in 'injection pressure' could cause a change in 'injection speed'.

To study the affects of changes in moulding conditions on the properties of injection moulded components one needs a machine in which changes in one machine setting do not affect other machine settings. At the present time a servo-valve controlled machine appears to offer the best way of achieving this desirable state of affairs.

Such a machine was used to study the effect of changes in blend preparation methods, and of changes in machine operating procedures on the properties of PP/EP(D)M samples (see Chapter 7).

5.3.1 PP/EP(D)M blends

PP/EP(D)M blends are now being used widely by the automobile industry to produce car bumpers and bumper covers. Such mouldings are large (e.g. 1·5 m long) and are produced in multiple-gated moulds (e.g. four gates may be used) from materials based on mixtures of PP and EP(D)M, or of mixtures of PP copolymers and EP(D)M.

The PP material may be tumble mixed with the EP(D)M and then fed to the injection machine (direct blend injection moulding method or DBIM method) or it may be melt compounded (e.g. in a twin-screw extruder) and regranulated before being fed to the moulding machine. Both systems are in use as although melt compounding gives superior properties the DBIM route has obvious attractions.

5.3.2 Weld strength results

Dumb-bell shaped samples, gated from one end (see Fig. 1), were produced from the base polypropylene material, a tumble blend of PP/EP(D)M and a melt-compounded blend of PP/EP(D)M; selected samples were strained at a rate of 100 mm min^{-1}.

No significant difference could be found between the samples produced

TABLE 7

TEST RESULTS OBTAINED FROM PP AND PP/EP(D)M BLENDS

Type of material	*Melt temp. (°C)*	*Elongation (%)*	*Yield stress ($N mm^{-2}$)*	*Breaking stress ($N mm^{-2}$)*	*Charpy impact strength ($kJ m^2$) (notched)*		*Vicat softening point (°C)*
					24°C	*8°C*	
Polypropylene (PP)	240–245	62	31·8	24·5	—	—	—
PP plus 10% Nordel 1500—tumble blended	240–245	64	28·7	23·0	—	—	—
As above but melt mixed on a twin-screw extruder	240–245	92	27·6	22·7	—	—	—
PP	268–273	301	29·3	24·6	23·6	12·3	134
PP plus 10% Nordel 1500—tumble mixed	268–273	128	26·2	21·6	44·1	26·5	121
PP plus 10% Nordell 1500—melt mixed on a twin-screw extruder	268–273	307	25·5	22·4	42·0	23·7	128

The rate of cavity filling was $\simeq 7\,g\,s^{-1}$.
Gate size for both cavities (length × width × depth) was 1·5 × 3·5 × 1·5 mm.
Shot weight was 13·8 g.

from cavity one as opposed to those produced from cavity two. There was a difference between the three materials and in general it was concluded that where high elongations at break are required that the materials should be melt compounded before being moulded at high moulding temperatures. The conclusion about high moulding temperatures should be treated with caution as reduced orientation, in the test direction, may be responsible for some of the high elongation at break results. Adding rubber to PP improves the impact strength (see Table 7) but the method of rubber incorporation does not have the same dramatic effect as it does for the elongation at break results.

Having established what was possible from each material identical gates were inserted into the other end of each cavity so that the flowing melt stream produced a weld at the centre of the dumb-bell shaped cavity. The strength of the weld was then assessed by measuring break stress and elongation at break (Table 8).

It was now found that there was a significant difference (see Table 8) between the cavities; the cavity lying closest to the ground gave the highest weld strength. The lower cavity may have filled first and so the material at the weld may have had more time to develop a strong bond (unless one has a specific interest in studying such effects, this test result highlights the wisdom of only producing one sample at a time—see Section 5, Chapter 6). The difference between the two cavities was most marked at a lower melt temperature (245 °C) and at slower filling speeds.

The results obtained were very surprising as the gates in each cavity were identical. What was even more surprising was that fast cavity filling followed by long dwell times using high melt temperatures did not give the expected large value of weld strength. The work seemed to indicate that:

1. High melt temperatures lengthen the cycle without necessarily giving very high values of weld strength
2. The available injection time should be devoted to mould filling, rather than fast mould filling followed by packing

It would therefore appear that useful weld strengths can be obtained from PP/EP(D)M blends by using slow filling speeds at a melt temperature of 250 °C and with little or no dwell pressure. Such conditions can give very high elongation at break values but more work is needed to clarify the differences between the two cavities. What is immediately obvious from this piece of work is that good compounding of the two materials is essential and that the rate of injection is important.

TABLE 8

WELD STRENGTH RESULTS FROM PP/EP(D)M BLENDS

Type of material	*Cavity number*	*Rate of fill*	*Elongation at break* (%)	*Range* (%)	*Average of 1 and 2*	*Breaking stress* ($N\,mm^2$)	*Range* ($N\,mm^2$)	*Average of 1 and 2*
		8a Melt temperature 240–245°C						
Polypropylene (PP)	1	Slow	345	162–520	—	21·8	21·4–22·4	—
	2	Slow	37	20–62	191	21·6	21·3–22·6	21·7
	1	Fast	276	20–540	—	21·6	21·1–22·1	—
	2	Fast	21	14–30	149	21·4	21·0–22·0	21·5
PP plus 10% EP(D)M—tumble mixed	1	Slow	34	28–46	—	14·8	14·2–15·9	—
	2	Slow	9	4–16	23	12·5	11·4–13·4	13·7
	1	Fast	4	4–4	—	12·0	10·9–13·0	—
	2	Fast	4	4–4	4	12·2	11·6–12·7	12·1
PP plus 10% EP(D)M—melt mixed	1	Slow	31	14–62	—	20·1	17·7–20·4	—
	2	Slow	24	13–36	28	18·6	17·8–19·3	19·4
	1	Fast	21	14–34	—	18·0	17·2–18·8	—
	2	Fast	16	14–18	19	18·1	17·2–18·8	18·1
		8b Melt temperature 268–273°C						
PP	1	Slow	30	11–50	—	20·2	19·7–20·6	—
	2	Slow	24	7–41	27	19·4	18·8 20·0	19·8
	1	Fast	17	16–18	—	19·8	19·6–19·9	—
	2	Fast	14	14–14	16	19·8	19·3–19·6	19·7
PP plus 10% EP(D)M—tumble mixed	1	Slow	14	3–24	—	13·5	10·8–16·1	—
	2	Slow	3	3·0–3·6	9	12·0	11·5–12·5	12·8
	1	Fast	4	3·4–4·4	—	11·0	10·1–11·9	—
	2	Fast	4	3·4–4·4	4	11·4	10·3–12·5	11·2
PP plus 10% EP(D)M—melt mixed	1	Slow	16	11·2–21·0	—	17·4	17·3–17·5	—
	2	Slow	12	11·4–12·6	14	17·8	17·6–18·0	17·6
	1	Fast	18	16·4–19·0	—	17·1	17·0–17·1	—
	2	Fast	16	15·4–15·6	17	17·1	17·0–17·2	17·1

Rate of filling: slow, 1·8 $g\,s^{-1}$; fast, 9·2 $g\,s^{-1}$. Shot weight was 15·5 g.

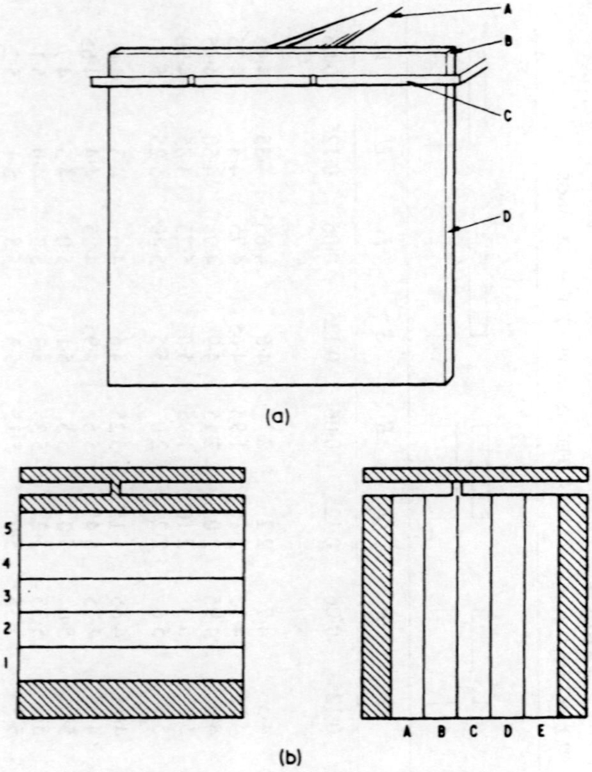

Fig. 4. (a) Schematic design of mould cavity for preparing plaques used in orientation experiments. Melt is fed by sprue A into a reservoir B. C is a removable gate section in which gates of varying dimensions (including length), shape, number and disposition may be used. D is the mould cavity. (b) Numbering and lettering of strips from a plaque mould describing results in Table 9. Shaded areas indicate discarded material (Brydson, 1970).

6 EFFECT OF CHANGING GATE SIZE AND POSITION

The effect of changing gate size and position on the properties of a given injection moulding are quite remarkable. This is mainly because one alters the amount of orientation and, perhaps even more importantly, the direction of orientation within that particular moulding.

The effect of changes in gate dimensions, and to a lesser extent gate position, can be studied using the variable gate mould shown in Fig. 4.

TABLE 9

EFFECT OF GATE TYPE, LENGTH AND POSITION ON ORIENTATION EFFECTS IN PLAQUE MOULDINGS

Gate type and position										
	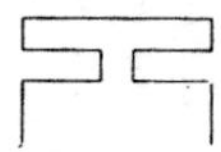		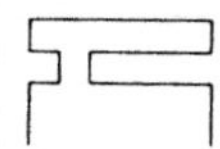						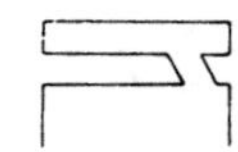	
Sample	*M*	*N*	*O*	*P*	*Q*	*R*	*S*	*T*	*U*	*V*
Gate length, in	0·125	0·06	0·125	0·06	0·125	0·06	0·125	0·06	0·125	0·06
Flexural load to break, kg										
1	2·55	2·55	4·5	4·7	2·2	2·1	4·6	4·65	4·35	4·45
2	2·4	2·3	4·75	4·75	2·1	1·95	4·65	4·75	4·4	4·25
3	2·4	2·4	4·9	4·95	2·05	2·15	4·9	4·9	4·50	4·65
4	2·4	2·4	5·25	5·3	2·10	1·95	5·0	5·25	4·85	4·90
5	2·65	2·05	5·4	5·6	2·2	2·0	5·5	5·40	5·25	5·3
A	6·1	6·1	4·65	4·65	5·15	5·25	4·6	4·6	4·3	4·4
B	6·3	6·3	4·75	4·75	5·45	5·5	4·95	4·85	4·4	4·65
C	6·3	6·2	5·0	5·0	5·3	5·5	5·1	5·0	4·5	4·7
D	6·25	6·1	5·3	5·35	5·25	5·35	5·3	5·3	4·9	5·1
E	5·9	5·8	5·3	5·35	5·05	5·15	5·3	5·5	5·1	5·2

6.1 The Variable Gate Mould

This mould produces a square shaped tile or plaque which measures 88·9 × 88·9 × 1·58 mm (3·5 × 3·5 × 0·062 in). This size was chosen as it suited the smallest machine which was available to the author at that particular time. The mould cavity is fed from one edge via a smaller chamber or cavity which is centrally filled via the sprue. Material therefore enters the larger cavity via a channel or gate which is machined into an insert. The gates are machined into bars of ground flat stock which are readily available in standard sizes at a reasonable price. As bars of up to 6 mm (0·25 in) thick can be accommodated a wide variation in gate length, width, thickness, angle, number and location can be studied.

Gate changing is very easy with this mould as all that needs to be done is to loosen two bolts and tighten two nuts. The gate changing block is thus slackened and the gate may be removed and another inserted.

If the tiles which this mould produces are cut into strips (see Fig. 4) then it can readily be shown that the material properties vary from one point to another within the moulding and that in general, orientation is highest in the flow direction (see Table 9). The strength of the material is therefore highest in that direction.

However it the strength increases in one direction then it falls in another; if a brittle material such as PS is being investigated then the 'weak direction' can be found by striking the moulding with a pointed metal tool. The crack thus produced generally follows a path which indicates the flow direction and its direction also indicates that molecular orientation is highest in that direction (Fig. 5).

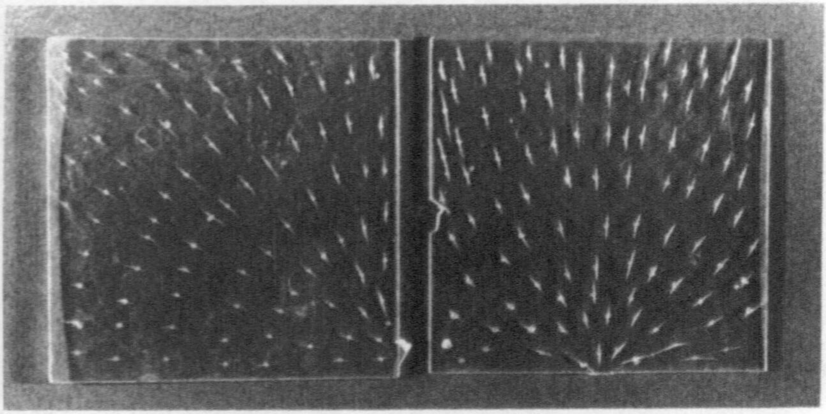

Fig. 5. Use of cracks to show the direction of orientation.

TABLE 10

FLEXURAL TESTING OF CENTRE AND CORNER GATED SAMPLES

(material general purpose polystyrene; results in kg cm^{-2})

(*a*) *Corner gated specimens*

Moulding no.	*Position*					*Grand average*
	A	*B*	*C*	*D*	*E*	
1	—	7·1	6·9	6·8	6·9	
2	6·9	6·6	6·2	6·2	6·0	
3	6·7	6·3	6·3	6·4	6·3	
4	7·0	6·9	6·6	6·3	5·9	
Average	6·9	6·7	6·5	6·4	6·3	6·6
	5	*4*	*3*	*2*	*1*	
5	8·0	7·1	6·7	4·7	6·0	
6	7·8	7·0	6·3	5·7	5·8	
7	8·0	7·0	6·4	6·2	5·0	
8	6·3	6·3	5·8	5·7	4·9	
Average	7·5	6·8	6·3	5·6	5·4	6·3

Ratio = 6·6/6·3 = 1:1.

(*b*) *Centre gated samples*

Moulding no.	*Position*					*Grand average*
	A	*B*	*C*	*D*	*E*	
1	6·7	6·6	6·9	6·9	6·9	
2	6·3	6·2	6·5	6·6	6·5	
3	6·9	7·1	6·9	6·7	6·7	
4	6·1	6·1	6·2	6·4	6·5	
Average	6·5	6·5	6·6	6·6	6·6	6·6
	5	*4*	*3*	*2*	*1*	
5	2·9	2·2	2·8	2·5	3·0	
6	3·4	2·3	2·9	2·5	3·2	
7	3·0	2·6	2·8	2·8	3·1	
8	2·5	2·2	2·1	2·7	2·5	
Average	3·0	2·3	2·6	2·6	3·0	2·7

Ratio = 6·6/2·7 = 2·4:1.

6.2 Effect of Changing Gate Position

It is commonly believed in the plastics industry that corner gating is better than centre gating as the mouldings have a reduced level of orientation. This may be true in some cases, but it is not automatically so, although experiments can be devised which appear to confirm the belief.

If, for example, tiles are produced using both a centre and a corner gate then the tiles may be examined to detect the presence of orientation. This may be done by cutting the tiles into 12 mm (0·5 in) wide strips and measuring the flexural load to break. By cutting the strips from the tiles in two directions at right angles, as shown in Fig. 4, the effect of orientation when using a material such as polystyrene may be shown (see Table 9).

The results from one typical experiment are shown in Table 10. By averaging all the results obtained from one direction and then comparing them (as a ratio) with the results obtained in the other direction, then a measure of anisotropy may be obtained. Judged in this way the corner gated specimens appeared to give more isotropic mouldings, i.e. the flexural strength was similar in both directions. However, the mouldings are not isotropic and this was proved by measuring the falling weight impact strength of the two moulding runs (Fig. 6). The falling weight impact strength of the two moulding runs was found to be virtually identical; all that was different was that the corner gated specimens failed in the diagonal direction.

6.3 Gate Size Changes

Gate cross-sectional area is important as this controls gate sealing. Material thickness at the gate is usually less than the part wall thickness and so when the cavity has filled, and the material stops flowing, the temperature of the material in the gate region changes most rapidly.

If a small gate is being used (with fast cavity fill rate, low packing pressures, short packing times and high mould temperatures) then parts with reduced orientation will result for many thermoplastics materials. This is because under these conditions a great deal of orientation relaxation is possible because of Brownian movement. Larger gates cool more slowly and because they take longer to seal they can increase the level of orientation within the moulding unless the packing pressures and times are reduced.

The length, or land, of a gate should be kept as short as possible as shorter landed gates provide least resistance to flow. The rate of mould fill is also higher with a short landed gate and because of this, decreasing land length will commonly result in the gate remaining open for a longer time—

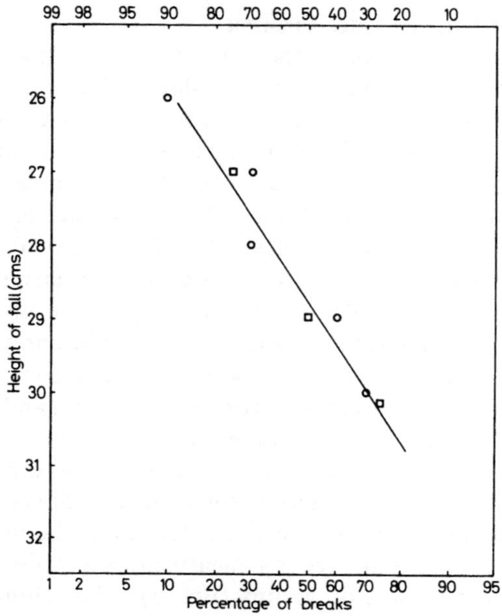

Fig. 6. Falling weight impact strength of centre (○) and corner (□) gated specimens.

this will affect the level of orientation as previously described. When the size of a gate is changed, then the properties of the moulding will also change because of changes in, for example, the level of orientation.

It must be remembered that orientation is not the only factor which should be considered when gate size and location are being decided. Moulding dimensions and therefore shrinkage are also dramatically effected by gate size and position. If the gate freezes too quickly then the affect of cooling cannot be compensated for and high shrinkages will result. As usual, therefore, one must make a compromise between various requirements.

7 EXPERIMENT DESIGN IN INJECTION MOULDING

If it is decided to investigate the effect of changes in processing conditions on properties it is suggested that after preliminary investigations an in-depth investigation should be performed on a restricted number of variables.

7.1 Restricting Experimentation

It is commonly found that processing affects the properties of injection mouldings. Properties such as orientation, crystallinity, filler particle size and shape, etc. are affected by melt temperature, injection speed, screw speed, etc.

If it is decided to investigate all the machine settings which it is thought influence properties, then a tremendous number of experiments would have to be performed. For example, if it is thought that eight machine settings are involved and if it is decided to investigate each of these at three levels (low, medium and high) then 3^8 experiments would have to be performed.

However, by choosing randomly it is found that significant trends can be observed with a much smaller number, e.g. 20 → 30. This approach to the problem was suggested by Bown and Robinson (1970) and the following scheme was derived from their suggested programme.

7.2 The Variables Investigated

Let us suppose that it has been decided to investigate the following eight variables at three different levels (L = low, M = medium, H = high) and it has been decided to restrict the number of experiments to 20. The eight chosen variables are:

1. Dwell or second-stage pressure
2. Screw rotational speed
3. Injection rate
4. Back pressure
5. Melt temperature
6. Dwell time
7. Mould temperature
8. Cooling time

Injection pressure has been omitted because most machines are now fitted with a two-stage pressure system; the first stage pressure is used to obtain the desired rate of mould filling and on many machines the set high pressure is not seen as the machine is adjusted so that just before the mould fills the pressure drops to the second-stage pressure. During the mould filling operation the resistance to flow is relatively small and so the injection line pressure never reaches the preset value.

Back pressure and screw rotational speed have been incorporated as these settings dramatically affect the amount of shear that is developed in the system. Cooling time is present, as apart from affecting production rate its variation also affects residence time in the machine.

Advice on measuring melt temperature has been given elsewhere but it must be emphasised that this must be done only after the operatives have been instructed on a 'safe procedure' as the melt can be shot from the nozzle at high rates. If a material's normal or usual running temperature is 200 °C and it is decided to investigate at 180 °C (low) and 220 °C (high), then appropriate barrel settings should be found which give these temperatures; these settings should then be used when it is necessary to vary the 'melt temperature' as detailed below.

The question may well be asked 'How do you know the value of low, medium, and high?' The answer is, usually, 'experience' because by the time it has been decided to investigate the effects of processing on properties usually some experiments have been done and experience accumulated. The values at which to set a machine control may, therefore, have been obtained from consulting the material manufacturer's literature or by experiment.

7.3 Level Allocation

Obtain 20 small file cards and on each draw a table which has 10 vertical columns and, after labelling, sufficient space for just one line (see Table 11). Insert the column headings and the initial experiment number.

TABLE 11

LAYOUT OF EXPERIMENTAL CARD

Final expt. number . . .

Initial expt. no.	Melt temp.	Dwell pressure	Screw speed	Injection rate	Back pressure	Dwell time	Mould temp.	Cooling time	Quality rating
1	M	L	M	L	L	H	M	L	

A die may then be used to decide whether a particular machine setting should be low, medium or high. For example, suitable values of melt temperature may have been decided as being 180 °C (L); 200 °C (M) and 220 °C (H). Let die faces 1 and 2 be L; faces 3 and 4 be M and faces 5 and 6 be H.

Roll the die and if a 3 comes up insert M in the column under melt temperature. Repeat for the other settings as shown in Table 11.

After all 20 cards have been completed in the above fashion it may be found that it is more convenient to change their order so that, for example, all the experiments at one mould temperature are done at the same time.

7.4 Performing the Experiments

Assign final experiment numbers to each card, rearrange the cards and perform the experiments. (For reference purposes it may be found more convenient to transfer the information to one large table on a single sheet of paper.)

Make certain that the machine has 'settled down' at each new setting before samples are taken, e.g. even on small machines when barrel temperature settings are changed the melt temperature may not equilibrate for 20 min.

Each moulding is then given a quality rating (or assessment) which is plotted on a graph against each variable in turn. If the variable does not influence quality the graph will be a straight line parallel to the axis but if it does have some influence the line will slope or will reach a peak. It may then be decided to investigate the 'significant' variables in greater depth.

7.5 An Examination in Greater Depth

The experiments discussed in the previous section may have indicated that three machine/melt variables should be investigated in greater depth. For example, the initial experiments may have shown that melt temperature, injection speed and cavity pressure should be investigated. Preliminary experimentation or other reasons, may also have indicated the maximum and minimum values for each of the three selected variables. For example, it may be known that a particular grade of polypropylene is normally processed over the range 230–250 °C, and that above a certain pressure the mouldings will be flashed. Initially, therefore, the temperature range will be approximately 20 °C and the maximum values of cavity pressure and injection rate will be limited to those which will not cause flashing. What is required are the extreme values for each of the three variables so that no matter what combination is chosen, reasonable mouldings can be produced. Some further 'preliminary' experimentation may be necessary in order to get this information.

If each value is identified as H, M and L (see Section 7.3) then each variable may be investigated as shown in Table 12.

If it is decided to investigate a different temperature region then another nine experiments may be conducted at say 250–270 °C; for this series of experiments a different cavity pressure range may be appropriate.

TABLE 12
EXPERIMENT DESIGN FOR THREE VARIABLES

Expt. no.	*Melt temperature*	*Cavity pressure*	*Injection rate*
1	M	M	M
2	H	H	H
3	L	L	L
4	L	L	H
5	L	H	L
6	H	L	L
7	L	H	H
8	H	H	L
9	H	L	H

7.6 Statistical Analysis

The results should be analysed so that the following information is obtained.

1. The average value
2. The experimental error
3. The effect of a variable on the property being investigated and its correlation coefficient
4. The interaction between the variables

For example, Allen and Van Putte (1974) used multiple regression analysis on their results and, in addition to the above, derived equations which expressed each property measured as a function of the process variables. The equations were used to create computer generated contour plots for some of the physical properties so that these plots could be subjected to visual analysis. The authors performed 27 carefully controlled experiments; they varied injection pressure and fill rate over three temperature ranges.

Such statistically designed experiments are very useful as they give the maximum amount of quantitative information from the least number of experiments. A number of variables may be worked with or handled at the same time and their separate and simultaneous responses may be investigated. The significance and reliability of the experiments may be obtained by calculating the standard deviation and by performing the 'F' test. If multiple correlation coefficients are calculated for each of the properties measured (e.g. shrinkage) then it should be possible to decide if the variables investigated are responsible for the changes encountered. If they are not, then other factors may need to be examined in detail so as to determine the machine parameters responsible for the change.

8 MOULDING QUALITY AND UNKNOWN CHANGES

Injection moulding should be operated so that consistent mouldings of the required quality are produced at the lowest possible cost.

It sounds a simple and straightforward statement but it is surprising how often it is not followed. Very often it is not followed because people do not appreciate what the statement implies. To put it bluntly, many machines and materials are not as good as the makers and/or users believe.

The properties of a particular moulding, if made from a given polymer/machine composition, can be altered dramatically. However, once the properties required in the moulding have been specified then it should be possible to produce mouldings of the required quality provided that machine performance is consistent. If consistent operation can be guaranteed operating costs should decrease as, for example, labour is not employed in checking and inspection, the reject rate decreases and time is not wasted on the production of over-specified mouldings.

It is commonly assumed that if all the settings of an injection moulding machine (e.g. barrel temperature, injection time, etc.) are held strictly constant then the mouldings which that machine is producing will also be constant. This is very often not true since the results of the injection moulding operation are dependent on many factors and most of these factors are not subject to direct control. For example, unknown changes may occur in:

1. The material. The material supplied by the manufacturer may be inconsistent or the ratio of virgin to reground material may be altered.
2. The environment. Air and cooling water temperatures may alter and/or the relative humidity may change thus introducing changes into the material or into the machine, e.g. due to oil temperature changes.
3. The machine. Voltage changes may cause barrel temperatures (and therefore material viscosity) to alter, the back-flow valve on the screw may function erratically, etc.
4. Machine operation. In a semi-automatic cycle the operator may introduce changes by varying the mould open time.

Better machine monitoring and control, using microprocessor based systems will help eliminate such variables and thus give more consistent mouldings.

REFERENCES

Allen, E. O. and Van Putte, D. A. Know all your moulding variables, *Plastics Engineering*, p. 37, July 1974.

Ballman, R. L. and Toor, H. L. Orientation in injection moulding, *Modern Plastics*, p. 113, Oct 1960.

Bown, J. and Robinson, J. D. *Injection and Transfer Moulding and Plastics Mould Design*, Business Books, London, 1970.

Brydson, J. A. *Flow Properties of Polymer Melts*. Published for The Plastics Institute by Iliffe, London, 1970.

Brydson, J. A. *The Effect of Processing on Properties*, an NCRT information document, 1977.

Cleerman, K. J. Injection moulding shapes of rotational symmetry with multi-axial orientation, *SPE J.*, p. 43, **23**, Oct 1967; Injection moulding shapes of rotational symmetry with multi-axial orientation, *SPE J.*, p. 55, **25**, Jan 1969; Rotation injection moulding. Part 3. Isotatic polystyrene, *SPE J.*, p. 44, **25**, Sept 1969.

Craft, J. L., Cox, R. and Govindasamy, S. Private communication, NCRT, 1980.

Han, Chang Dae, *Rheology in Polymer Processing*, Academic Press, London, 1976.

ICI. *The Variation of Moulding Behaviour and Physical Properties of Polythene with Melt Flow Index and Density*, IS 649, ICI, Herts, 1960.

Jensen, M. and Whisson, R. R. *Determination of Frozen-in-stresses in Thermoplastic Injection Mouldings*, RAPRA Research Report, p. 197, 1971.

Rubin, I. I. Injection moulding theory and practice, *SPE Monograph*, John Wiley and Sons Inc., New York, 1972.

Sharples, A. *Introduction to Polymer Crystallinity*, Arnold, London, 1966.

Shearer, N. H., Guillet, J. E. and Coover, H. W. Relations between physical properties and molecular structure of polypropylene. *SPE J.*, p. 83, **17**, Jan 1961.

Thomas, K. and Meyer, D. E. Anisotropy arising from processing of reinforced plastics. In *Composites—Standards, Testing and Design*, IPC, Guildford, p. 131, 1974.

Thomas, K. and Meyer, D. E. Ultrasonic measurement of reproducibility and anisotropy of processed polymers, *Plastics and Rubber: Materials and Applications*, p. 136, Jan 1976.

Thomas, K. and Meyer, D. E., Anisotrophy and impact properties of injection moulded glass-fibre reinforced thermoplastics. Paper presented at the Fifth European Conference on Plastics and Rubbers, Paris, 12–15 June 1978. Available from the National Physical Laboratory, Teddington, UK.

Treloar, L. R. G. *Introduction to Polymer Science*, Wykeham Publications, London, 1970.

Wheelans, M. A. *Injection Moulding of Rubber*, Newnes Butterworths, London, 1974.

White, J. L. and Dee, H. B. Flow visualisation for injection moulding of polyethylene and polystyrene melts. *Polym. Eng. Sci.*, p. 212, **14**, 1974.

CHAPTER 6

Testing and Quality Control

1 THE OBJECT OF TESTING

Testing is very often a time consuming and expensive business and before one can discuss specific tests, and test specimen production, it is reasonable to decide the purpose or object of testing in general.

Testing is not something new, as from the earliest of times, man has sought to compare materials of the same type and/or materials of different types. Over the years, therefore, many tests have been devised in order to enable judgements to be made as to:

(a) The quality and consistency of raw materials
(b) The quality and consistency of associated products
(c) The merits, or otherwise, of new or modified materials
(d) The suitability of a particular design

The need to carry out adequate fitness-for-purpose and quality control testing has been increased greatly in recent years by legislation on liability for product performance. New legislation is making the risk of punitive damage claims for product liability a very real possibility which manufacturers and suppliers must face and guard against; particularly by ensuring that their materials and products are not in any way defective. Hence the need for adequate testing (Dewey, 1980).

It must be remembered that within the family of plastics and rubbers, properties vary widely and, furthermore, the properties of an individual polymer can be dramatically altered by compounding with other ingredients or by altering the processing conditions. Consequently, the measurement of physical properties, i.e. physical testing of plastics and rubbers, is extremely important in the development of the science and technology of these materials. Specialist books on this subject are listed at the end of this chapter (Ives *et al.*, 1971; Turner, 1973; Scott, 1965).

TABLE 1

LIST OF INTERNATIONAL STANDARDS (ABBREVIATED)[a]

ISO 178-1975	Plastics—Determination of flexural properties of rigid plastics
ISO/R 179-1961	Plastics—Determination of the Charpy impact resistance of rigid plastics (Charpy impact flexural test)
ISO/R 180-1961	Plastics—Determination of the Izod impact resistance of rigid plastics (Izod impact flexural tests)
ISO/R 292-1967	Plastics—Determination of the melt flow index of polyethylene and polyethylene compounds
ISO 293-1974	Plastics—Compression moulding test specimens of thermoplastic materials
ISO 294-1975	Plastics—Injection moulding test specimens of thermoplastic materials
ISO 295-1974	Plastics—Compression moulding test specimens of thermosetting materials
ISO 305-1976	Plastics—Determination of thermal stability of polyvinyl chloride, related chlorine-containing polymers and copolymers and their compounds—Discoloration method
ISO 306-1974	Plastics—Determination of the Vicat softening temperature of thermoplastics
ISO/R 527-1966	Plastics—Determination of tensile properties
ISO 868-1978	Plastics—Determination of indentation hardness by means of a durometer (Shore hardness)
ISO/R 899-1968	Determination of tensile creep of plastics
ISO/R 960-1969	Plastics—Determination of the water content in polyamides
ISO/R 1183-1970	Plastics—Methods for determining the density and relative density (specific gravity) of plastics excluding cellular plastics
ISO 2557-1976	Plastics—Amorphous thermoplastic moulding materials—Preparation of test specimens with a defined level of shrinkage
ISO 2557/2-1979	Plastics—Amorphous thermoplastic moulding materials—Part 2: Preparation of test specimen with a defined level of shrinkage in the form of rectangular plates by injection moulding

[a] Copies of the above standards may be obtained from the Standards organisation in each country—in the UK this would be The British Standards Institute.

2 STANDARDISATION

The tests that have been devised fall into two main categories; those which are suitable for quality control and specification purposes, and those which yield data suitable for design purposes. The former tend to be called single point tests, i.e. ones which are carried out at a single test temperature and/or speed, whilst the latter take into consideration the marked effect that changes in time, temperature and environmental conditions have on the properties of polymers.

Whether a test is used to determine an absolute value of a property, i.e. for research or whether its purpose is routine quality control, a major requirement of the test is that it should be repeatable. People in the same laboratory or in different laboratories should be able to get the same result on the same material. Since there are a large number of factors, such as test machine design, test piece size, shape and method or production, temperature of test, etc., which will influence the result obtained, it is fairly obvious that for agreement to be possible standardisation of the test methods is essential.

To achieve this, various organisations have evolved in many countries and each organisation prepares and issues standards suitable for testing a wide range of materials (e.g. The British Standards Institute (BSI), The American Society for Testing and Materials (ASTM), The International Organisation for Standardisation (ISO)). Copies of standard test methods relevant to plastics and rubbers can be obtained from these organisations, for example, BSI and ISO standards are available in the UK from BSI, Newton House, Pentonville Road, London N1 9ND. Some ISO standards are listed in Table 1. No matter which test method is employed it is, important to ensure that the samples have received the same conditioning before testing. For many materials this conditioning is simply storing the specimens at a standard temperature before testing. Despite its simplicity such physical ageing is important as postmoulding changes commonly occur.

3 COMMON SHORT-TERM TESTS

In Section 1 it was pointed out that there are a number of good and valid reasons why one should test. The tests may be grouped into two main categories; those which are suitable for quality control and specification purposes, and those which yield data for design purposes. Within the first category are the most commonly performed tests; these are sometimes known as short-term, mechanical tests.

Despite their limitations, short-term tests are the most widely used as they are relatively cheap to perform and, provided account is taken of their limitations, they give useful information, e.g. on material uniformity.

In this category of tests are numbered tensile testing, impact testing and flexural testing. These tests are roughly listed in order of importance. Such tests are commonly performed on specially prepared, standard test specimens.

3.1 Tensile Testing

Tensile testing is probably the most widely used short-term mechanical test of all. This is because it is relatively easy to perform, gives reproducible results and yields a great deal of information. From this one test one can obtain not only tensile strength, but also elongation and modulus; the same basic machine may also be used to perform a number of other tests (e.g. flexural strength) with relatively simple modifications.

Such machines are usually constructed so that the test sample can be deformed at a standard rate; the resistance to deformation and the amount of elongation is measured. In order to ensure that deformation and breaking occur in a specified region the sample is usually dumb-bell shaped or waisted.

The sample dimensions in the waist region are first measured with a micrometer or dial gauge and then the sample is firmly gripped in the jaws of the machine. An extensometer may then be clipped to the sample and one jaw of the machine is drawn away from the other (usually by an electric motor) at the specified speed.

The application of this tensile stress causes the specimen to stretch from its original length (l_o) to a longer length (l_1). Tensile strain (Σ_l) commonly referred to as 'strain' may then be defined as:

$$\Sigma_l = \frac{\text{change in length}}{\text{original length}} = \frac{l_1 - l_o}{l_o}$$

If sufficient stress is applied then the sample will break. The elongation at break (E_B) is usually expressed as a percentage of the original length and this, therefore may be expressed as:

$$E_B = \frac{(l_1 - l_o) \times 100}{l_o}$$

The tensile strength at break, i.e. the ultimate tensile stress, may be obtained by dividing the force necessary to cause failure by the cross-sectional area at break. As this area is difficult to measure, the tensile strength that is commonly quoted is that which is given by:

$$\frac{F}{A_o}$$

where F = the force which causes failure and A_o = the original cross sectional area. (Yield strength may be calculated from the load at which the specimen continues to elongate without additional load.) The stress and strain obtained by using the original sample dimensions are sometimes referred to as 'nominal' or 'engineering' stress and/or strain.

The ratio of stress/strain, if stress is proportional to strain, is called the Young's modulus, and this figure is commonly quoted for many materials as it gives an appreciation of the stiffness of a particular material. For materials such as steel the value of Young's modulus, obtained from the slope of a stress-strain plot (up to the yield point), does not depend markedly on the test conditions and as such has come to be regarded as a material constant. Such is not the case for polymeric materials (see Section 2); for test specimen procedure see ISO recommendation R 527 or ASTM D 638.

3.2 Rigidity and Flexural Strength

In general the rigidity of plastics materials is very low and this has hindered their applications in many areas. It must also not be forgotten that a material which is suitable for use at one temperature will not be suitable for use at another. With some rubber modified thermoplastics, excessive stiffening may occur at low temperatures and for some applications this can be considered a disadvantage, e.g. car bumpers.

Flexural properties are usually measured by applying a stress at the centre of a rectangular bar which is supported at two other points, i.e. three-point loading. During the test the force applied and the deflection which results is measured and the test may be easily performed on what has been previously described as a tensile strength machine.

A useful specimen measures 50 mm long by 6 mm wide by 4 mm thick (1·95 × 0·23 × 0·16 in) and it is called 'useful' because it can be used for another test, i.e. the Charpy impact test (see Section 3.3.2). The two supports are set at 40 mm (1·56 in) apart and the crosshead moves at a rate of 10 mm min^{-1} (0·39 in min^{-1}). When a deflection of 6 mm (0·23 in) is

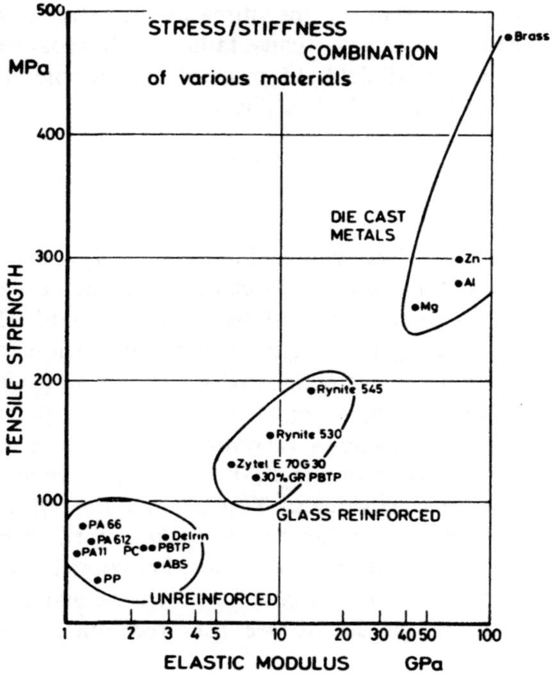

Fig. 1. Strength/stiffness combination of varied materials. Rynite = polyethylene terephthalate (PETP); Rynite 530 = 30 % glass fibre; Rynite 545 = 45 % glass fibre; Zytel E = PA66 = 30 % glass fibre (after Anon., 1980).

observed, the force reading is measured. The test procedure is described in DIN 53452.

The ASTM standard (D 790) uses a sample which measures 127 × 12·7 × 3·2 mm ($5 \times \frac{1}{2} \times \frac{1}{8}$ in) and the span width is 50·8 mm (2 in). The flexural strength in MN m^{-2} or psi is calculated from the load recorded at the breaking point, the cross-sectional area of the bar and the distance between the bar supports. If the material does not break, then the flexural yield strength is obtained from the force recorded at 5 % strain. From these data, the flexural modulus of elasticity may be calculated: this is the ratio of flexural stress to flexural strain before permanent deformation has occurred. The modulus had the same units as those used for pressure (e.g. MN m^{-2} or psi) and if the modulus has a high value, then the material is regarded as stiff; conversely if a low value is recorded, then the material is regarded as being flexible. At room temperature, high density polyethylene could have a value of 210 MN m^{-2} (1500 psi) whereas polystyrene could have a value of 1400 MN m^{-2} (10 000 psi).

Glass fibres are commonly added to plastics materials in order to improve their stiffness and the effect on strength and stiffness of glass fibre addition, may be obtained from Fig. 1. This figure is based on one which was published (Anon., 1980) and it shows how the strength and stiffness values for thermoplastics have been increased in recent years.

3.3 Impact Testing

During an impact test, one is measuring the ability of the plastics material to resist a rapidly applied stress. The test is most commonly performed by striking a plastics bar with a pendulum or hammer at a specified speed and then measuring the load to break. It is usual to test more than one specimen, e.g. 5.

3.3.1 Izod impact strength

The specimen required for an Izod impact test consists of a rectangular bar which can be, for example, 63·5 × 12·7 × 6·3 mm (2·5 × 0·5 × 0·25 in). A standard v-shaped notch is machined or moulded into one face of the bar and then one end of the bar is gripped or clamped in a vice. The other end of the bar is then struck, at a specified height and speed, with a swinging pendulum: the energy expended in breaking the specimen is obtained. The results may be expressed in various ways, e.g. foot-pounds per inch of notch; in this method the energy to break is multiplied by that number which is obtained when the notched face width is inverted, i.e. 4 if $\frac{1}{4}$ in samples were used. Nowadays the result is often reported at kJ m^{-2} (see Fig. 2).

It is difficult to imagine how this test ever became a standard as the samples are not very representative of plastics parts. For example, they are very thick and they contain notches. Both are unusual in plastics technology. Good specimen production (e.g. free from voids) is always a problem in Izod impact testing and such considerations may explain why the test appears to have declined in popularity in recent years. ISO recommendation R 180-1961 E for example, explains how the Izod impact strength of rigid plastics is determined (see also ASTM D 256-72a; BS 2782, 1970; DIN 53453).

3.3.2 Charpy impact testing

The Charpy impact test uses a rectangular bar which is struck by a pendulum or hammer (sometimes also referred to as a tup). The sample may

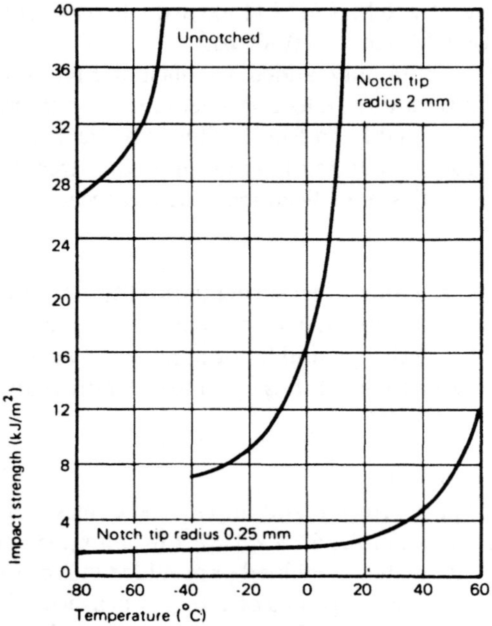

Fig. 2. Impact strength/temperature curves for a rigid PVC. Notch depth 2·75 mm (after ICI, 1979).

be notched or unnotched and the results may be expressed as (a) the amount of energy needed to break a specified bar, (b) in terms of foot-pounds per inch, or (c) Joules per metre. The test is markedly different to the Izod test as:

1. The sample is supported at each end
2. The sample is not clamped
3. The sample is struck at the centre
4. The notch may be of a different shape to that used in the Izod and the notch is on the side which is not struck.

The standards (e.g. DIN 53453; ASTM 256-72a; BS 2782, 1970; ISO 179-1961 E) specify a range of samples, the smallest of which is much smaller than the Izod impact specimen mentioned previously. If this size of sample is adopted, then the sample preparation is easier and in this respect the test is superior to the Izod. However, the test does not appear to be very useful for assessing the impact strength of very flexible materials as these samples can jump from the anvil on which they rest, this is because they are not clamped.

3.3.3 Tensile impact

This test is in effect a high speed tensile test and is generally performed on a modified Charpy test rig. One end of the sample has a large protrusion or jig attached to it and the other end of the sample is bolted to the pendulum of the testing machine. When the pendulum swings, then the sample tries to pass through the jaws of two stationary anvils but the protrustion prevents this happening. The sample is thus broken by this blow which is delivered along the longitudinal axis of the sample; the energy required to do this is recorded (e.g. ASTM D 1922).

3.3.4 Instrumented falling weight

This test is not commonly encountered as the equipment is not readily available as it is new and expensive. It is used to impact test actual components and uses a very heavy weight or striker heavy enough to break the components being tested. Behind the 'nose' of the striker is a transducer and this measures the force involved in destroying the moulding as the striker falls. If a high speed recorder is used then the information on the break may be played back through a slow speed recorder. This test provides a way of determining the impact properties of the actual mouldings and by recording the force/deformation signal it is possible to derive comparative data from a small number of samples. In 1980 there was no British Standard, there is, however, a DIN recommendation on this subject. (It is expected that there will be a British Standard within the near future.)

3.4 Thermal and Flow Properties

In this section the following commonly performed tests will be described: Vicat softening point, heat distortion temperature and melt flow index (MFI). High shear rheometry will also be mentioned as it is a more useful test than the more commonly employed MFI test.

3.4.1 Vicat softening point

In this test a plastics specimen is heated at a constant rate (e.g. 50 °C h^{-1} in an oil bath) whilst being subjected to a load applied via a needle or indentor. The load applied is specified (e.g. a 1 kg weight is applied to the indentor) and the end of the indentor in contact with the plastic specimen has a cross-sectional area of 1 mm^2. The Vicat softening point is the temperature at which this flat-ended needle sinks into the plastics material to a depth of 1 mm (0·040 in).

To ensure that the sample is heat conditioned, the test should be commenced with the oil bath at a fairly low temperature, e.g. at least 50 °C below the expected Vicat softening point. Sample preparation is very important in Vicat softening point determinations as if a sample with small holes or voids is used, then a low result will be obtained. For this reason, injection moulding samples of between 3 and 5 mm are desirable.

The test is widely used nowadays to assess polypropylene and rubber-modified polypropylene as it indicates the maximum short-term temperatures to which parts made from those plastics may be subjected to, when no load is being applied to them. The test is therefore commonly used to assess a material's short-term heat stability so that materials of the same type may be ranked in order of increasing heat resistance.

3.4.2 Heat distortion temperature

This test is also known as heat distortion point and/or the deflection temperature under load. This last term is the one now favoured as it describes the test quite succinctly. The test is used mainly for routine control purposes and one of its great advantages is that the apparatus required is relatively simple. The same oil bath (used to raise the sample temperature) may also be used for Vicat softening point measurements.

In this test a rectangular plastics bar is supported at two points which are a specified distance apart, e.g. in ASTM D 648 the distance is approximately 100 mm (4 in). A load is placed on the centre of the bar (e.g. to produce either 9 or 37 MN m^{-2} (66 or 264 psi) outer fibre stress) and then the temperature of the bath is raised at a rate of 2 °C min^{-1}. The deflection temperature of the plastics material under load is reported as being that temperature recorded when the bar has deflected by 0·25 mm (0·01 in). This test is therefore a measure of the flexural modulus of the material under a specific set of test conditions.

3.4.3 Melt flow index

This test is widely known as MFI and is normally used to measure and compare the flow properties of polyethylene.

The material to be tested is placed inside a heated barrel whose temperature is regulated to give a specified temperature, e.g. 190 °C. After a prescribed waiting period, a loaded piston is applied on top of the material and the load applied forces the heat softened material to extrude through a die. After a predetermined time, the extruded material is cut and weighed and the flow rate or MFI value is reported as the rate of extrusion in grammes of material per 10 min.

The test therefore measures extrusion rate, through a die or orifice of a specified length and diameter, under standardised conditions of temperature and pressure. Its main advantage is its relative simplicity but it must not be forgotten that it measures the flow rate under conditions which are not typical of those encountered in injection moulding machines. The shear rates encountered in injection moulding machines are far higher than those encountered in melt flow index experiments.

3.4.4 High shear rheometer

In this test the material is placed inside a heated barrel and after a selected heating time (e.g. 20 min) a piston is brought against the material at a specified rate. The resistance to flow is then measured and the speed is increased. After a short 'settling down' period the resistance to flow is again measured at the new speed. This procedure is repeated for a number of speeds, e.g. over the range 2–160 mm min^{-1}. The rheological behaviour of the material over a range of shear rates may then be expressed as a flow curve in which shear stress is plotted against shear rate. Such curves are useful as they show how the material will respond to changes in rate.

If the shear rate in a particular process can be calculated then the high shear rheometer can be used to measure viscosity at that particular shear rate. Material viscosities could therefore be compared at rates which are appropriate to a particular process—if the temperature of the barrel is systematically changed then plots of viscosity v. temperature may be constructed (see Fig. 3). The viscosity (strictly speaking the apparent viscosity) may be calculated by dividing shear stress by shear rate.

Shear stress (τ), at a particular rheometer speed, may be obtained from the following formula:

$$\tau = \frac{Pr}{2L}$$

where P = measured rheometer pressure, r = radius of the rheometer die, and L = length.

For a circular die orifice shear rate ($\dot{\gamma}$) is

$$\dot{\gamma} = \frac{4Q}{\pi r^3}$$

where Q = volumetric flow rate in $cm^3\ s^{-1}$ and r = the radius of the orifice in cms.

3.4.5 Spiral flow testing

This test is not a standard test although it is widely known in the plastics moulding industry. The test uses an injection moulding machine fitted with

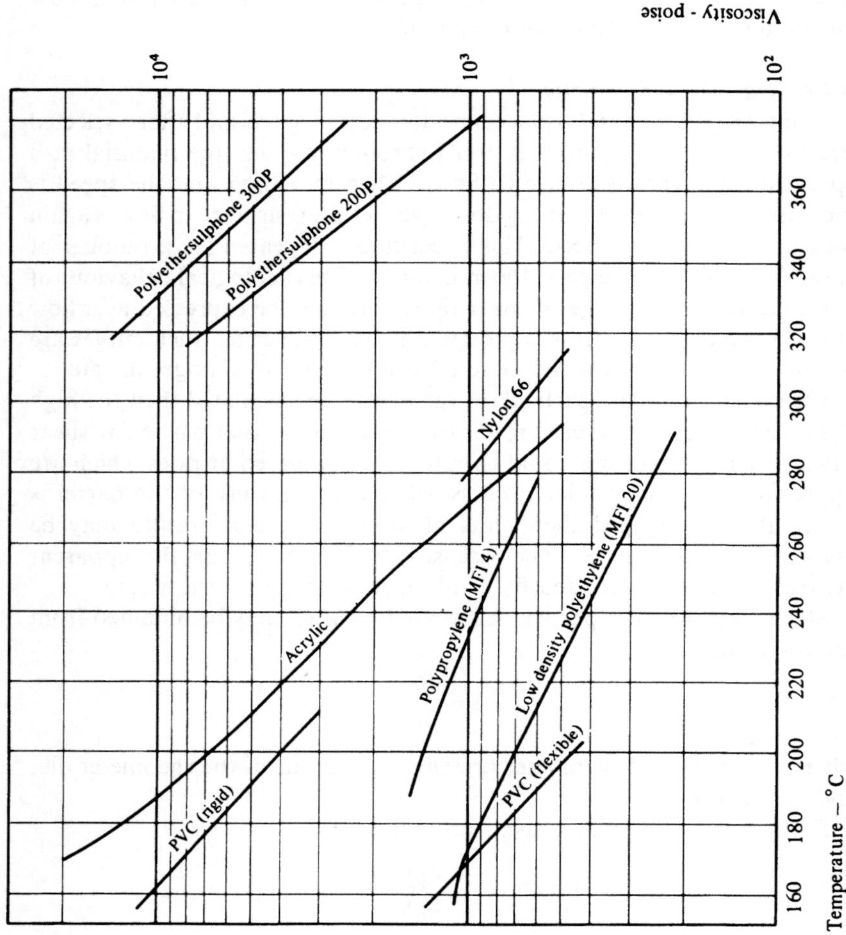

Fig. 3. Viscosity/temperature curves for various plastics moulding materials at 1000 s^{-1} (after ICI, 1979)

a mould which has an unusual cavity cut into one mould half. The cavity consists of an Archimedean spiral which is fed from the centre via the sprue. As material is forced through the sprue and so into the spiral, flow continues until the material sets or cools. The test is not therefore a straightforward rheological test as the hot material is flowing into a cool mould. After ejection the length or weight of the spiral is measured and reported. When this test is being performed, it is important to maintain the screw cushion size at a constant value whilst changing another machine setting, e.g. temperature or flow rate. If the affect of temperature is being investigated, then the material injection rate, or flow rate, should be held constant and vice versa.

3.4.6 Flow path: Wall thickness ratio (FR)

This test is not a standard test although the FR ratio is widely quoted for thermoplastics, e.g. 150:1. Like the spiral-flow test, measurements are made on an injection moulding machine but unlike the spiral flow test a special mould is not used. The average thickness of the cavity is measured and then the maximum flow path length is determined—this is the furthest distance that the material has to travel when filling the cavity. The test is commonly performed with a range of moulds so that the affect of flow channel thickness on the ease of flow can be shown; thicknesses of 1, 2 and 3 mm are used and a range of temperatures may also be employed. The other settings on the machine are adjusted to give average, or normal, values appropriate for that material.

The FR values so obtained are intended to convey to the moulder how the material may be expected to flow during moulding.

3.5 Other Commonly Performed Tests

3.5.1 Density measurement

Density determinations may be performed by weighing the sample first in air and then in water. By using the following formula the density may then be calculated

$$D = \frac{W_a}{W_a - W_w} d$$

where W_a = Weight in air, W_w = Weight in water and d = density of water. Because of the relationship that exists between density and crystallinity, accurate density determinations are most commonly needed with crystalline materials such as the polyethylenes. These materials float in water and as the density is required to be determined very accurately, the following procedure is recommended.

A couple of small pieces of the material, e.g. granules, are placed in a beaker which also contains a water/alcohol mixture. The density of the water/alcohol mixture should be approximately 0·9 g cm^{-3}. If, for example, 35 ml of methylated spirits was mixed with 25 ml of distilled water then at 23 °C the mixture would have a density of somewhat less than 0·9 cm^{-3}. Additional water may then be slowly added to the beaker. When the density of the liquid is the same as the density of the polymer, then the pieces of polyethylene will remain suspended, i.e. neither sinking nor floating. It is important to add the additional water slowly and to keep the solution well stirred, these precautions ensure that the liquid density is uniform and that the heat of mixing does not cause the temperature to rise to an unacceptable level. The density of the water/alcohol mixture may then be determined by a weighing bottle technique (Weston, 1959).

Because a material has a density less than 1, it must not be assumed that it is a polyethylene, polypropylene or an ionomer; it could be a cellular or expanded plastics material.

3.5.2 Shrinkage

Shrinkage measurements are sometimes performed on plastics test specimens as the information obtained is useful for mould design purposes. In the laboratory mould shrinkage is most easily measured by having two lines scribed (on the surface of the mould) which are a reasonable distance apart. Care should be taken to ensure that these lines do not destroy a sensitive test surface. By measuring the line separation on both the mould and the specimen, the moulding shrinkage may be calculated and expressed, for example, as shrinkage per unit length, e.g. 0·006 mm mm^{-1} or 0·006 in in^{-1}.

If this information is used to compute cavity dimensions, then it is important to ensure that the correct formula is used. The suggested formula (Rubin, 1972) is as follows:

$$D_c = D_p + D_pS + D_pS^2$$

where D_c = the cavity dimension (in or mm), D_p = the dimension of the moulded part (in or mm) and S = shrinkage (in/in or mm/mm).

3.5.3 Hardness

Hardness ratings are obtained by measuring how far a steel ball or indentor sinks into a surface when a specified load is applied to the indentor.

The two most common types of hardness tests that are employed in the polymer industry are the Rockwell (e.g. ASTM D 785) and the Shore durometer tests (e.g. ASTM D 2240).

3.5.3.1 Rockwell hardness. This test is commonly performed on the harder types of plastics material and the hardness reading is obtained by applying a specified load to a steel ball of specified diameter. The hardness reading is read from a scale and to cover the range of hardnesses commonly encountered, four scales have been specified; these are (in order of increasing hardness): R, L, M and E.

It is usual to use the scale which gives the highest reading but considerable overlap between the scales is possible. A wide range of results may therefore be quoted under the generic name of Rockwell hardness and because of this the subject is rather confused.

3.5.3.2 Durometer hardness. Shore durometer hardness is obtained from the indentation of a probe into the polymeric material. If the probe has a pointed tip, then the hardness test is referred to as Shore D and if a blunt probe is used, then the hardness is quoted as Shore A. The Shore A scale is used for softer materials.

Neither of these hardness tests, i.e. the Rockwell or the Shore, correlates with the scratch resistance of the material and the tests should not therefore be used to assess this property.

4 LONG-TERM TESTING

Tensile testing, although widely used, is not capable of giving information which could be used, for example, in the design of a continuously stressed component in a particular environment. This is because the conditions of test may not, for example, reflect the way in which the component is to be loaded.

4.1 Constant Load Testing

Many articles or components operate under conditions where full loads, or deformations, are applied within a short space of time and then maintained constant. In service such loads may be externally applied or they may be self-weight loads. If such a load is applied, and left applied, then the polymer sample will keep on deforming, i.e. the sample will 'creep' at comparatively low loads. The effect of applying a fluctuating stress or strain may also have to be considered in some applications.

Creep tests are often performed by loading a plastics sample, e.g. with a weight, whilst keeping it at a specified temperature. At periodic intervals,

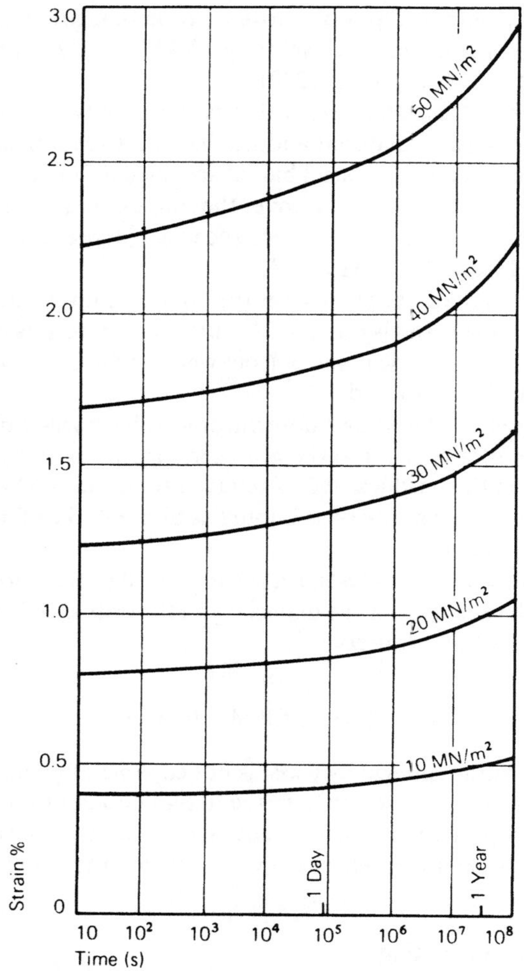

Fig. 4. Polyethersulphone 300P. Creep in tension of 20 °C (after ICI, 1979).

measurements of deflection or extension are taken; the time span may extend over several years. Such tests indicate that quite small loads applied for long periods of time can cause significant deformations of plastics materials. Such deformation is not, however, necessarily permanent.

A knowledge of a particular component's response to a constantly applied load is therefore important and must be appreciated by the designer.

4.2 Creep Curves

If a load is applied to a plastics component (under specified conditions of temperature, humidity, etc.) then a deformation or strain will result. Creep is defined as the total strain which is time dependent resulting from the applied stress (load).

The results may be expressed as a 'creep curve' in which the strain (resulting from a constant stress) is plotted against time. By using different loads a family of curves may be assembled; in order to get a more convenient scale, strain is plotted against 'log time' (see Fig. 4). Such an assembly of curves provides a relationship between stress, strain and time under specified conditions. It will be seen that as the stress increases the strain increases and that the strain is time dependent.

4.3 Isometric Stress–Log Time Curves

The same information may be rearranged and displayed in a different way. For example, if a line is drawn parallel to the time axis then it will intersect, or cross, the family of curves. At each intercept the stress required to produce that constant strain (i.e. an isometric stress) may be noted and then used to construct an isometric stress–log time curve (Fig. 5).

Such curves help to provide design information; for a given component it could be that the design limitation specified is a maximum deformation or distortion. The graph may be used to obtain the maximum stress which can be tolerated for a specified time.

4.4 Isochronous Stress–Strain Curves

If a line is drawn on the family of creep curves (see Section 4.2) parallel to the 'strain axis' then an isochronous stress-strain curve may be constructed. 'Isochronous' means 'uniform in time' and the line may be drawn at say the 100 s mark. This procedure is commonly used to illustrate the effect of changes in crystallinity, water content, etc. on properties (see Fig. 5).

4.5 Tensile Creep Modulus

A third representation of the basic creep data is a modified form of the isometric stress–log time curves; a family of tensile creep modulus curves as a function of strain and log time, may be plotted. The creep modulus may be substituted into classical elastic design formula (see Fig. 5).

4.6 Other Data Required

The representations of the data contained in some of the creep curves does not convey any new information nor do such representations tell a

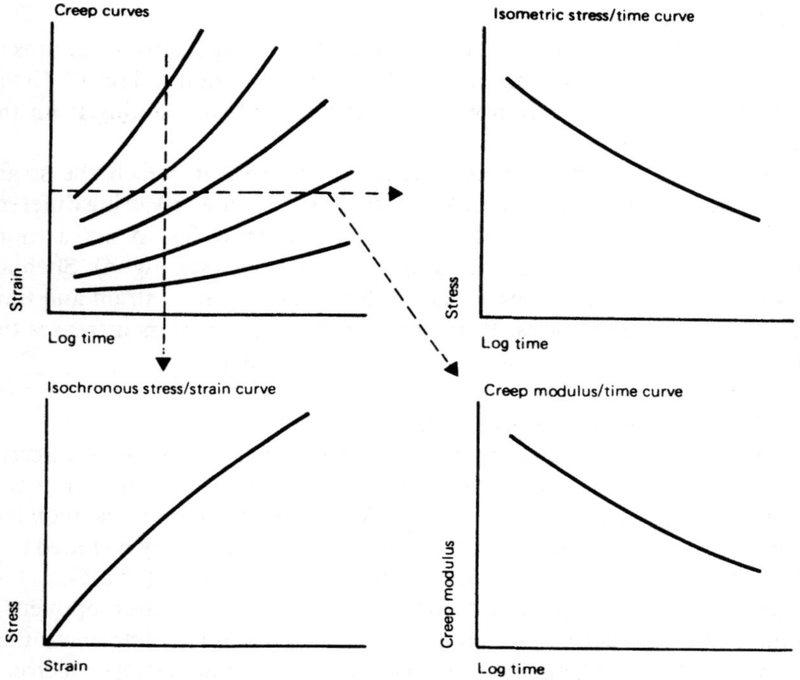

Fig. 5. Presentation of creep data (after ICI, 1979).

designer all that he needs to know about the long-term behaviour of plastics under load. Many other factors need to be investigated. For example, the following should be studied:

1. Effect of temperature. The behaviour of thermoplastics to an applied load is temperature dependent
2. Water concentration. Water can have a dramatic effect on materials such as nylon 66 as the water acts as a plasticiser
3. Testing for longer times. In order to keep testing to a minimum and to get information reasonably quickly, the long-term behaviour may be obtained by extrapolation—this has obvious disadvantages
4. Dynamic load application. This is important as many engineering components are loaded intermittently
5. Effect of processing on properties. The machine settings and the operating sequence used have a tremendous effect on properties
6. The effect of additives. Even quite small additions (e.g. of pigment) may modify the crystallinity behaviour of some polymers

5 INJECTION MOULDING OF PLASTICS TEST SPECIMENS

5.1 Codes of Practice

In order to satisfy standards or specifications many moulders, end-users or raw material suppliers must produce test specimens. Although such specimens should ideally be cut from the proposed component it is very often not practical or convenient to do so. In such cases the specimens are usually produced by moulding techniques such as compression or injection moulding. Because of the production speeds possible injection moulding is very often employed (Whelan and Craft, 1978a).

However, it can be readily shown that test specimens moulded under different injection moulding conditions can give markedly different properties. Such differences have led, for example, to disputes between a moulder and a raw material manufacturer and/or an end-user. It is therefore proposed to offer this code of moulding practice, which should minimise the effect of moulding variables on properties and which will assist in the setting of the injection unit so as to give results which are representative for a particular material.

It has been recognised for a long time that the use of injection moulding machines to produce test specimens introduces problems and for this reason, organisations such as the ASTM (American Society for Testing and Materials) issue codes of moulding practice. These help to improve the reproducibility of results if different materials are being compared or, if the properties of a given material are being checked to see if they are as specified. Two such codes of practice are ASTM D 1897-72 and D 1130-63. These are both standard recommended practices and both refer to the injection moulding of plastics test specimens. The intention behind such codes is to offer advice on how to achieve temperatures, pressures, cycle times, etc. which will give representative samples; emphasis is placed on the recording of the machine settings. Such recording is necessary, of course, for future reference.

However, neither of the standards refers to the need for controlling the screw cushion, i.e. the amount of material left in front of the screw after injection has been completed. This could be very important as if, for example, there is no screw cushion then the pressure being applied to the material in the mould will be zero. As injection pressure dramatically affects many properties, e.g. moulding weight, tensile strength, etc. it is recommended that this setting should also be adjusted to a constant value and the setting recorded. It is also recommended that the moulding weight should be monitored and checked against previous findings as consistent shot weight is a reliable indication of moulding cycle reproducibility.

The setting of screw speed also needs some discussion as it is known that high screw speeds can result in high, but nonuniform, melt temperatures. Screw speed must also therefore be monitored and recorded.

The procedure outlined below is not a short cut to perfect mouldings; it is one which will use relatively large amounts of materials and sample production will take quite a long time. However, if it is followed specimens whose properties are representative for that material when processed into test specimens by injection moulding will be produced.

5.2 Suggested Equipment

A suitable machine for commonly moulded specimens (e.g. tensile dumb-bells, impact bars, etc.) is a small, reciprocating screw-machine of about 40 to 50 tons lock. Such a machine, if operated correctly, can give well conditioned melt (i.e. in terms of temperature) and can handle a wide range of materials. The machine must be capable of giving a shot volume of material which is greater (e.g. by 50%) than the volume of the total moulding, i.e. specimens plus the runners and sprue. This is because if the machine is operated at the limit of its capacity then melt variations could occur because the screw length available for melt conditioning decreases as shot volume is increased. Injection speed and pressure must be capable of being independently varied and at least two 'injection pressures' (see Section 5.4.3) should be available.

The mould must be capable of producing mouldings to within the specified dimensional tolerances. Mould lay-outs for test specimen production are specified in standards such as ASTM D 647. Such test specimen moulds for plastics are of multicavity construction and when using such moulds it is best to concentrate on just one type of cavity or specimen at a time (see Section 5.4). This can result in material wastage as the other test specimens are thrown away (unless the other cavities can be blocked off). In order to save time and money an insert mould could be used with a single cavity machined into each insert. (Such a mould was designed by J. E. Nightingale and made by Springate Engineering Ltd and is in current use at the National College of Rubber Technology.) It is extremely important that the mould be fitted with temperature control devices and that the temperature of the two mould halves can be checked periodically, e.g. by means of thermocouples connected to an electronic thermometer. This electronic thermometer can also be used to check melt temperatures if fitted with an appropriate needle probe (see Section 5.1.2).

A balance, capable of reading to 0·01 g is also extremely useful as it can give valuable information (see Section 5.4.3). The injection machine must have fitted a line pressure gauge and a means of measuring screw position and therefore displacement or speed. Cycle timing is extremely important and even if the machine is fitted with cycle timers it is recommended that a separate clock or watch be available which is capable of measuring time accurately, e.g. to within 0·1 s. This may be used to check timer settings, injection speeds, mould open times, etc.

5.3 Definitions

Because of their importance some important machine conditions or settings are defined:

5.3.1 *Time*

All the following times should be recorded in seconds.

5.3.1.1 Starting time. The injection cycle time starts the instant the injection ram or screw moves forward.

5.3.1.2 Screw forward time also commences when the screw starts to move forward and finishes the moment the screw starts to move back. This portion of the cycle therefore covers a number of activities, for example:

1. Screw advance. The screw is pushing the melt along but there is little resistance to movement. This will happen if melt decompression has been used to stop drooling from the nozzle.
2. Melt compaction. Before the melt will flow into the mould cavity, pressure must be generated on the material and this will only happen when there is a build-up of pressure caused by resistance to flow.
3. Initial mould filling. This term refers to the time taken to fill the cavity with melt. One way of measuring this is to measure the time from the instant the injection gauge first shows pressure until the moment that the screw stops its rapid forward movement. Without special equipment it is difficult to measure on fast cycling machines as mould filling is commonly completed in less than 1 s. However, when test pieces are being produced it is recommended that mould filling be done slowly so as to produce good-surfaced mouldings with 'representative' properties. Such slow filling is possible as test specimens are usually of heavy cross-section.
4. Dwell time. This time commences when initial mould fillings stops and lasts until the screw injection pressure is removed. Dwell time

and dwell pressure can have a tremendous affect on part properties as during this period additional material may be packed into the cavity to compensate for shrinkage. Whether this is desirable is debatable as the action of forcing layers of partially set material past each other can introduce considerable stress into the component. To avoid this the dwell time, and the dwell pressure, should be kept as low as possible. This part of the cycle may be referred to as final mould filling.

5.3.1.3 Screw recovery time. This starts when the dwell time finishes and ends when the screw stops rotating. For a given material the screw recovery time is dictated by screw rotational speed and back pressure. Screw recovery rate may be obtained by dividing the linear distance moved, during screw recovery, by the time taken.

5.3.1.4 Cooling time. This commences with the lapse of dwell time and finishes when the mould starts to open. The screw returns during this part of the cycle and melt decompression or 'suck-back' may then occur on some machines, i.e. the screw is pulled back within the barrel to 'suck-back' the melt away from the nozzle.

5.3.1.5 Mould closed time. This is the total time for which the moulds are held closed. On many machines when the two mould halves touch the signal is given to the screw to start injection.

5.3.1.6 Mould open time. This is capable of being set on many machines and it commences the instant the mould faces start to part and finishes when the mould faces touch to commence the next cycle.

5.3.1.7 Total cycle time. This is the sum of the mould open time and the mould closed time.

5.3.2 Pressure and speed

5.3.2.1 Injection line pressure. Most machines are fitted with vane pumps and these can commonly generate pressures of $10{\cdot}5\ \mathrm{MN\,m^{-2}}$ (1500 psi). By setting the relief valve at an appropriate value any pressure up to this set value may be obtained if there is sufficient resistance to flow, i.e. the line pressure so selected is an upper limit and not necessarily what is used. It is commonly found that it is only when the mould is full that the pressure builds up to the preset value.

5.3.2.2 Injection pressure. This is related to injection line pressure by the following formula:

$$\frac{\text{line pressure} \times \text{area of actuating piston}}{\text{area of screw}}$$

It is therefore the pressure immediately in front of the screw and as this is difficult to measure (as the screw position changes constantly during injection) nozzle pressure is sometimes used in its place. There is a direct relationship between line pressure and nozzle pressure on most machines. Because of pressure losses the nozzle pressure is approximately 10% lower than the injection line pressure.

5.3.2.3 First- and second-stage pressure. On many machines two injection line pressure control valves are provided. These may be set at two different levels to suit two different parts of the cycle, i.e. initial mould filling, done with the first valve set at a high value, e.g. 7 MN m^{-2} (1000 psi) followed by mould packing, where the second valve is set at a lower value, e.g. 3·5 MN m^{-2} (500 psi). The change from one setting to another may be done in various ways, e.g. by means of a striker (attached to the screw) and a microswitch. During the initial mould filling stage (first stage) the pressure usually builds up gradually as the resistance to flow increases and it may, or may not, reach the preset line pressure.

As line pressure and injection speed are interrelated it is probably best to divide the injection part of the cycle in the following way:

(a) Initial mould filling. This is performed at the required rate with the screw acting as a ram. If possible this should be recorded, e.g. in mm s^{-1}, and the pressure used to maintain this flow rate should also be recorded.

(b) Final mould filling. In this part of the cycle the speed of movement is very slow as extra material is being packed into the mould to compensate for shrinkage. This should be recorded as a pressure, e.g. 3·5 MN m^{-2} (500 psi).

The overall injection time should also be recorded and if possible the division between the initial mould filling and final mould filling times should be noted.

5.3.2.4 Back pressure. This is the pressure that the screw must overcome before it can move back during the plasticising part of the cycle. It is usually read from the same gauge as injection line pressure and so has the same units but is of much smaller magnitude, e.g. up to 20% of full line pressure. The use of back pressure may help mixing and reduce shrinkage but it can also lengthen the cycle and can cause excessive shear.

5.3.2.5 Clamping pressure. This should be as high as possible in order to stop the mould opening during the injection. Test specimens normally have a large area and for this reason high clamping pressures are required, e.g. up to 49 MN m^{-2} (7000 psi) of a moulding's projected area.

5.3.3 Temperature settings

All machines have instruments for controlling the barrel temperature. It is important to realise that this temperature is not necessarily the temperature of the melt nor of the barrel close to the melt. On screw machines considerable heat can be generated by friction, or shear, and as machines are not normally fitted with cooling devices, melt and barrel temperatures can easily deviate from those set on the instruments if, for example, screw speeds are very high and/or high back pressures are employed.

It is therefore recommended that the screw rotational speed, and back pressure, be kept low and that melt temperatures are measured and recorded. It is important that the zones of the barrel are clearly marked for recording purposes, e.g. nozzle, barrel front, barrel middle and barrel rear would seem reasonable definitions.

The mould used must be capable of being held at a specified temperature (e.g. by a hot water circulating unit) and both mould halves should contain a temperature sensor (see Sections 5.4.1 and 5.4.2). Hydraulic oil temperature should also be controlled and monitored, e.g. at 40 °C $\pm$1 °C.

5.4 The Recommended Procedure

If using a multicavity mould decide which type of specimen you wish to produce and then ignore the others. Repeat the following procedure for all the types of specimen required even if two of the cavities produce 'identical' components. If similar components are capable of being produced from the same mould (but from different cavities) ensure that each component is clearly identified with a cavity number.

5.4.1 Moulding temperature

Consult the material supplier's literature or your own records (see Chapter 7) so as to obtain suitable setting temperatures for both the injection unit and the mould. After the set-point temperatures have been reached (i) allow at least 20 min for the heat to penetrate completely through the screw and barrel assembly before moulding is commenced, and (ii) check that the screw turns freely before the hopper-throat slide is opened.

Set the screw speed to a low value and if the screw does not turn freely then the machine should be allowed to 'heat-soak' for a longer time—if necessary at a higher temperature. Periodically check the temperature of the melt, moulds, hydraulic oil and hopper block. An electronic thermometer may be used for this purpose (Whelan and Craft, 1978b).

Because of the importance of the melt temperature the method of measuring it with an electronic thermometer will be described in detail. (The hopper block temperature should be checked because this ensures that the cooling water has been turned on and that it is circulating at the required rate.)

5.4.2 Melt temperature measurement

5.4.2.1 Air shot technique. Allow the machine to operate under 'typical' operating conditions until you are satisfied that the machine has 'settled down'. The hydraulic oil temperature may only reach its operating temperature after 1 h but it is important that it does so as otherwise erratic machine performance may result.

Commence moulding and then interrupt the cycle when the screw or ram has retracted (after refilling the barrel) and switch to 'hand operating' if necessary.

Withdraw the injection carriage away from the mould and ensure that no other people are in the immediate vicinity of the injection unit. Keep your hands well clear of the barrel and nozzle area, shoot the material into the air and collect the melt on a piece of board.

Remove the board from the nozzle area and insert the heated probe into the melt.

Stir gently and record the highest temperature observed. (Before any temperature is measured it is best to preheat the probe to the temperature that will be measured.) Repeat this procedure and adjust the temperature controllers until the required melt temperature is obtained.

Gases may emerge violently with the melt, therefore keep the hands and face away from the nozzle and barrel area.

Ensure that no other person is in the immediate area who may be endangered by your actions.

Avoid reaching over or around the mould and wear insulating gloves and safety glasses.

Use only *brass* or *copper* scrapers to clean the nozzle and mould before bringing the barrel back into contact with the mould.

5.4.2.2 The SPI nozzle. Various nozzles are available which have melt thermocouples incorporated. A very useful design is one suggested by the SPI (SPI, 1967) (see Chapter 3).

5.4.3 Simple machine setting

It is probable that the machine employed for test specimen production will be an in-line screw machine with two injection pressure settings (see

Section 5.2) and it is this type of machine which is meant by the term 'simple machine'.

5.4.3.1 Initial cycle time setting. If the mould has been used to produce a particular component previously then those settings can be reset (providing that the material under test is similar). The problems start when a new type of specimen is being moulded but luckily suggestions have been made (e.g. Bown, 1979) which enable the various parts of the cycle to be calculated. The answers obtained from such calculations should only be treated as guide-lines and should be modified as the occasion demands.

5.4.3.2 Mould open time. The mould open time may çonsist of the following operations:

1. Slow initial mould separation
2. Fast mould separation
3. Final mould separation stage (slow)
4. Part ejection
5. Mould pause time. After the mould has finally ceased moving the guard may need to be opened and closed if the machine is being operated semi-automatically
6. Fast mould closing
7. Slow mould closing
8. Mould clamping

Most of these stages are set to the maximum rate possible, or appropriate, for that particular stage; the manufacturer's handbook will give the speeds possible for each stage.

It will probably be found that approximately 4 s will need to be allowed for this series of stages as slow part ejection, using a hydraulic ejector (if possible), should be used so as to avoid component damage or distortion.

5.4.3.3 Injection time calculation. As the thermoplastics material enters the mould it contacts the relatively cold walls and then it starts to solidify by cooling—this reduces the gap through which the material must flow if it is to reach the extremities of the mould. Each material is offered in a range of grades which, for example, differ in their flow path:wall thickness ratio. If the flow path:wall thickness ratio is 175:1 then this means that the material will flow for 175 mm if the wall thickness of the moulding is 1 mm. (The figure quoted should be regarded as a maximum figure.)

It has been suggested that the maximum time available for filling can be calculated (Bown, 1979) by using the following formula:

$$t = \frac{(d_w p_f / p_{fr})^3}{8[(T_x - T_m)/(T_c - T_m)]^3}$$

where t = filling time in seconds, d_w = minimum wall thickness of the moulding in mm, p_f = maximum flow path in mm, p_{fr} = flow path:wall thickness ratio, T_x = heat distortion temperature in °C, T_m = mould temperature in °C and T_c = maximum cylinder temperature. If the melt temperature is known then this may be used in place of the maximum cylinder temperature.

Bown (1979) has suggested the following values for the expression $(T_x - T_m)/(T_c - T_m)$:

LDPE	0·3
PP	0·32
PS	0·35
PA66	0·52
PA66 (glass filled)	0·65
ABS	0·17
POM	0·32

As an example the maximum filling time for a bar, 150 mm long × 1·5 mm thick (6 in × 0·06 in), moulded in PS with a p_f of 175:1 would be equal to:

$$\frac{\left(\dfrac{1{\cdot}5 \times 150}{175}\right)^3}{8 \times (0{\cdot}35)^3} = \frac{1{\cdot}286^3}{8 \times 0{\cdot}043} = \frac{2{\cdot}127}{0{\cdot}343} = 6\,\mathrm{s}$$

Test specimens should be free from voids and have a good surface finish—both these conditions can be best met by using a slow filling rate. On the other hand, test specimens should not contain excessive orientation which such a slow filling rate could induce. If a mould is filled quickly then, providing the gate seals rapidly, the orientation will be low as relaxation by Brownian motion will occur; however in this case the shrinkage may well be unacceptable.

To get low uniform shrinkage, make the gate (see Section 5.3.3.4) the full depth of the cavity; the width should be twice the depth. The land length should be as short as possible (e.g. 1 mm) so as to minimise the pressure drop which occurs when the material flows through this narrow restriction.

Adjust the filling time so that the cavity is filled slowly but steadily in the calculated time (if necessary enlarge the feed system if the mould cannot be filled in this time) and when the mould is full drop the pressure to the lowest possible value for the shortest possible dwell time.

5.4.3.4 Dwell time. Test cavities are commonly filled via rectangular side gates as such gates are easy to machine in one mould half. Such a gate has six sides, only four of which make any significant contribution to cooling. The other two faces, or sides, consist of polymer—one faces the cavity and the other faces into the runner.

During the mould filling time it is unlikely that material will accumulate in the gate; the steady flow of melt will keep it open. Once the melt stops moving, then the temperature of the material in the gate will drop rapidly as the cross-sectional area is relatively small and the material is in contact with four metal surfaces.

If the gate freeze time is calculated using the formula suggested for the cooling time (see Section 5.4.3.5) then the answers obtained will be too high; the heat transfer equations on which that solution is based refer to relatively long cooling times and to situations in which the object or moulding is cooled from the top and bottom surfaces only, i.e. the sides are ignored.

In a paper by Olson and Schultz (1942) a method was suggested whereby temperature changes could be calculated for a number of component shapes. For this approach to be successful it is necessary to know the thermal diffusivity of the material, the material temperature and the temperature at which flow becomes impossible. This information is fed into a formula, tables of figures are consulted and an approximate answer is obtained; the approximate answer is then used to obtain a better answer. Some freeze time calculations for polystyrene are shown in Table 2 (Weston, private communication, 1980).

Using this approach it is possible to construct graphs in which, for

TABLE 2
FREEZE TIME CALCULATIONS FOR POLYSTYRENE (Weston, 1980)

Gate number	*Gate dimensions (mm)* *Depth*	*Width*	*Time to cool to 110°C (s)*
1	1	1	0·6
2	1	2	0·9
3	1	3	0·95
4	1	Infinite	1·1
5	2	1	0·9
6	2	2	2·3
7	2	3	3·1
8	2	4	3·5
9	2	Infinite	4·4

Assumptions: Mould temperature 40 °C
Melt temperature 200 °C
Temperature at which flow ceases 110 °C
Thermal diffusivity of polystyrene $1{\cdot}172 \times 10^{-7}\ \mathrm{m^2\,s^{-1}}$.

$$\left(\text{thermal diffusivity} = \frac{\text{thermal conductivity}}{\text{density} \times \text{specific heat}}\right)$$

example, freeze time is plotted against mould temperature for each gate size. Separate plots would be needed for each melt temperature and for different gate shapes.

Such graphical solutions are very easy to use; a useful collection of references on this subject is to be found in the work by Rubin (1972) on page 200.

If the specimen mentioned in Section 5.4.3.3 was being moulded via a gate which measured 2 mm wide by 1 mm deep (0·08 × 0·04 in) then from Table 2 the freeze time would be 0·9 s.

5.4.3.5 Cooling time. This is very often the longest part of the cycle; the approximate cooling time can be obtained by squaring the maximum part thickness and then multiplying this product by a factor (Bown, 1979). With PS the factor suggested is 3·40 and for a bar with a wall thickness of 1·5 mm (0·06 in) the calculated cooling time would be $3{\cdot}40 \times 1{\cdot}5 \times 1{\cdot}5 \simeq 8$ s.

Bown (1979) gives a list of factors for some of the more commonly used thermoplastics, e.g. ABS, 1·90; POM, 3·05; PA66, 1·40; LDPE, 2·00; and PP, 3·38. Despite the high moulding temperatures required for PA66, it can be seen that the cooling times required for this material can be quite short and can justify using a comparatively expensive material for everyday jobs; for the test bar the cooling time would be approximately 3 s.

The runner dimensions will be very much larger than the maximum part thickness and so, on ejection, the runner will distort as, on average, it will be above the glass transition temperature. This does not matter unless a high runner temperature causes problems on ejection, e.g. sticking to the component, etc. If this happens the cooling time will have to be lengthened.

5.4.3.6 Machine purging. Set the calculated times on the machine controls (in the example given 19 s) and then reduce the injection line pressure setting, and the injection speed control, to low values. After the machine has reached the operating temperature of the material in the barrel introduce the new material into the hopper. If the new material is unstable, or of too low a viscosity at these temperatures, then a buffer material may need to be used.

For example, do not mould acetals directly after PVC (or vice versa) as serious and dangerous decomposition may occur. If another polymer contains a halogen then this restriction also applies; flame retardants commonly contain halogen compounds and so compounds containing such materials cannot be used before, or after, acetals.

Bring the carriage away from the barrel and purge the old material out—ideally this should have been done when the machine was shut-down after the last run. Introduce the new material into the barrel and rapidly run it through the barrel so that the incoming material 'scrapes' the hotter, old material from the barrel walls. Purge the barrel for 3 min in this way and then shut-off the feed supply from the hopper: purge clean and allow the empty barrel to stand for 5 min.

Introduce more fresh material and repurge for a further 3 min. Start changing the temperatures to those required for the new material. Repeat the purging procedure as often as is necessary. About 4 kg (9 lbs) of material should be sufficient for purging on a small machine, e.g. with a 27 mm (1 in) diameter screw. Great care must be taken during the purging step as otherwise a serious accident can easily occur, e.g. burning. During the purging operation note the fully forward position of the screw—this is sometimes known as 'screw bottom'.

5.4.3.7 Machine operation. Allow the machine to settle down at its correct operating temperatures and then using the suggested times commence moulding, using if possible a fully automatic cycle but with the feed setting reduced to a low value, i.e. insufficient to fill the mould.

Continue moulding and increase the feed setting until the mould is almost full. Adjust the line pressure (first stage) and the injection speed until the required rate of injection is obtained. At this stage the moulding should be slightly short but the surface finish should be that which is required. At the end of the injection stroke, or time, the screw should 'bottom'.

Set the final mould filling (second stage, injection pressure) to come into operation at this point—use a line pressure which is approximately a quarter that of the first-stage value.

Increase the amount of feed (i.e. allow the screw to rotate for a longer time) until the screw is about 3 mm (0·125 in) away from its maximum forward position when injection is complete. If necessary increase the second-stage pressure until the mouldings are free from sinks or voids and then weigh the moulding.

Increase the second-stage pressure until the moulding weight no longer increases or until the moulding starts binding or sticking in the mould. It may prove necessary to extend the injection time if the moulding weight does not settle down to a steady value. However, the temptation to overpack the mould must be avoided at all costs and the screw feed should be adjusted to maintain a constant cushion of material in front of the screw.

After each adjustment allow the machine to settle down, e.g. for six shots, before making any further alterations.

Adjust the cooling time until the component can be ejected without distortion; the cooling time should also be long enough to allow relatively slow (e.g. 100 rpm) screw rotation and recovery.

5.4.3.8 Sample production. Allow the machine to run, for say six shots, after the last adjustment and then collect the required number of samples. Label each sample with a shot number then check the melt temperature as previously described. (The screw should be allowed to refill and the material should be shot into the air at the time it would have been injected into the mould.)

Record this temperature as the melt temperature used to produce the samples and make certain that all other settings have been recorded. A standard form should be developed for this purpose and then machine settings such as the following should be recorded.

5.4.3.9 Reporting the moulding procedure. The report shall include:

1. Type and description of material used
2. Pretreatment of material (if any)
3. Mould used
4. Description of specimen
5. Gate size and description
6. Make and model of machine used
7. Type of machine control used, e.g. automatic or hand
8. Screw size and brief description
9. Rated heating capacity of heating chamber
10. Weight of moulding plus sprues
11. Set temperatures
12. Melt temperature
13. Mould surface temperature (cavity and core)
14. Hopper block temperature
15. Oil temperature
16. Set injection line pressures; first and second stage
17. Back pressure
18. Clamping pressure
19. Mould open time
20. Initial mould filling time
21. Final mould filling time
22. Cooling time
23. Screw recovery time
24. Sprue break times
25. Overall cycle time
26. Injection or mould filling speed
27. Screw rotational speed
28. Screw cushion size
29. Length of feed stroke
30. Number of mouldings made

If possible the pressure required to maintain the specified rate of flow should be recorded as this can provide a useful indication of the ease of flow of the material.

5.4.4 Multistep machines

Machines are now available on which a number of injection speeds and second stage dwell pressures can be preselected, e.g. by digitally set thumbwheels. Such machines appear more complicated than those previously described but in practice they are easier to set or use.

Many hydraulic control units are based on closed-loop, servovalve systems in which the injection stroke can be divided into a number of steps, e.g. five. Each step, which is a preselected percentage of the stroke, can be performed at a different speed.

5.4.4.1 Multistep setting. The injection moulding produced consists of two main parts—these are the feed system and the moulded component. Of these two parts it is only the moulded component which is of immediate interest. The feed system will be regranulated and so should not influence the moulding cycle unless, for example, high speed filling causes degradation.

Set the first injection speed to a high value for a small percentage of the shot and set the other four speed settings to zero. After the screw has travelled a short distance (equivalent to the small percentage setting) it will stop. Make a series of mouldings and adjust the percentage setting of the first stage until the flow of material almost reaches the gate.

Set the second stage speed to approximately half that of the first stage (so that the system can respond) and adjust the percentage setting until the gate is just reached. Bring in the third stage at a low speed and percentage setting so that the material just crosses the gate and forms a button on the cavity side of the gate—this should avoid 'jetting' into the cavity.

The cavity may now be almost filled at the desired speed, using the fourth-speed step. The percentage setting should be adjusted until the cavity is almost full, i.e. it is a 'short'. At this point a very slow fifth stage may be brought in to complete the cavity filling, i.e. within the calculated time available for mould filling.

Dwell pressure or final mould filling may then be applied using, if possible, cavity pressure control (CPC).

By following this procedure high speed filling of the sprue is possible and yet the cavity may be filled at whatever speed is most appropriate. Jetting is avoided by reducing the speed as the material reaches the gate and a two-stage reduction is recommended at this point so as to try and ensure that the

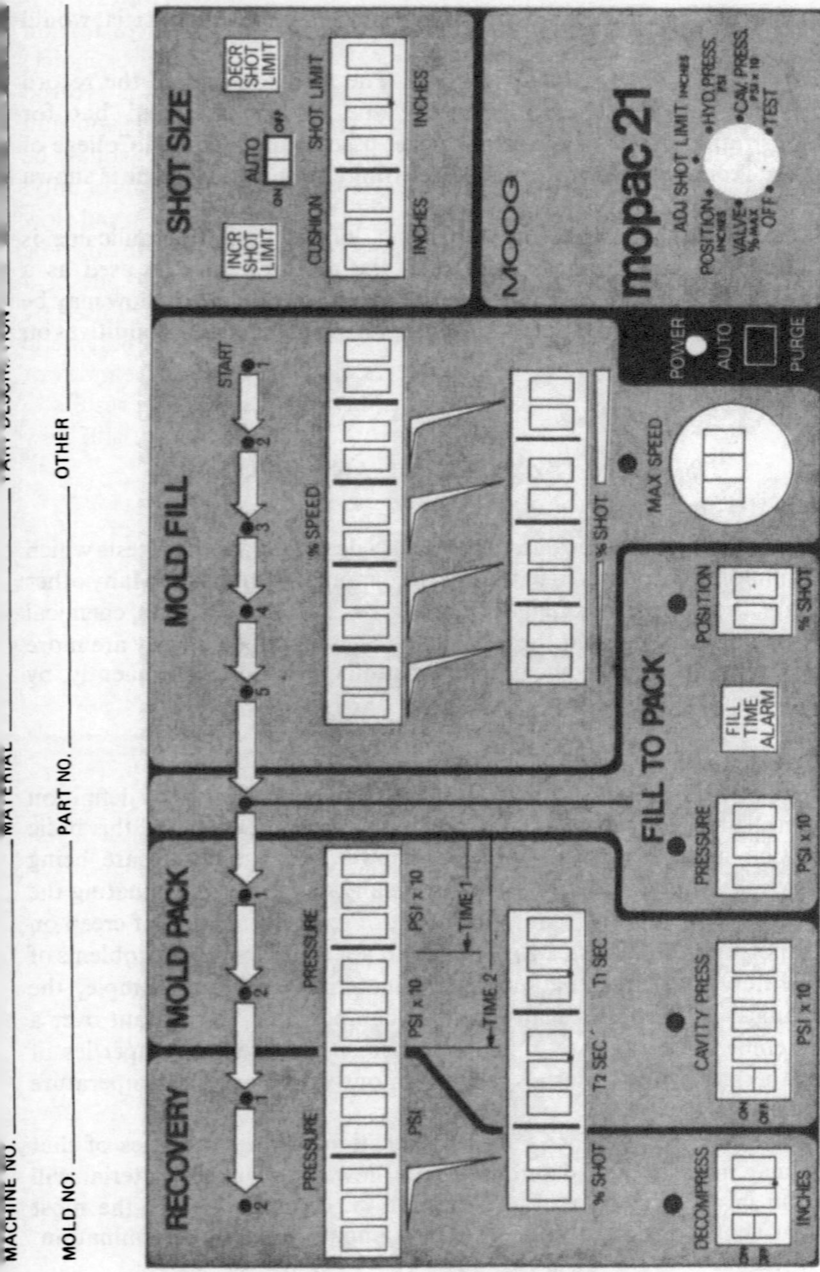

Fig. 6. Record sheet.

system could respond, i.e. so that over-shooting, because of inertia, would not occur.

5.4.4.2 Recording multistep settings. The actual layout of the record sheet will obviously depend on the type of machine used but for demonstration purposes the record sheet used at the National College of Rubber Technology for recording the settings on such a machine is shown (see Fig. 6).

If the output from a pressure transducer, located in the hydraulic line, is connected to a recording system then the machine may be used as a rheometer—the pressure required to maintain a certain rate of flow may be used to characterise materials and/or to investigate the effect of additives on flow properties.

6 QUALITY CONTROL AND TESTING

The short-term, mechanical tests previously described are those tests which are commonly encountered in the injection moulding industry. Many other tests are also performed on plastics materials, e.g. electrical tests, chemical tests, ageing tests, etc. but these have not been described as they are more commonly performed by the polymer manufacturer or, less frequently, by an end-user.

6.1 Limitations of Tests

The short-term tests described all suffer from limitations as by definition they measure a property under a specified set of conditions and this basic limitation must always be borne in mind when test results are being interpreted. Many of the tests were originally designed for evaluating the properties of metals and ignore the affect of temperature and of creep on many polymeric materials. Most metals do not suffer from the problems of temperature and time that polymeric materials do. For example, the mechanical properties of many metallic components are constant over a wide temperature range (e.g. 500 °C) whereas the strength properties of many thermoplastics materials vary enormously over a 100 °C temperature change.

Test values are therefore only an indication of the properties of that particular material and do not indicate how a particular material will perform in a particular end-use application. Consider one of the most popular short-term tests of all—the Izod impact strength determination.

Some of the defects associated with this test are as follows:

1. Notch sensitivity. Very few plastic components are designed with notches or steps, which could act as stress-raisers, and so the value of the information obtained is debatable
2. Range limitations. The test is only performed at one testing speed and usually at only one test temperature. If either or both of these were varied, then different results would be obtained
3. Poor part correlation. The Izod test cannot predict how well a part will perform in service as, for example, the design or shape of a particular component will greatly influence its impact strength
4. Low modulus materials. If a material is soft or has a low modulus, then the Izod sample will bend markedly before breaking and a very high impact strength will be recorded

This test should only be used as a ranking, or 'pecking order' test and used to assess the impact strengths of different grades of the same type of material. The only real way of judging the material's suitability for a particular application is to manufacture that product by the selected processing technique and then assess fitness for use by testing the product under anticipated service conditions. Such procedures are expensive to operate and difficult to initiate and for these and other reasons, many moulders only employ 'fit and finish' testing. However, the value of a more elaborate, component test programme should not be under-estimated in certain circumstances (see Section 6.3).

6.2 Fit and Finish Inspection

In many moulding shops little or no formal testing is performed on the mouldings or on the material from which the mouldings were made. In many of the applications for which plastics materials are used, it is often only necessary to ensure that the surface finish of the moulding is of the required quality and that it will fit, or mate, with other components in an assembly.

However, before a mould is put into service, it makes sense to produce a trial batch of mouldings so as to ensure that each moulding will meet the component specification. The complete moulding spray, or moulding shot, should be measured and recorded very carefully so that in the event of a subsequent change the precise nature of the change in dimensions may be determined accurately. For the purposes of comparison, it is also recommended that these mouldings are kept in a cool, dry, dark place, so as to ensure that changes (in colour, dimensions, etc.) are minimised.

By checking on every dimension of a multicavity tool before production starts, it should be possible to confirm whether or not each individual moulding from that tool is acceptable. By considering the dimensions, and after inspecting these initial mouldings, it should also be possible to decide which of the moulding's dimensions are critical so that appropriate gauges may be constructed. For everyday use gauges or jigs are acceptable, as they are easier to use than relatively complex micrometers.

Once the mould is in production, e.g. on an automatic production line, then samples of the mouldings may be taken periodically (e.g. every 2 h) by a quality control inspector. To assist that person in his task, acceptable examples of the mouldings under discussion are often attached to the side of the machine for comparison purposes. Also attached to the machine is a record which the inspector must sign after completing a standardised procedure. He is required to enter the number and type of components being produced, the number of cavities in use, the faults found, the time of inspection, etc. By adopting this procedure, and by giving batches of mouldings their own identification, the chances of faulty mouldings finding their way into the assembly system are minimised as if, for example, the inspector finds that the mouldings produced since the last inspection are faulty, then the faulty batch can be traced and scrapped if necessary.

At the end of the production run, a sample of the mouldings should again be taken and the dimensions of these should be checked against the originals. Before the mould is put into store it should be brought back up to standard (e.g. by replacing worn inserts) cleaned and lightly greased. The water cooling channels should be drained and cleaned so that rust build-up is minimised. The mould should then be clearly labelled (with the date, job description, machine on which used, etc.) and a sample of the mouldings, which it produces, attached firmly to the outside. Once this has been done, then the mould should be placed in a clean, dry store and its particulars entered in an appropriate record.

6.3 Component and Material Testing

Finished part testing, or component testing, is done in an effort to maintain component quality; when used properly it can also reduce machine down-time and lead to a reduction in the number of reject mouldings produced. Viewed in this way quality control can reduce moulding costs. However, to do this it is necessary to eliminate unnecessary testing and this means that test programmes should be reviewed periodically so that the value of the information obtained can be assessed.

The tests described below are probably those which are most commonly used when precision components are being produced, e.g. based on engineering thermoplastics. Some of the most useful tests for engineering thermoplastics are weight, density, flow tendency, moisture, oven-heating, photoelastic inspection, impact test and end-use testing (Moritz, 1980). Such tests are run alongside a fit and finish programme.

6.3.1 *Weight*

Of all the tests which can be performed alongside an injection moulding machine, component weight is now the easiest one to perform. A modern, digital electronic balance needs little if any setting-up and component weight can be determined very easily and rapidly. Such a determination indicates how well a mould or a particular cavity within a multicavity mould, has filled. Component weight can be readily correlated with the dimensions of a particular component and if required, the digital balance can be fitted with an electronic calculator so that the readings obtained may be statistically analysed.

If density is specified instead of part weight as a control test, then this is more difficult to perform. Probably the easiest way of performing the measurement is to use two containers each containing a solution of the appropriate density. These solutions are selected so that the moulding will sink in one and float in the other.

6.3.2 *Flow tendency*

The flow tendency, or ease of flow, of a material may be assessed in various ways such as melt flow index, high shear rheometry and spiral flow testing. However, such tests require special equipment to be set aside and as such are not commonly employed in the moulding production industry. It is obviously easier to perform the flow test during the moulding operation.

One method of assessing a materials flow behaviour is by means of a flow tab. This consists of a thin, graduated channel which is say, located at the end of the runner system. If the flow tab length alters, then this is due to either a change in machine settings, operating procedure or material properties.

6.3.3 *Moisture content*

With many plastics materials the level of moisture in the material, fed to the hopper, must be kept below quite small amounts as otherwise the properties of the material will alter; properties such as ease of flow, density and strength may all be affected.

Because the moisture content is so small and because the moisture is often held tightly by the plastics material, heating and weighing techniques are slow and troublesome. Instruments are now available which can give a rapid and even automatic indication of moisture content within a particular material. A sample of the material may be introduced into a cell within the instrument and the electrical conductivity of that sample measured. As conductivity alters with water concentration, the change in conductivity may be used to measure moisture content.

6.3.4 Oven heating

Many moulded components are used at elevated temperatures and for this reason their dimensions may be checked after an oven-heating period. The upper surface temperature should be ascertained and unless otherwise specified, it is suggested that the oven temperature be set 25 °C higher than that of the upper surface temperature. The length of time for which the component is in the oven will depend upon its thickness but for most components (e.g. of less than 3 mm cross-section) a heating period of 30 min is sufficient. Formulas used for calculating cooling times may be used or, the time may be measured experimentally by inserting a thermocouple in the thickest part of the moulding and then heating the assembly.

Oven heating is also used to determine whether or not a component has excessive moulded-in stress. If the orientation stresses within a component change, then its resistance to oven heating will also change. As the orientation stresses increase, then the heat distortion temperature reduces.

6.3.5 Photoelastic inspection

For transparent materials the easiest way of detecting frozen-in stresses is to examine the component through polarised light. Such light is produced when ordinary light passes through certain materials. The crystals which cause polarisation are usually carried in sheets of plastic and for the purpose of moulding examination, two such sheets are required. One sheet is placed over a light source and the second sheet is placed on top of the first. The second sheet is rotated until the light passing through the assembly is at a minimum. If the moulding is now placed between the two sheets, then a series of coloured fringes will be seen; if the number of fringes increases, from one moulding to another, then the stress level is also increasing.

6.3.6 Impact testing

The ability to withstand an impact, commonly referred to as toughness, is a very important requirement for many plastics mouldings and one which is not often measured by the standard type of test. For this reason it has been suggested (Moritz, 1980) that a pendulum type tester be modified so that it can either be used to test small components directly or so that it may be used to test a moulded tab in the case of a large component.

6.3.6.1 Testing of small components. In this test, the energy absorbed in breaking the component is measured on a pendulum type impact tester by allowing a sufficiently massive pendulum to strike against the component held in a vice or fixture in a standardised or regulated way. This is because the point of impact will affect the test value. In order to get more consistent results, it is also recommended that the time between moulding and testing be held constant, i.e. at 1 h.

A single-impact tester may be used on a wide range of parts by using interchangeable pendulum weights, adjustable vices and individual part fixtures.

6.3.6.2 Testing of large mouldings. Using the apparatus described above, sections from large mouldings may be tested on similar equipment and in such cases it is important to ensure that the method of test specimen preparation, as well as the method of test specimen mounting, is held constant. As each part of the moulding will probably give a different energy to break value, it is important to ensure that test samples are always cut from the same position within a moulding. To avoid cutting larger mouldings it is suggested that an impact test tab should be located off the main runner; the gate which feeds this runner should be sized so that this tab will pack out in a similar fashion to the moulding. It should therefore give a realistic energy-to-break reading. The tab should be end-gated and the tab thickness should be the same as the part wall thickness; the tab width should be approximately three times its thickness.

6.3.7 Simulated end-use testing

For components which must meet very critical specifications, recourse is sometimes had to a test which simulates the sort of treatment that a component will experience in use. Such a test may be a dynamic test in which, for example, the component is repeatedly flexed so as to duplicate what the moulding will experience in service. Such a test may be performed at two extreme temperatures so that performance may be guaranteed.

Quality assurance testing is extremely expensive to perform and as such is only performed when absolutely necessary. Quality control exists as a production-support function within the manufacturing enterprise solely to answer three questions about the product (Price, 1976). The three questions are:

1. Can we make it OK?
2. Are we making it OK?
3. Have we made it OK?

The first question to answer therefore is whether or not the moulding machine is capable of producing mouldings whose dimensions fall within the specification limits.

6.4 Production Capability

In design specifications, limits on dimensions are usually set in the form of a nominal value and a plus or minus tolerance, for example 10·00 ± 0·02 mm. It should be noted that the plus and minus deviations need not be equal. Such limits ensure that the items manufactured are serviceable and can be assembled together with other parts. These are the specification limits. In the actual manufacture of the items a natural variation in dimensions occurs due to chance errors (Neville and Kennedy, 1964). Before specification limits are agreed it would be advisable to check whether or not a particular injection moulding machine is capable of producing mouldings whose dimensions fall within these specification limits. To do this it may be necessary to construct a pilot cavity so that a limited number of mouldings may be produced on the machine in question and then appropriate measurements may be made on these mouldings. From these measurements, it is possible to predict the process capability, or the natural tolerance limits, for that particular machine using simple statistics.

6.4.1 The normal distribution curve

In the early days of statistics it was found that many groups of data could be represented by an approximately symmetrical curve which was obtained by plotting the frequency (the number of times a particular value of a dimension was recorded) against the property being measured. This curve was called the normal curve or the normal frequency distribution. To define the frequency distribution it is necessary to know (a) Its location or level, which is usually expressed by the average ($\bar{x}$), and (b) The spread or dispersion, which is usually expressed by the standard deviation (σ). The standard deviation is defined as 'the square root of the mean of the squares of the individual deviations from the average'. That is, it may be represented by the following formula

$$\sqrt{\frac{\Sigma(x-\bar{x})^2}{n}}$$

On many pocket calculators the standard deviation and the average of a group of numbers may be obtained simply by pushing a button.

6.4.2 Use of the normal distribution curve

If a very large number of mouldings were produced and a particular dimension (e.g. their diameter) was measured, then a normal distribution curve could be constructed. If a large enough number of samples were taken, then in the ideal case, a bell-shaped curve would result which would be symmetrical about the average value. Approximately 68 % of all the measurements would lie within ±1 standard deviation of the average, 95 % would lie within ±2 standard deviations, and 99·7 % would lie within ±3 standard deviations of the average.

From a comparatively small number of measurements on a sample run it is possible to estimate the standard deviation and so in turn estimate what the total spread of dimensions will be on the production run. To ensure that a normal distribution curve is followed, the sampling moulding run should be divided into small lots so that averages (of say, four individual measurements) are used as it is known that the distribution of averages tends to follow the normal distribution curve (the central limit theory).

6.4.3 Estimation of total range

Allow the moulding machine to 'settle down' and then produce a series of mouldings, e.g. 32. The larger the number of mouldings, then the more accurate will the result be, however, the length of this sampling run is often controlled because of practical considerations. As each moulding comes from the machine, give it a number and put it to one side to stabilise, e.g. for 20 min. Group the mouldings in batches of four. Perform the appropriate measurements (e.g. diameter) on each moulding and for each group record the average value and the range. The range is obtained for each group by subtracting the lowest value from the highest.

The average value determined on this small sampling run should be the same as the average value on the much larger production run. From the range an estimate can be simply produced which will indicate what the total spread of diameters will be.

The normal distribution curve predicts that 99·7 % of all production will lie within 6 standard deviations of the average —that is, 3 standard deviations above and 3 standard deviations below. An easy way of obtaining these upper and lower limits is to multiply the range by 1·5. If this figure is added to the previously obtained average value, then the upper

TABLE 3
CALCULATION OF NATURAL TOLERANCE LIMITS
(E.G. FOR DIAMETER)

Set No.	*Measured value, e.g. diameter (mm)* *Average or mean* ($\bar{x}$)	*Range* (R)
1	10·20	0·17
2	10·15	0·12
3	9·98	0·20
4	9·80	0·03
5	9·85	0·15
6	10·02	0·07
7	10·13	0·20
8	9·87	0·02
Total	80·00 ($\Sigma\bar{x}$)	0·96 (ΣR)
Number (of sets)	8 (n)	8 (n)
$\frac{\Sigma\bar{x}}{n} = \bar{\bar{x}}$	10·00	—
$\frac{\Sigma R}{n} = \bar{R}$	—	0·12

Upper natural tolerance limit $= \bar{\bar{x}} + 1{\cdot}5\bar{R} = 10{\cdot}00 + 0{\cdot}18 = 10{\cdot}18$ mm.
Lower natural tolerance limit $= \bar{\bar{x}} - 1{\cdot}5\bar{R} = 10{\cdot}00 - 0{\cdot}18 = 9{\cdot}82$ mm.
Each set is based on the average diameter (and the range between) four mouldings. Both these limits must lie within the specification limits.

limit will be obtained; if it is subtracted from the average value, then the corresponding lower limit will be obtained (see Table 3).

This procedure (Price, 1976) gives a quick way of establishing the natural tolerance limits of the production process but if an estimate of the standard deviation for the production run is required, then this may be obtained by multiplying the range by 0·4857. Multiplying this figure by 3 gives the factor suggested by Price. If required, the estimated standard deviation figure may be used to construct control charts.

It should be noted that if groups other than 4 are employed during the sampling run, then different factors should be applied. Such factors may be obtained from text books on statistics and in, for example, Neville and Kennedy, these factors are listed under heading of 'range coefficient *d*'.

When the number of samples within each set is 5, the range coefficient is 0·4299.

6.5 Monitoring Production

If required, the statistical properties measured previously may be used to extend the scope of quality control so that changes or drifts in the pattern of production may be readily observed.

6.5.1 Quality control charts

The first step in constructing a quality control chart is to define limits or boundaries within which production is acceptable. If the limits are exceeded, then the moulding process has drifted out of control or, it is in danger of doing so.

A graph is drawn in which the appropriate frequency distribution (e.g. diameter) is plotted on the vertical scale; the horizontal scale is marked off with an appropriate time interval. This is chosen to suit the testing period, e.g. 1 h. A horizontal line is drawn at the average value—for the example previously shown this would be 10·00 mm.

Two pairs of control limits are now placed on the chart. The first pair represents warning or inner limits and are placed so that each line is 1·96 standard deviations away from the average value. One is the upper warning limit and the other is the lower warning limit. 1·96 times the standard deviation gives a 2·5 % probability of a sample mean having a value below the lower limit and a 2·5 % probability of its having a value above the upper limit. These are also referred to as the 1 in 40 limits.

Action or outer limits may be drawn at a distance of 3·09 times the standard deviation away from the average value, i.e. the distance between them is 6·18 standard deviations and this means that 1 in 1000 of all measurements should fall above the upper limit and 1 in 1000 should fall below the lower limit.

To ensure that the results follow probability theory individual results are not plotted but rather small sets of results (e.g. 4) are averaged and plotted.

6.5.2 Sample gauging

In some factories, gauges are used to judge the product and the gauges may be set so that they conform to, say, the inner limits. The laws of chance dictate that periodically a sample will be outside the gauge limits and in such an event, another sample should be immediately taken. If everything is going well, then the probability of finding two consecutive samples which are outside the limits are very small and so if two consecutive samples are faulty, then action should be taken. However, if the second sample is within the gauge limits, then the first result may be ignored.

Many moulds are of the multi-impression type and in order to cut down on the amount of work associated with monitoring the output, it may be possible to test only one cavity. It is important to choose a cavity which is representative of the whole shot so that the fluctuations in its properties represent the fluctuations of the other mouldings. The fit and finish of this representative cavity are therefore checked at specified times and if necessary appropriate physical tests are also performed on the moulding.

6.6 Finished Goods Inspection

In the example considered in previous sections, quality control was defined only in terms of a specified dimension. However, mouldings must be free from other defects, such as contamination, burning, surface marking, etc.

6.6.1 Acceptable quality limits

The specification agreed between the moulder and his customer may specify that the supply mouldings shall not contain more than a certain number of defects. If more than this number of defects are found, then a particular lot of mouldings may be rejected. This figure may be set at say 99 % so that each lot or batch of mouldings must not contain more than 1 % of defective mouldings. The defects should, of course, be precisely defined in the specification.

In most cases it is impractical to inspect every moulding and for this reason samples of the production run are taken, inspected and the number of defective items counted. If this number exceeds preset limits, then the batch is assumed to be faulty. To save the production of work which will subsequently be rejected, Price (1976) has suggested that the sampling may be done by the operator so that if the number of defective mouldings becomes excessively large, production may be stopped. The example quoted by Price is given in the next section.

6.6.2 Random sampling

If the injection moulding machine is producing 90 000 mouldings per day, then the moulder would like to know how many mouldings he must visually inspect and how many defective mouldings he is allowed to find in the sample.

Various organisations (e.g. the ASTM and HMSO) publish tables which answer these questions. In the HMSO publication (Defence Specification DEF 131A) it is recommended that for a sample size of 90 000 a random sample of 500 items should be chosen. With an acceptable quality limit of 1 % the whole 90 000 mouldings are deemed to be satisfactory if the sample 500 contain 10 or fewer defective mouldings.

There is no need to wait to produce the total batch before performing the random sampling; if required, quality control charts may be constructed which if filled in as the run progresses will indicate whether or not the output is going to be acceptable. Further details can be obtained from books on statistics (e.g. Neville and Kennedy, 1964).

6.7 Total Inspection

Robot handling systems are now playing an increasingly important part in automatic handling systems. In the field of injection moulding their big advantage is that they do not tire like a human operator and this means that moulding cycles are more consistent. Robot arms have been mounted on injection moulding machines and these arms or manipulators reach into the mould after each shot and pull out the part. The arm may then place the moulding on a table or moving belt so that finishing (e.g. degating, paint spraying, etc.) may be performed by operators and/or robots.

With the advent of the microprocessor it is possible to program very complex instructions for such machines and before long such machines will receive their instructions in English. By adding a TV camera into the system it is possible to produce an automatic inspection system. Such a system would therefore consist of a computer (usually a microprocessor), a video camera and a robot.

The inspection process can proceed in the following sequence (Blomm, 1980):

1. Find the part
2. Orient the part with a master image
3. Compare the part with the master
4. Evaluate the part
5. Decide on 'go' or 'no go'

At the present time such a system is probably not economically feasible but as the cost of manual inspection rises and the price of automatic systems falls, then computer–video systems will become more popular. Already systems are being discussed which will incorporate laser measuring systems, within the computer–video system, so that dimensional tolerances may be checked accurately and swiftly.

REFERENCES

Anon. How metal prices are favouring plastics. *Plastics and Rubber Weekly*, p. 13, 31 May 1980.

ASTM. Injection moulding of specimens of thermoplastics materials, ASTM D 1130-63 (reapproved 1969).

ASTM. Injection moulding of specimens of thermoplastic moulding and extrusion materials, ASTM D 1897, 1972.

Blomm, B. Computer + video camera + robot = automatic inspection system. *Plastics Eng.*, p. 55, Nov 1980.

Bown, J. *Injection Moulding of Plastics Components*, McGraw-Hill Book Co., New York, 1979.

Dewey, M. A. P. Testing: Reduce your liability risk. *British Plastics and Rubber*, p. 61, Sept. 1981.

ICI. *Presentation and use of Data on the Mechanical Properties of Thermoplastics.* Technical Service Note G 123, ICI Plastics Division, Herts, 1979.

Ives, C. G., Mead, J. A. and Riley, M. M. *Handbook of Plastics Test Methods*, Published for the PRI by Iliffe, London, 1971.

Moritz, J. A. Practical ways to improve moulding Q.C. *Plastics World*, p. 80, Sept 1980.

Neville, A. M. and Kennedy, J. B. *Basic Statistical Methods*, Intertext Books, London, 1964.

Olson, F. C. W. and Schultz, O. T. Temperatures in solids during heating or cooling, *Industrial and Engineering Chemistry*, p. 874, July 1942.

Price, F. A poor mans guide to quality control, *Plastics and Rubber Weekly*, p. 10, 9 Jan 1976; Can we make it? *Ibid*, p. 13, 16 Jan 1976; Quality control by constellations, *Ibid*, p. 11, 30 Jan 1976; Know your shortcomings before you sign, *Ibid*, p. 16, 13 Feb 1976; Putting some back in the pot, *Ibid*, p. 16, 20 Feb 1976.

Rubin, I. I. Injection moulding theory and practice, *SPE Monograph*, John Wiley and Son Inc, New York, 1972.

Scott, J. R. *Physical Testing of Rubbers*, McLaren and Sons Ltd, London, 1965.

SPI. *Standard Test Procedure for Determining Plasticising (Plasticating) Performance of Screw Injection Machines*, Soc. Plast. Ind., New York 1967.

Turner, S. *Mechanical Testing of Plastics*, Published for the Plastics Institute by Iliffe, London, 1973.

Weston, D. Some special test methods applicable to polyethene. *Plastics*, p. 465, **24**, Nov 1959.

Whelan, A. and Craft, J. L. *The Injection Moulding of Test Specimens*, NCRT, London, 1978a.

Whelan, A. and Craft, J. L. Machine ailments—low cost diagnosis, *British Plastics and Rubber*, p. 56, June 1978b.

CHAPTER 7

Material Recommendations for Injection Moulding

1 POLYSTYRENE

1 *Common name:* Polystyrene.

2 *Abbreviation:* PS or GPPS, i.e. general purpose polystyrene.

3 *Systematic chemical name:* Poly(1-phenylethylene).

4 *Some suppliers:*	5 *Trade names or trade marks:*
Ato Chimie	Lacqrene
BASF	Polystyrol
BP	—
CDF Chimie	Gedex
Dow	Styron
Hoescht	Hostyren N.
Monsanto	Lustrex
Montedison	Edistir
Shell	Carinex
Sterling	Sternite

6 *Material properties:* A hard, rigid material which in its natural form has a high gloss, sparkle and transparency. A wide range of colours is readily available and as the mouldings can also be decorated by a large number of techniques, attractive components can be easily produced. Low water absorption and good electrical insulation properties are other desirable characteristics. However, the polymer is brittle, burns easily and has poor outdoor weathering properties. It is easy to process by injection moulding but this apparent ease can be deceptive as unless care is taken parts that stress-crack in service may result. The density is about $1{\cdot}05\,g\,cm^{-3}$.

7 *Ease of flow:* Relatively easy flow; special easy flow grades available.

8 *Shrinkage:* Because of the low shrinkage of this amorphous material

parts with good dimensional accuracy can be obtained. Shrinkage is about 0·5% (about 0·004 in/in).

9 Resistant to the following: Resistant to acids (except oxidising acids), alkalis, mineral oils, detergents and the lower molecular weight alcohols. Relatively unaffected by exposure to water and to high energy radiation.

10 Not resistant to: High temperatures and a wide range of organic solvents, e.g. aromatic and chlorinated hydrocarbons, esters and ketones. Liquids such as white spirit can cause stress-cracking and can therefore be used to detect strain. Decomposed by prolonged contact with oxidising agents such as concentrated sulphuric acid.

11 Material detection or identification: Sinks slowly in water. Burns easily with a yellow, sooty flame and produces a lot of smell, smoke and soot. Mouldings are brittle and emit a metallic ringing noise when dropped on a hard surface. Does not cut easily or cleanly with a knife. Dissolves in carbon tetrachloride.

12 Colouring: Can be readily coloured, e.g., dry colour; masterbatch—tumble mixing has proved extremely successful. Normally percentage of colourant/masterbatch is 1–4% dependent upon base colour of styrene and tintorial power of colourant. Liquid colourant is readily accepted but can cause problems, e.g. with respect to processing and properties of the end product—development taking place.

13 Material and component handling: Does not absorb water (sufficient to upset processing) normally. Store in a clean, dry area. Predrying not usually necessary but when required heat at 70 °C for up to 3 h.

GPPS has very low impact strength and low abrasion resistance and these factors should be taken into account when mouldings are handled.

14 Mould and gate considerations: Can be moulded with gates of small cross-sectional area. Pinpoint, tab, insulated runner and hot runner are widely used.

15 Flow path: wall thickness ratio: The material has a maximum flow path:wall thickness ratio which is intermediate between that of PP and ABS, e.g. at 1 mm wall thickness PP, 170:1; PS, 150:1; ABS, 140:1.

16 Projected area considerations: Depends on grade, temperature, etc. Generally the clamping pressure required is of the order of 1–2 tsi of projected area (15–30 $MN\,m^{-2}$) for GPPS. For thin walled mouldings 3–4 tsi (45–60 $MN\,m^{-2}$) may be required. With easy flow materials only 1–2 tsi may be needed even for thin walled components.

17 Cylinder equipment: The cylinder is usually equipped with a shut-off nozzle when decompression is not available; a valve, to prevent back-flow, is usually fitted to the screw.

18 Screw cushion: About 3 mm.

19 Shot capacity: Because of this material's good heat stability the shot volume can be as little as 5% of the rated capacity if required; the maximum rating of the barrel can also be utilised.

20 Melt temperature—as measured in the nozzle or by an air-shot technique: 220–260 °C (428–500 °F).

21 Barrel residence time: Good resistance to heat normally. Grades which contain ultra-violet additives may tend to yellow if left to 'cook' in the barrel.

22 Temperature settings: Please note that it is the melt temperature which is important; those in Table 1 are only suggested, initial settings. The temperature of the hydraulic oil and of the material in the hopper should not vary excessively.

TABLE 1

Zone No.	*Location*	*Temperatures °C* *From*	*To*	*Temperatures °F* *From*	*To*
1	Barrel rear (near the hopper)	150	180	302	356
2	Barrel middle	180	230	356	446
3	Barrel middle	210	230	410	446
4	Barrel front	210	280	410	536
5	Nozzle	210	280	410	536
	Mould	10	80	50	176

23 Injection speed—mould filling speed: As high as possible consistent with other requirements, e.g. surface finish. To get high speeds it is not necessary to use very high pressures.

24 Injection pressure: The machine should be capable of giving the following: First stage: <1500 bars; 150 MN m^{-2}; 21 400 psi. Second stage (dwell or follow-up pressure): <750 bars; 75 MN m^{-2}; 10 700 psi.

25 Screw rotational speed (rpm): High screw speeds possible but better to run at slow speeds.

26 Back pressure: <150 bars; 15 MN m^{-2}; 2140 psi.

27 Shutting down: No need to purge with another material; empty the hopper and purge the barrel clean.

28 Reprocessing: Up to 100% regrind may be run.

29 Finishing: Degating is relatively easy, e.g. by hand using a sharp knife or scalpel. Drilling or tapping, to produce a thread, may prove difficult or unsatisfactory in service, so inserts are used. Mouldings can be readily decorated, e.g. by (i) silk screening, (ii) printing, (iii) hot foiling (coloured or metallic), or (iv) painting. Components are readily joined by ultrasonic welding or solvent welding.

30 Other comments: This material readily attracts dust and should therefore be kept covered at all times.

31 Typical components: Toys, containers, tape cassettes, disposable tumblers, etc. Usually parts that require a combination of colour, clarity, stiffness, low cost and good appearance but which do not require high heat, solvent and impact resistance. When suitably light stabilised the material is used in light fitting applications, e.g. diffusion louvres. The addition of glass fibres improves stiffness, impact strength and heat distortion temperature thus opening up new markets. Rubber modification and/or copolymerisation are other techniques widely employed (see Styrene Acrylonitrile and Toughened Polystyrene).

2 STYRENE ACRYLONITRILE

1 Common name: Styrene acrylonitrile.

2 Abbreviation: SAN.

3 Systematic chemical name: Not known but the following has been suggested: poly(1-phenylethylene-*co*-1-cyanoethylene).

4 Some suppliers:	*5 Trade names or trade marks:*
BASF	Luran
Dow	Tyril
Monsanto	Lustran SAN
Montedison	Kostil

6 Material properties: When styrene is copolymerised with acrylonitrile (20–30%) a material with a higher softening point, rigidity and solvent resistance (compared to PS) results. This polymer is transparent but not water white and clear mouldings have a yellow tint (which can be offset by a blue dye). The material has better stress crack resistance than PS and a higher impact strength. Resistance to staining, e.g. by food, is good. The densities of SAN and PS are similar but SAN has a higher resistance to creep under load; maintains its impact strength from -40 to $+50$ °C without change and has a higher heat distortion temperature.

7 Ease of flow: In general it is a stiffer flow material than PS, therefore more pressure is needed to get equivalent rates of flow.

8 Shrinkage: An amorphous material with low shrinkage, e.g. about 0·5% ($\simeq$0·004 in/in).

9 Resistant to the following: Because of the polar nature of the acrylonitrile molecule these copolymers have better resistance to hydrocarbons, oils and greases than PS. It is resistant to acids, alkalis, detergents, fats and oils, petrol and salt solutions.

10 Not resistant to: Swollen by aromatics, dissolved by ketones and methylene chloride (used for glueing). The resistance to alcohols is only fair and its resistance to chlorinated hydrocarbons is poor. Weathering resistance is not very good but is improved by stabilisers.

11 Material detection or identification: Sinks in water and cannot be cut easily with a knife. Usually found as clear mouldings which are classed as brittle. Not readily soluble in carbon tetrachloride and when burnt gives off styrene as well as a bitter or acrid odour; burns without forming drips.

12 Colouring: Widely used colours are translucent greys (smoke), brown, opal and clear. Dry colouring is not recommended except for small mouldings whose colour is relatively unimportant. Compounded material works best because the percentage of colour added is relatively small. If this small amount is added at the moulding stage then batch to batch inconsistencies may arise with dry colour and large mouldings.

13 Material and component handling: SAN is usually used in premium mouldings and it must therefore be stored under conditions which will not result in contamination, e.g. by dust or water. Predrying is not normally required but can be performed by heating at 70 °C for up to 3 h. To stop the surface becoming marred or marked some components are individually wrapped.

14 Mould and gate considerations: Gate sizes are larger than for PS. To get good optical properties use tab, deep fan and edge gates. The surface finish of the tool should be of a very high standard (e.g. no imperfections and chrome plated) because of the applications of this material.

15 Flow path: wall thickness ratio: A range of grades is available (as for most polymers) so that, for example, one grade of PS may be stiffer flow than a grade of SAN. In general however SAN is stiffer flow than PS with a flow path: wall thickness ratio of approximately 140:1 at 1 mm (0·04 in) wall thickness, i.e. similar to ABS.

16 Projected area considerations: This material is usually moulded into components which have relatively thick sections; resistance to flow is therefore low and this can reduce the need for very high clamping pressure when moulding items such as jugs, tumblers, etc. Clamping pressures may be of the order of 2 tsi (30 MN m^{-2}) for such components. Pressures of 4 tsi (60 MN m^{-2}) have been used for mouldings which have large surface areas, e.g. music centre covers.

17 Cylinder equipment: The cylinder is usually fitted with a shut-off nozzle when decompression is not available. A valve, to prevent back-flow, is usually fitted to the screw.

18 Screw cushion: About 4 mm.

19 Shot capacity: Under-rate machine by approximately 20% (compared to polystyrene).

20 Melt temperature: About 30 °C higher than for PS but do not overheat (i.e. above 270 °C) as the material may degrade. Tighter control is required for this material compared to PS.

21 Barrel residence time: This can be critical and can result in colour changes. At high temperatures (approaching 270 °C) the barrel residence time may be quite low, e.g. 5 min.

22 Temperature settings: As in Table 2. High mould temperatures give parts with good surface finish and fewer moulded-in stresses therefore mouldings with improved craze resistance and toughness result; they can be further improved by annealing, e.g. in a water bath at 70 °C.

TABLE 2

Zone No.	*Location*	*Temperatures °C*		*Temperatures °F*	
		From	*To*	*From*	*To*
1	Barrel rear (near the hopper)	150	180	302	356
2	Barrel middle	180	230	356	446
3	Barrel middle	210	250	410	482
4	Barrel front	210	260	410	500
5	Nozzle	210	250	410	482
	Mould	30	85	86	185

23 Injection speed: As high as possible consistent with other requirements, e.g. surface gloss, etc.

24 Injection pressure: The machine should be capable of giving the following: First stage: up to 1500 bars; 150 MN m^{-2}; 21 400 psi. Second stage (dwell or follow-up pressure): up to 750 bars; 75 MN m^{-2}; 10 700 psi.

25 Screw rotational speed (rpm): Use slightly slower screw speeds than for PS.

26 Back pressure: Up to 150 bars; 15 MN m^{-2}; 2140 psi.

27 Shutting down: No need to purge with another material. Empty the hopper and purge the material clean.

28 Reprocessing: The water absorption of these materials is 10 times that of PS, i.e. similar to that of PMMA; they should therefore be stored carefully after regranulating and will probably have to be predried.

29 Finishing: Can withstand machining operations that would crack PS. Gates may be removed by machining (milling, routing, etc.) and/or cropping—gate scars are sometimes covered by self-adhesive labels. The components may be decorated by processes such as hot stamping and silk screening; the appearance may be improved by buffing or polishing.

30 Other comments: Even though this material is claimed to have twice the impact strength of PS it must not be imagined that it is very tough—it is a brittle, notch-sensitive material like other amorphous thermoplastics.

31 Typical components: Cups, tumblers, dishes, picnic items, trays, disposable cutlery, houseware, tape casette storage racks, dials, cosmetic containers, etc. In general this type of material is specified when PS is not good enough. It is usually cheaper and tougher than PMMA but does not possess the weathering resistance and excellent transparency of this polymer.

3 TOUGHENED POLYSTYRENE

1 Common name: Toughened polystyrene or high impact polystyrene.

2 Abbreviation: HIPS or TPS or SB.

3 Systematic chemical name: Not known—one suggestion is toughened poly(1-phenylethylene).

4 Some suppliers:	*5 Trade names or trade marks:*
Ato Chimie	Lacqrene
BASF	Polystyrol
BP	—
Chemische Werke Hüls	Vestyron
Dow	Styron
Hoescht	Hostyren S
Monsanto	Lustrex
Montedison	Edistir
Union Carbide	—

6 Material properties: A semirigid material produced by polymerising styrene in the presence of a rubber, e.g. polybutadiene. The fine rubber particles are chemically grafted on to the PS chains and this gives materials with a high impact strength but with lower strength, stiffness, hardness and softening point (cf. PS). The transparency and surface gloss of PS is also lost by the addition of the rubber. Normally the resistance to light is poor and the material has a poorer resistance to chemicals than PS. As the rubber concentration is increased the impact strength is also increased but the stiffness, tensile strength, density and softening point are decreased. A very high impact grade will have a density of 1·02 g cm^{-3}. In general, HIPS is not such a good electrical insulator as GPPS.

7 Ease of flow: Similar to PS. Both general purpose and easy flow grades available.

8 Shrinkage: Similar to PS—about 0·005 in/in.

9 Resistant to the following: Can resist machining operations, e.g. drilling, punching, etc. which would shatter PS. Will resist acids, alkalis, fats, oils and salt solutions.

10 Not resistant to: A wide range of aromatic and chlorinated hydrocarbons; poorer chemical resistance than PS.

11 Material detection or identification: Clear mouldings not possible (cf. PS). Cuts more easily than PS and when burnt gives off an odour of burning rubber as well as styrene. Sinks very slowly in water and softens at a lower temperature than PS.

12 Colouring: Dry colouring techniques (by tumble blending) are widely used and give satisfactory results; with large colourant concentrations adhesion of the colouring system to the granules may be improved by first tumbling the polymer with butyl stearate. Masterbatches are also used, e.g. for specific purposes such as with fluorescent pigments. Liquid colours have been satisfactorily used with HIPS.

13 Material and component handling: More prone to water absorption than PS, however, predrying is not usually necessary, although heat at 70 °C for up to 3 h may be used. Normal packaging procedures adopted for components.

14 Mould and gate considerations: Similar to PS. Often moulded with gates of small cross-section (e.g. for disposable cups). Pinpoint, tab, insulated runner and hot runner are widely used. Submarine-ejector type gating used, e.g. for fascia mouldings.

15 Flow path: wall thickness ratio: Available in a range of grades. At a wall thickness of 1 mm standard grades can have a maximum flow path: wall thickness ratio of approximately 130:1. Easy flow materials available but these may have a lower softening point.

16 Projected area: The clamping force requirements range from 1 to 3 tsi (15 to 45 MN m^{-2}). Easy flow materials can usually be moulded using low clamping forces.

17 Cylinder equipment: The cylinder is usually equipped with a shut-off nozzle when decompression is not available. A valve to prevent back-flow, is usually fitted to the screw.

18 Screw cushion: Usually about 4 mm, however, on some jobs a screw cushion is not used, e.g. disposable cups. Special machines have been built to mould these items and some manufacturers recommend that the screw cushion should be eliminated; accurate and reproducible dosing is necessary on such machines.

19 Shot capacity: 90% of the machines rated capacity may be used.

20 Melt temperature—as measured in the nozzle or by an air–shot technique: 180–250 °C (356–482 °F). Surface gloss improved by high melt and mould temperatures. To obtain the uniformly polished matt finish (associated with a satisfactory HIPS moulding) the melt temperature may need to be fairly high.

21 Barrel residence time: Not critical although it is good practice not to let the material 'cook' in the barrel.

22 Temperature settings: Please note that it is the melt temperature which is important; those in Table 3 are only suggested initial settings. The temperature of the hydraulic oil and of the material in the hopper should not vary excessively.

TABLE 3

Zone No.	*Location*	*Temperatures °C*		*Temperatures °F*	
		From	*To*	*From*	*To*
1	Barrel rear (near the hopper)	160	190	320	374
2	Barrel middle	170	210	338	410
3	Barrel middle	180	230	356	446
4	Barrel front	190	250	374	482
5	Nozzle	180	240	356	464
	Mould	10	80	50	176

23 Injection speed—mould filling speed: Should be as high as possible but consistent with other requirements, e.g. surface gloss, jetting, etc. High pressures not needed to get fast flow rates.

24 Injection pressure: First stage: up to 1500 bars; 150 MN m^{-2}; 21 400 psi. Second stage (dwell or follow-up pressure): up to 750 bars; 75 MN m^{-2}; 10 700 psi.

25 Screw rotational speed (rpm): Fast screw speeds are possible but not necessarily desirable as their use can result in temperature fluctuations.

26 Back pressure: Up to 150 bars; 15 MN m^{-2}; 2140 psi.

27 Shutting down: No need to purge with another material.

28 Reprocessing: Up to 100% regrind may be run.

29 Finishing: A wide variety of techniques are used to degate mouldings; the components may be separated by twisting, cutting (scalpel or knife) or by automatic systems. Cropping of edge gates is widely used. Components may be decorated by a variety of techniques (i) hot stamping (ii) painting or spraying (iii) vacuum metallising (iv) silk screening, etc.

30 Other comments: Moulding quality can be judged by surface finish: in general the better the finish, the higher is the quality for a given grade.

31 Typical components: Household appliances, toys, furniture components, disposable cups, appliance housings, etc. Applications where a certain degree of impact resistance is required and where the relatively poor chemical resistance of this material is unimportant or can be put to good use, e.g. to solvent weld components together. Scale model kits are a good example of this application.

4 ACRYLONITRILE-BUTADIENE STYRENE

1 Common name: Acrylonitrile-butadiene styrene.

2 Abbreviation: ABS.

3 Systematic chemical name: Not known—one suggestion is poly(1-butenylene-*g*-(1-phenylethylene-*co*-1-cyanoethylene).

4 Some suppliers:	*5 Trade names or trade marks:*
Abson	Abtec
Anic SpA	Ravikral
BASF	Terluran
Bayer	Novodur
Borg Warner	Cycolac
Dow	Dow-ABS
DSM	Ronfalin
Goodrich	Abson
Monsanto	Lustran
Montedison	Urtal
Polymon	Formid
Rhone Poulenc	Afcoryl
SIR	Restiran
Sterling	Sternite ABS
Ugine Kuhlmann	Ugikral

6 Material properties: A hard, tough material with good resistance to impact even at low temperatures. The material has low water absorption and this means that its electrical insulation properties, which are good, are relatively unaffected by humidity changes. It is normally available in translucent or opaque colours and the resultant mouldings can have a high gloss. The surface is resistant to scuffing and marking but the unmodified material has not got very good weathering properties. Superior heat resistance and impact strength compared to HIPS. Density about 1·05 but can vary depending on grade, i.e. whether medium impact, self-extinguishing, glass filled, etc.

7 Ease of flow: Relatively easy flowing. The ease of flow will depend on the grade selected but will, broadly speaking, be similar to PS.

8 Shrinkage: About 0·6 %, i.e. about 0·005 in/in.

9 Resistant to the following: Resistant to staining and to alkalis, acids (not concentrated oxidising acids), oils and fats; the majority of alcohols and hydrocarbons.

10 Not resistant to: Acetone, carbon tetrachloride, ether, ethylene dichloride, ethyl acetate, toluene, trichloromethane, i.e. fairly low molecular weight organic materials such as ketones, esters, aldehydes and some chlorinated hydrocarbons.

11 Material detection or identification: Cuts cleanly with a knife and cuts have smooth edges; ranges from semibrittle to tough. Burns with a yellow, sooty flame and gives off an acrid odour which also smells of rubber. As the material has a density of 1·05 $g\,cm^{-3}$ it will sink slowly in water. Resists carbon tetrachloride better than HIPS.

12 Colouring: Natural colour is an ivory shade so a wide range of colours is possible; dark colours give the best resistance to ageing. Problems may arise, however, due to the high processing temperatures employed and also due to leaching. Dry colours and masterbatches are used but may cause problems if an exact colour match is required. Where such matching is required the normal procedure is to purchase the coloured polymer from a specialist compounder or a raw material manufacturer. Base polymer colour affects the final colour and should therefore be controlled.

13 Material and component handling: Material requires moisture removal before processing in order to achieve optimum properties. Strict control of storage can help to eliminate extensive drying operations. Drying temperatures of 70–80 °C for 2–4 h may be used.

Many ABS mouldings are electroplated and in order to minimise surface marking the mouldings should be handled as little as possible and when handling is necessary clean gloves (changed regularly) should be worn. The components should be wrapped, e.g. in polyethylene bags.

Absorbs water more than PS or HIPS so predrying may be necessary if mouldings are blistered or streaked. If a vented machine is not available predry for about 4 h at 80 °C. Predrying is necessary if the material has absorbed more than 0·2 % moisture during storage.

14 Mould and gate considerations: Similar to PS; pinpoint, hot runner or insulated runner may be used. Where electroplating or metallising of the components is to be performed the feed systems should be designed so that high levels of stress are avoided, e.g. by using large gates and runners.

Where unavoidable surface defects are obtained these may be masked by the use of grained, mould surfaces, such grained surfaces also help to mask any marring which occurs in service.

15 Flow path: wall thickness ratio: Similar to PS. At 1 mm wall thickness the flow path: wall thickness ratio will be about 150:1 for PS and 140:1 for ABS. Easy flow grades of ABS are however available and such grades will give a much higher value.

16 Projected area: The clamping pressure required is usually of the order of 2 tsi of the projected area (30 MN m^{-2}).

17 Cylinder equipment: Because of water absorption a vented barrel can be advantageously used with this material. The cylinder is usually fitted with an open nozzle as shut-off nozzles may cause burn marks; a valve, to prevent back-flow, is usually fitted to the screw.

18 Screw cushion: About 4 mm.

19 Shot capacity: Up to 90 % of the rated (PS) capacity of the machine may be used.

20 Melt temperature—as measured in the nozzle or by an air-shot technique: Surface gloss is improved by high mould and melt temperatures, e.g. 220–260 °C. Do not exceed 270 °C as discoloration will occur.

21 Barrel residence time: If high melt temperatures are being used then the residence time in the barrel will be low as otherwise discoloration or changes in colour may occur. Different barrel residence times on different machines can cause colour changes.

22 Temperature settings: Please note that it is the melt temperature which is important; those in Table 4 are only suggested, initial settings. The temperature of the hydraulic oil and of the material in the hopper should not vary excessively.

23 Injection speed—mould filling speed: Very high pressures are not needed to get high speeds; to get the best surface finish (e.g. for electroplating) use injection velocity control. Surface finish is improved by using well dried material, high melt and mould temperatures and slow filling speeds.

24 Injection pressure: The machine should be capable of giving the

TABLE 4

Zone No.	*Location*	*Temperatures °C* *From*	*To*	*Temperatures °F* *From*	*To*
1	Barrel rear (near the hopper)	150	180	302	356
2	Barrel middle	180	230	356	446
3	Barrel middle	210	250	410	482
4	Barrel front	210	260	410	500
5	Nozzle	210	250	410	482
	Mould	60	90	140	194

following: First stage: up to 1500 bars; 150 MN m^{-2} 21 400 psi. Second stage (dwell or follow-up pressure): up to 750 bars; 75 MN m^{-2}; 10 700 psi. It is not necessary to have long dwell times or high packing pressures.

25 Screw rotational speed (rpm): High screw speeds can be used with this material but it is better to run at slow speeds if possible, e.g. 100 rpm.

26 Back pressure: Up to 250 bars; 25 MN m^{-2}; 3570 psi. At high temperatures and with insufficient back pressure air can cause scorching of the polymer.

27 Shutting down: Purging with a different material is not necessary. Note that ABS and PS are not compatible.

28 Reprocessing: When regrind is being used it is blended with virgin material in relatively small quantities, e.g. ratio of virgin to regrind is 75:25. Only good quality regrind should be used, e.g. free from oxidation, contamination, etc. Do not use regrind for electroplating applications.

29 Finishing: A wide variety of techniques are used to degate small mouldings, e.g. cropping, cutting (scalpel or knife) or automatic systems which are operated by the mould opening action. Large gates may be removed by routing, sawing, etc. Mouldings may be decorated by printing, embossing and metallising. Assembly into large components may be made by using interference fits, gluing and ultrasonic welding. Static electricity may interfere with printing and surface charges should be removed. If the mouldings are degreased prior to electroplating then the degreasing chemical must be carefully chosen, e.g. to avoid stress cracking. Mouldings should be carefully handled and/or individually wrapped as they scratch easily, e.g. the swarf can cause scratching during machining.

30 Other comments: A very wide range of grades of ABS is possible as, for example, the material can be made by two main methods: (i) melt mixing of nitrile rubber with SAN, or (ii) polymerising styrene and acrylonitrile in the presence of a rubber (polybutadiene or styrene-butadiene rubber). The second procedure is the commercial procedure as this gives polymers with the best balance of properties. A 'typical' composition could contain 20–30% acrylonitrile, 20–30% butadiene and 40–60% styrene.

31 Typical components: Luggage, appliance housings, radio and TV cabinets, vacuum cleaner cases, bathroom cabinets, complete instrument panel surrounds on cars, telephone handsets, grilles, safety helmets, etc. When used in automotive applications the material is commonly electroplated as, by chemical etching techniques, good adhesion occurs between the metal and the moulding. In such applications a good surface finish on the mouldings is essential as the plating will emphasise (not hide) surface marks.

5 ACRYLIC

1 Common name: Polymethyl methacrylate.

2 Abbreviation: PMMA or acrylic.

3 Systematic chemical name: Poly[1-(methoxycarbonyl)-1-methylethylene].

4 Some suppliers:	*5 Trade names or trade marks:*
Altulor	Altulite
Du Pont	Lucite
ICI	Diakon
Montedison	Vedril
Resart	Resarit
Rohm GMBH	Plexiglas

6 Material properties: PMMA is a hard, rigid material which is renowned for its clarity (transmits about 85% of daylight) weathering resistance and depth of colour. A wide range of colours is possible and the mouldings can be readily decorated by a wide variety of techniques (e.g. metallising) so that attractive components can be readily produced. Low water absorption and good dimensional stability are other desirable attributes. The polymer also appears to be unaffected by contact with human tissue. Copolymerisation of methyl methacrylate (MMA) with other monomers results in copolymers whose properties are intermediate between those of the parent materials. For example, copolymers of MMA and styrene have properties intermediate between those of PS and PMMA.

7 Ease of flow: Stiffer flow than ABS or PS and therefore higher pressures may be required. For thick parts, fill slowly—a variable speed injection facility would be useful.

8 Shrinkage: This amorphous material shrinks by $<0{\cdot}8\%$ (0·002–0·007 in/in).

9 Resistant to the following: Dilute acids and alkalis, fats, oils, aliphatic hydrocarbons (e.g. white spirit and paraffin), dilute alcohols and detergents. Swollen by alcohols, phenols, ether and carbon tetrachloride. Relatively unaffected by strong alkalis (e.g. ammonium hydroxide) and concentrated hydrochloric acid.

10 Not resistant to: Strong acids and alcoholic alkalis, chlorinated hydrocarbons, esters, low molecular weight ketones and aromatic hydrocarbons. Ethyl acetate can be used to detect strain.

11 Material detection or identification: As the density is $1{\cdot}18\,\mathrm{g\,cm^{-3}}$ the material sinks in water. It does not cut easily with a knife and is usually clear, hard and brittle. Upon ignition the material burns with a blue flame which is tipped or edged with yellow; a sweet fruity odour is emitted but the amount of smoke evolved is low. It continues to burn when the source of ignition is removed and the flame crackles; frothing may occur (e.g. when the flame is extinguished). There is very little residue left after burning. Softens in the region of 100 °C.

12 Colouring: Normally supplied already coloured by the manufacturer (colour matched if necessary). Masterbatches are used in preference to other systems when in-house colouring is being employed. Only a few commercial masterbatches are used, e.g. red/amber for the automotive industry and opal for the lighting industry. Liquid colouring does not appear to be widely used due to dispersion and processing problems.

13 Material and component handling: The material will absorb water if stored incorrectly, e.g. condensation may occur on the surface. Long storage is to be avoided and a first-in first-out system should be rigorously adhered to. Components susceptible to static are normally individually wrapped, e.g. in tissue. To protect surfaces, handle carefully, e.g. by wearing gloves.

Normally suitable for moulding without any preliminary drying operations if stored correctly. The use of a vented barrel may prove very advantageous if the material is damp. Slight traces of moisture result in frothing (in the cylinder) and splash marks (in the mouldings) therefore dry at 70 °C for 4 h.

14 Mould and gate considerations: Sprues and runners need to be larger than those used for PS as the material does not flow as well. If cavities can be filled quickly then small gates may be used, e.g. pingates on low wall thicknesses. However, to maintain the optical properties it is usually necessary to fill slowly and therefore short, thick gates are used. Tab gating helps to prevent problems caused by jetting.

15 Flow path: wall thickness ratio: Depends on grade of material, part thickness, temperature, etc. At the same wall thickness, easy flow, medium flow and stiff flow materials will give ratios of 150:1, 120:1 and 100:1, respectively. Compared to PS this material has much stiffer flow; the maximum flow path: wall thickness ratio for PMMA will be 100:1 whereas for PS it could be 150:1 (at 1 mm wall thickness).

16 Projected area: For most products approximately 3 tsi (46 $MN\,m^{-2}$) of projected area is used; for lenses and parts with long flow section 4 tsi (62 $MN\,m^{-2}$) may be needed to achieve satisfactory clamping.

17 Cylinder equipment: A check ring is usually fitted to the screw (which can be of general purpose design). Both valved and open nozzles may be used—if excessive decompression is used then splash marks may be seen in the moulding.

18 Screw cushion: 2–6 mm.

19 Shot capacity: 20–85% of the cylinders volumetric capacity may be used.

20 Melt temperature—as measured in the nozzle or by an air-shot technique: The melt viscosity of acrylic moulding powders is more temperature dependent than PS or PP, therefore accurate temperature control is necessary. Melt temperatures are usually in the region of 225 °C; for thin sectioned parts, melt temperature may reach 260 °C.

21 Barrel residence time: If PMMA is subject to an excessive heat treatment (which is dependent on both time and temperature) then it will depolymerise to give methyl methacrylate (MMA) and therefore bubbling; 13 min at 275 °C may cause such degradation.

22 Temperature settings: Please note that it is the melt temperature which is important; those in Table 5 are only suggested, initial settings. The temperature of the hydraulic oil and of the material in the hopper should not vary excessively.

23 Injection speed—mould filling speed: Usually low so as to avoid jetting and to maintain optical properties.

24 Injection pressure: The machine should be capable of giving the following: First stage: <1750 bars; 175 $MN\,m^{-2}$; 25 000 psi. Second stage (dwell or follow-up pressure): <1050 bars; 105 $MN\,m^{-2}$; 15 000 psi. In the case of lenses pressure may need to be applied for a long time, e.g. 3 min.

25 Screw rotational speed (rpm): Low. Usually acrylic mouldings are fairly thick so the screw speed should be adjusted to suit the cooling cycle, e.g. less than 100 rpm to avoid over-heating. The screw should be driven by a high torque, low speed motor.

26 Back pressure: <400 bars; 40 MN m^{-2}; 5700 psi. Too low a back pressure may result in bubbles in the mouldings.

27 Shutting down: Purging with a different material is not necessary.

28 Reprocessing: Regrind must be thoroughly dried—usually used in opaque moulding as reground material impairs optical properties. Where large amounts of scrap are being made then, after regranulation, impact modifiers are blended in, and the mixture used instead of impact styrene for some applications.

TABLE 5

Zone No.	*Location*	*Temperatures °C From*	*To*	*Temperatures °F From*	*To*
1	Barrel rear (near the hopper)	150	180	302	356
2	Barrel middle	180	210	356	410
3	Barrel middle	200	230	392	446
4	Barrel front	220	240	428	464
5	Nozzle	225	245	437	473
	Mould	40	80	104	176

29 Finishing: Threaded inserts are used instead of drilling and tapping; components may be ultrasonically welded. A high polish may be obtained on cut edges by buffing. Components decorated by silk screening and vacuum metallisation. A wide variety of degating techniques is employed, e.g. (i) simple (using a scalpel), (ii) hot wire, (iii) routing, (iv) end milling and (v) peripheral deflashing using sticks cut from PF, cloth laminates (Tufnol). Components can be joined by painting the edges with chloroform or with a solution made from PMMA in methylene chloride.

30 Other comments: The addition of lubricants, to improve flow, usually results in a loss of clarity. To minimise problems caused when changing from other materials some moulders keep a cylinder just for acrylic.

31 Typical components: PMMA's excellent optical properties, depth of colour and good resistance to weathering result in its use for streetlight fixtures, lenses and light-covers on cars. Also used for instrument panels, dials and vending machine housings. Mouldings are not subject to microbiological attack.

6 UNPLASTICISED POLYVINYL CHLORIDE

1 Common name: Unplasticised polyvinyl chloride.

2 Abbreviation: UPVC, hard PVC, or rigid PVC.

3 Systematic chemical name: Poly(1-chloroethylene). As the term UPVC simply means that the compound is based on PVC then the compounds may be known under the trade name of the parent polymer or, under the trade name of a proprietary compound.

4 Some suppliers:	*5 Trade names or trade marks:*
Anic SpA	Fimat
Ato Chimie	Lacquyl
BASF	Vinoflex
BP	Geon
CdF Chimie	Gedevyl
Chemische Werke Hüls	Vestolit
Diamond Shamrock	Dacovin
Dynamit Nobel	Trosiplast
Hoescht	Hostalit
ICI	Corvic or Welvic
Lonza	Lonza compound
Montedison	Sicron
Solvay	Solvic
Wacker Chemie	Vinnol

6 Material properties: Unplasticised PVC has many desirable properties, e.g. it is strong, stiff, transparent, resistant to oils, etc. However it is difficult to process, can have a low impact strength and a low heat distortion temperature. These deficiencies can be minimised by additive choice, machine design and operation. On heating, hydrochloric acid is evolved and this attacks both the machine and the operators. Without the addition of stabilisers (such as tribasic lead sulphate) processing would be impossible; it is also made easier by the use of lubricants (e.g. 2 phr of butyl stearate), processing aids (3 phr of acrylic processing aid) and low molecular weight resins (i.e. of low *k* value). Impact strength may be increased significantly by mixing with other plastics (up to 15 phr of a copolymer—ABS or MBS); heat distortion temperature may be raised by using fillers (e.g. 10 phr of china clay) and large concentrations of polymeric modifiers (e.g. 45 parts of a modified ABS). Unless the refractive index of the additive matches that of the PVC, an opaque composition results.

7 Ease of flow: At its processing temperature this is a very stiff flow material; high temperatures (e.g. greater than 200 °C) cannot be used because of decomposition.

8 Shrinkage: This is an amorphous material so shrinkage is low, approximately 0·6 %, i.e. about 0·004 in/in.

9 Resistant to the following: UPVC has excellent resistance to mineral acids (dilute and concentrated), alkalis and detergents; its resistance to alcohols and oils is good. Chemical and heat resistance is generally made worse by the addition of impact modifiers.

10 Not resistant to: Attacked by ketones, esters and aromatic hydrocarbons (e.g. benzene). Can have a high impact strength but is notch sensitive. Resistance to ultra-violet can be improved, e.g. by use of carbon black. Methylene chloride can be used to detect an under-gelled compound.

11 Material detection or identification: Compounds based on PVC have a high density (e.g. 1·3–1·6 g cm^{-3}) and therefore they sink quickly in water. It is a self-extinguishing material which when burnt in a flame, colours the body of the flame yellow and tips the edges green. A pungent, irritating odour of hydrochloric acid results; if a piece of the material is heated on a copper wire then the flame is coloured green. It is commonly seen in the form of opaque mouldings, e.g. grey or brown pipe fittings.

12 Colouring: Pigments may be added at the high speed mixing stage; when this is done pigments of fine particle size should be used (sometimes referred to as micropigments). For some jobs (where the colour required must be held accurately) compounded material is used.

13 Material and component handling: Ingredients usually mixed together in a high speed mixer. For example: (i) PVC added to a mixer together with other solid ingredients, (ii) mixed at high speed until the melting point of the lubricant is reached, e.g. 100 °C, (iii) lubricant added followed by acrylic processing and (iv) at 125 °C dumped into a cooler/ blender. A large ribbon mixer is used for this purpose and this cools the mix, reduces batch-to-batch homogeneity and also helps reduce electrostatic charges on the powder (thus stabilising bulk density). The powder may be used directly or it may be pelleted (using an extruder). Pelleting improves dispersion, adds to the cost and reduces thermal stability. Pellets are much easier to handle than the powder.

14 Mould and gate considerations: The mould, and ideally all machine parts such as the platens, should be acid resistant. Sprue taper should be generous (e.g. 5°) if ease of release is to be achieved. To minimise pressure losses the runner system should be short and of full-round cross-section, e.g. 5–10 mm in diameter. The gates should also be round and the edges of the gate should be radiused towards the component; gate thickness should be approximately three-quarters of the part thickness. Pin gates (>1 mm) can sometimes be used for small components. Gate lands should be kept small. When pipe fittings are being moulded (and cores are being employed) it is important to ensure that the locking unit can generate a large opening force as the contracting material may bind the mould together. Moulds should be hardened and may be chrome plated so as to improve acid resistance.

15 Flow path: wall thickness ratio: This is not an easy flow material; the flow path: wall thickness ratio (at 1 mm wall thickness) would typically be of the order of 60:1.

16 Projected area: The clamping pressure should be of the order of 2–2·5 tsi (30–34 MN m^{-2}).

17 Cylinder equipment: To reduce the risk of stagnation and subsequent degradation, the screw should not be fitted with a nonreturn valve. A plain screw tip (streamlined or shaped to match the internal nozzle shape) or one with a 'flying spiral' formation should be used. An open nozzle is also recommended together with a screw designed for UPVC processing (e.g. compression ratio 2:1). The screw must be tough and acid resistant, e.g. achieved by chrome plating a screw machined from tough steel. The screw drive system must be adequately powered.

18 Screw cushion: This should be as small as possible (e.g. 2 mm) so as to reduce the risk of material stagnation.

19 Shot capacity: If the screw has been designed for UPVC then the full shot capacity may be used.

20 Melt temperature: This should be adjusted so that the material is capable of being slowly purged from the nozzle; the melt should have a smooth, glossy appearance and should be uniformly plasticised. As the processing temperature and the decomposition temperature are very close, accurate control is essential; a temperature of between 180–200 °C should be used. Do not exceed 210 °C.

21 Barrel residence time: Keep as short as possible as otherwise degradation will start; once it starts it will rapidly spread. In the event of a stoppage, reduce the set temperatures and purge the barrel clean. At the recommended melt temperatures the material should be capable of being heated for about 20–25 min without serious degradation. Do not attempt to process UPVC in a ram or plunger machine.

22 Temperature settings: See Table 6.

23 Injection speed—mould filling speed: Excessive shear can cause

TABLE 6

Zone No.	*Location*	*Temperatures °C From*	*To*	*Temperatures °F From*	*To*
1	Barrel rear (near the hopper)	140	160	284	320
2	Barrel middle	160	170	320	338
3	Barrel middle	170	180	338	356
4	Barrel front	180	190	356	374
5	Nozzle	190	210	374	410
6	Mould	20	60	60	140

degradation (burning) and this can be caused by forcing the material, at high speed, through the gate. Adequate mould venting must be provided and a programmable injection speed unit is useful.

24 *Injection pressure:* The machine should be capable of giving up to the following: First stage: 1750 bars; 175 MN m^{-2}; 25 000 psi. Second stage (dwell or follow-up pressure): up to 1000 bars; 100 MN m^{-2}; 14 500 psi.

25 *Screw rotational speed (rpm):* To minimise degradation keep the screw speed low, e.g. below 80 rpm—speeds are approximately a third of those used for PS. As it is surface speed that is important, speed decreases as screw diameter increases.

26 *Back pressure:* Up to 50 bars; 5 MN m^{-2}; 700 psi. Back pressure control is very important when UPVC is being processed; it should be capable of being precisely set and must be reproducible so as to obtain an accurate heat input. Usual back pressure is 1 MN m^{-2} (150 psi).

27 *Shutting down:* Turn off the heaters and purge out slowly until the barrel is empty. Purging compounds are available which help to reduce degradation by keeping the barrel clean. Do not follow PVC with acetal (or vice versa); purge with PS first. It is better to keep a cylinder and screw just for PVC if at all possible.

28 *Reprocessing:* Each time UPVC is heated its processing life is shortened and once degradation starts it rapidly spreads. Therefore, ensure that no degraded material is fed back into the machine. If degradation does occur, it will first be seen as a dirty patch or streak in the component. When this happens, purge until clear and try and reduce the heat input to the material.

29 *Finishing:* Gates are removed by end-milling, drilling, cropping, etc. Mouldings may be decorated by hot stamping and silk screening.

30 Other comments: If PVC is to be successfully moulded then the equipment must be acid resistant and accurate control (over time, speed, temperature, pressure and material) is important.

31 Typical components: The mechanical properties of UPVC range between those of PP and PA; because of its chlorine content it does not burn like these two materials. Pipe is manufactured from this material and injection moulding is used to make the pipe fittings. Jointing can be by solving welding or it can be achieved by using 'interference fit' couplings. The pipe system is used for rainwater, soil pipes, guttering, etc. Other parts of such plumbing systems are also made from UPVC, e.g. gullies, grilles, traps, etc.

7 PLASTICISED POLYVINYL CHLORIDE

1 Common name: Plasticised polyvinyl chloride.

2 Abbreviation: Plasticised or soft PVC, or PPVC.

3 Systematic chemical name: Plasticised poly(1-chloroethylene) has been suggested.

4 Some suppliers:	*5 Trade names or trade marks:*
Chemische Werke Hüls	Vestolit
Dynamit Nobel	Trosiplast
Lonza	Lonza-compound
Solvay	Solvic
Wacker Chemie	Vinnol

(See also unplasticised polyvinyl chloride.)

6 Material properties: PVC is a very versatile material as its properties can be varied over a very wide range by the use of additives, e.g. fillers, stabilisers, plasticisers, lubricants, etc. The term 'plasticised' means a plasticiser has been added in order to produce a soft, flexible product and/or in order to make processing easier. Plasticisers are usually organic liquids called esters and the type and concentration used dramatically affects the properties of the finished product—properties such as cost, temperature resistance (low and high), bacteria and stain resistance, fire resistance, etc. In order to get a balance of properties, blends of plasticisers are commonly used.

7 Ease of flow: Increasing the plasticiser concentration gives a more easy flowing material.

8 Shrinkage: This is an amorphous material whose shrinkage varies from 0·01 to 0·05 in/in, depending on the formulation.

9 Resistant to the following: Excellent resistance to dilute mineral acids and good resistance to concentrated acids and to alkalis. If correctly formulated resistance to aromatic hydrocarbons, detergents and greases can be reasonable. Resistance to weathering is generally good; improved by the use of ultra-violet stabilisers and/or carbon black.

10 Not resistant to: Chlorinated hydrocarbons, alcohols and ketones. Some plasticisers may be attacked by bacteria—to minimise this use bactericides and/or linear esters (based on straight-chain alcohols). Plasticisers may be lost by exposure to high temperatures, water and solvents but the correct choice will help to minimise such losses.

11 Material detection or identification: Compound density is usually greater than 1 $g\,cm^{-3}$ (up to $1 \rightarrow 2\ g\,cm^{-3}$) so the material sinks in water. It may be cut easily and cleanly with a knife and the cuts have smooth edges. Transparent mouldings are possible although most injection mouldings are opaque. Provided that the plasticiser level is low, PVC compounds burn slowly and tend to be self-extinguishing. When heated in a flame on a copper wire, a green coloration is produced.

12 Colouring: Masterbatching can be used but problems may arise due to nonuniformity of the final colour. Pigments of fine particles are used for dry colouring but for colour consistency a compounded material may be preferred.

13 Material and component handling: Ingredients may be blended in a high speed mixer. Suspension polymers are usually used—when the plasticiser is added to the high speed mixer, rapid absorption occurs. The mix is then either cooled and injection moulded, or cooled, extruder pelletised and then injection moulded. It is not usually necessary to predry the material.

14 Mould and gate considerations: PVC is used to mould shoe soles and the moulds have been made from Kirksite-type alloys. These are cast to the size required and each casting contains water-cooling copper tube, each casting can then be mounted in a water-cooled steel bolster. Fast and frequent mould changing is therefore possible.

Brass and aluminium moulds have also been used. Small gates may be used, e.g. pin-gates with a diameter of 0·5 mm (0·02 in). Edge gates of equivalent cross-sectional area are also used on appropriate mouldings. For long running jobs moulds made from tool steel are the most popular.

15 Flow path:wall thickness ratio: Depends upon the formulation, e.g. plasticiser type concentration, etc. but flow path:wall thickness ratio (at 1 mm wall thickness) can be up to 180:1.

16 Projected area: The clamping pressure can be up to 2·5 tsi (34 $MN\,m^{-2}$) but is dependent on formulation, product design, etc.

TABLE 7

Zone No.	Location	Temperatures °C From	Temperatures °C To	Temperatures °F From	Temperatures °F To
1	Barrel rear (near the hopper)	145	155	293	311
2	Barrel middle	155	165	311	329
3	Barrel middle	165	175	329	347
4	Barrel front	175	180	347	356
5	Nozzle	180	185	356	365
6	Mould	20	30	68	80

17 Cylinder equipment: A conventional screw and barrel arrangement may be used, e.g. a back-flow valve on the screw and either decompression or, a nozzle shut-off valve to prevent drooling. Small components are still produced on ram machines without thermal degradation or component property changes.

18 Screw cushion: About 4 mm.

19 Shot capacity: Up to 80 % of the machines rated capacity on PS may be used.

20 Melt temperature—as measured in the nozzle or by an air-shot technique: 150–190 °C.

21 Barrel residence time: Depends on the method of operation, formulation, etc. When using very high back pressures the critical residence time may be reduced to 5 min.

22 Temperature settings: Please note that it is the melt temperature which is important; those in Table 7 are only suggested, initial settings. The temperature of the hydraulic oil and of the material in the hopper should not vary excessively.

23 Injection speed—mould filling speed: The speed should be as high as possible and consistent with other requirements. High speeds can cause shear separation of the components in some formulations and may therefore have to be reduced.

24 Injection pressure: The machine should be capable of giving up to the following: First stage: up to 1200 bars; 120 MN m^{-2}; 17 400 psi. Second stage (dwell or follow-up pressure): up to 800 bars; 80 MN m^{-2}; 11 500 psi.

25 Screw rotational speed (rpm): High screw speeds are possible but not necessarily desirable. Adjust the screw speed to suit the time available, e.g. 80–200 rpm.

26 Back pressure: 250 bars; 25 MN m^{-2}; 3570 psi. Back pressures higher than this have been used, e.g. when large proportions of scrap are being run.

27 Shutting down: When a particular run has finished, turn off the heaters and purge the barrel empty. If the run has been completed purge with PS or PE until there is no trace of PVC then turn off the heaters; special purging compounds are available. Do not purge with PE if the machine is to be restarted on PVC as traces of PE may be difficult to remove and delamination may occur subsequently.

28 Reprocessing: Up to 100% regrind has been run but the amount incorporated depends upon formulation and end-use application.

29 Finishing: Hand degating (using scalpels, side cutters, etc.) is common; automatic degating is used, e.g. with those compounds containing lower concentrations of plasticiser.

30 Other comments: In order to get soft, flexible products which are immune to plasticiser loss, nitrile rubber (in the form of powder) may be added to the PVC—up to 50 phr (based on the PVC) may be used.

31 Typical components: PVC can be formulated to be a soft, resilient material and as such it is used for washers, grommets, electrical flex-ends, etc. A major use for this material is in footwear and a large number of PVC soles and heels are now manufactured by injection moulding (using specially developed moulds and machinery).

8 CELLULOSICS

1 Common name: Cellulosics, e.g. cellulose acetate, cellulose acetate butyrate, cellulose propionoate.

2 Abbreviation: Depends on type, e.g. CA, CAB, CP.

3 Systematic chemical name: Not known—the following have been suggested: cellulose ethanoate (CA), cellulose ethanoate-butanoate (CAB), cellulose propanoate (CP).

4 Some suppliers:	*5 Trade names or trade marks:*
CA	
Bayer	Cellidor
Bergmann	Bergocell
British Celanese	Dexel
Eastman Chemical	Tenite
CAB	
Bayer	Cellidor
Eastman Chemical	Tenite
CP	
Bayer	Cellidor
Eastman Chemical	Tenite

6 Material properties: Cellulose is a naturally occurring high polymer which cannot be moulded unless it is modified by a combination of the following: (i) esterification, (ii) molecular weight reduction and (iii) plasticisation. CA is made by esterifying (or reacting) cellulose with acetic acid; CAB is formed when cellulose is reacted with a mixture of acetic plus butyric acid and CP is made by reacting cellulose with propionic acid. All are hard, stiff materials which have good gloss, excellent toughness and a wide colour range (e.g. from clear to black). Compared to the major synthetics CA has poorer electrical insulation properties, heat, water, chemical and weathering resistance. CAB offers good weathering properties, ease of moulding (particularly for thick sections) and good appearance and toughness (especially at low temperatures). CP does not have the rancid odour associated with CAB, but is not so weather resistant; in other respects it is similar to CAB.

7 Ease of flow: By adding plasticiser (e.g. dimethyl phthalate or DMP) a wide range of grades is available. Higher plasticiser contents give softer, tougher materials which flow more easily but which have lower heat distortion temperatures. The hard grades of these cellulose esters have similar flow properties to that of PS. Easy flow grades of all three are possible; CAB is available in softer flows than CP.

8 Shrinkage: Each ester may give the following shrinkages: CA from 0·005 to 0·008 in/in; CAB from 0·003 in/in and CP from 0·003 to 0·006 in/in (about 0·5 %). The postmoulding shrinkage is negligible.

9 Resistant to the following: Aliphatic hydrocarbons, petrol, detergents, oils and greases. CA has the best resistance to aromatic and chlorinated hydrocarbons; CAB has excellent resistance to perspiration.

10 Not resistant to: Affected by weak acids and alkalis and not resistant to strong acids and alkalis. Attacked by esters of the low molecular weight alcohols and by low molecular weight ketones and alcohols. Cellulose acetate absorbs up to 3 % water (and changes its dimensions accordingly) whereas CAB and CP absorb about 1 %. All are soluble in ethyl acetate, methylene chloride, acetone and *o*-chlorophenol.

11 Material detection or identification: Mouldings are usually fairly hard and can be clear and tough. Little smoke is produced when burnt. All three are soluble in acetone and trichloromethane. CA has the highest density at 1·26–1·30 $g\,cm^{-3}$; CAB and CP are similar with densities of 1·16–1·22 $g\,cm^{-3}$. CA burns slowly with a yellow flame and a vinegary odour; CAB burns with a blue flame and smells like rancid butter; CP has a blue flame tipped with yellow and the smell is quite pleasant.

12 Colouring: All three materials are available as crystal clear materials and can be purchased in a full range of compounded colours (i.e. transparent, translucent and opaque). Mottled effects may be produced by adding a harder flow grade (of the same polymer); colour bleeding (transfer of colour from one plastic to another) can occur with some colours. Cellulosics can, in general, be coloured by a variety of techniques.

13 Material and component handling: Care must be taken during storage to ensure that moisture contamination does not occur—particularly with CA. Long storage is to be avoided and a first-in first-out system should be adopted. The moisture will improve the flow but will also reduce moulding quality—a high water content is shown by silver streaking. Predry for 3 h at 70 °C but be careful not to keep the material in the oven too long as material flow and component quality will suffer due to excess volatile loss. Do not predry with other materials as they may become contaminated with plasticiser lost from the cellulosic. Predrying is not always necessary but is common.

14 Mould and gate considerations: Due to overheating (e.g. because of long residence times or frictional heating) volatiles may be evolved; these can cause burning and hinder melt fusion (necessary if the melt stream has been split). Try to avoid dividing the melt, ensure that the mould is vented and do not over-heat the material by using too small a gate at too high a speed. Pinpoint gates are now commonly used and jetting may be minimised by siting the gate (less than 3 mm ($\frac{1}{8}$ in)) opposite a cavity wall. To minimise component weaknesses, such gates are often used in the form of tunnel gates as this may move the area of weakness to a less critical place. Box-shaped components are best gated from one side using, if possible, a flash gate. On thick components, jetting may be prevented (if small gates are used) by the use of movable or spring-loaded cores. Ribs and inserts do not usually cause problems (e.g. sinking and cracking). Due to the resilience of these materials components may be made to jump considerable undercuts. Hot-runner moulds may be used to overcome long runners and/or to minimise scrap production.

15 Flow path: wall thickness ratio: CAB and CP will flow into thinner sections than CA. At 1 mm wall thickness the flow path: wall thickness ratio can be very high for some formulations, e.g. 300:1 is possible for easy flow formulations.

16 Projected area: Clamping pressure required is usually of the order of $1 \rightarrow 2$ tsi ($15 \rightarrow 30\ \mathrm{MN\,m^{-2}}$).

17 Cylinder equipment: Properly dried material at the correct temperature will not drool from the nozzle—some sealing type nozzles may cause degradation. Use an open nozzle if possible and if necessary use decompression. The screw should be equipped with a back-flow valve. (These materials are still processed on ram-type machines because of their ease of flow.)

18 Screw cushion: About 3 mm.

19 Shot capacity: Can be similar to that for PS but usually about 80 % of the barrel's rated capacity in PS.

20 Melt temperature—as measured in the nozzle or by an air-shot technique: Depends on the type of ester and on the amount and type of plasticiser used, for example:

CA 170–250 °C (338–480 °F)
CAB 170–230 °C (338–446 °F)
CP 170–250 °C (338–480 °F)

21 Barrel residence time: On purging the melt should be smooth, glossy, free from bubbles or gassing and with a honey-like viscosity. If the material is overheated then the surface appearance may be poor (e.g. matt finish), the flow properties may change (flashing may become a problem), distortion and colour changes may occur and excessive fumes will be emitted. Therefore do not allow the material to stagnate or 'cook' in the heated barrel.

22 Temperature settings: Please note that it is the melt temperature which is important, those in Table 8 are only suggested, initial settings. The temperature of the hydraulic oil and of the material in the hopper should not vary excessively.

The temperature settings are for CAB; CA will be lower by about 10 °C—for CP use higher temperatures (e.g. by 20 °C). When moulded under the correct conditions the depth of colour and gloss of CAB and CP will be similar to that of PMMA. Be careful not to process at too low a temperature as components with poor strength may result. High mould

TABLE 8

Zone No.	*Location*	*Temperatures °C*		*Temperatures °F*	
		From	*To*	*From*	*To*
1	Barrel rear (near the hopper)	140	160	280	320
2	Barrel middle	155	175	311	347
3	Barrel middle	170	190	338	374
4	Barrel front	185	205	365	401
5	Nozzle	180	200	356	392
	Mould	40	80	104	176

temperatures are recommended when thick-sectioned parts (e.g. tool handles) are being moulded as such a high temperature helps to stop void formation.

23 Injection speed—mould filling speed: For thin walled components high injection speeds may be possible—at high speeds the viscosity of cellulosics may be increased. As this point is increased by raising the temperature (and as moulding strength is also improved by higher temperatures) work at the highest temperature possible, i.e. without colour changes, degradation, etc. Fill thick moulds slowly; a programmable injection speed facility is very useful.

24 Injection pressure: The machine should be capable of giving the following: First stage: up to 1500 bars; 150 MN m^{-2}; 21 400 psi. Second stage (dwell or follow-up pressure): up to 300 bars; 30 MN m^{-2}; 4300 psi. Be careful not to overpack the mould. Adjust hold pressure and dwell time until sink marks just cease to be visible.

25 Screw rotational speed (rpm): Avoid overheating and get better temperature uniformity by using slow screw speeds.

26 Back pressure: Up to 150 bars; 15 MN m^{-2}; 2140 psi. Ensure that too high a back pressure is not being used as this could cause degradation; too high a pressure could also cause degradation due to air trapping.

27 Shutting down: Do not purge with another material as this could cause problems (such as haziness). Switch off the heaters, withdraw the carriage away from the mould and slowly purge the barrel until empty.

28 Reprocessing: On granulation and re-use the material may have different flow characteristics; predry, blend with fresh material (up to 20 %) and keep a check on the number of times of re-use. Do not use regrind on light or transparent compounds. Ensure that the regrind is kept perfectly clean, e.g. by stripping and cleaning the grinder before use.

29 Finishing: Easy to cut, drill, tap and polish; may be joined by cementing (that is, if the two pieces are of the same type of cellulosic, e.g. both CAB). Hot stamping, vacuum metallising, etc. are used but surface cleanliness is essential if good adhesion is to be obtained.

30 Other comments: If other plastics have been processed using the same screw and barrel then before the cellulosic is used, the whole assembly must be cleaned very thoroughly, e.g. stripped and cleaned mechanically. CA may be counted as a 'another plastic' to CP and CAB.

31 Typical components: CA is chosen when a reasonably tough, clear material is required and/or when special colour effects are required; it has a slight odour but can be made self-extinguishing. Examples of use are toys, appliance housings, pens, brushes, buttons, ornaments, etc.

CAB has good toughness, is tough at low temperature, has good dimensional stability and excellent outdoor weathering characteristics. It is probably the most widely used cellulosic moulding material and is used in outdoor signs and letters, tool handles, lenses, steering wheels, keys on calculators, etc.

CP has higher hardness, stiffness and tensile strength than CAB and is used for brush handles, steering wheels, toys, tool handles, pen barrels, seats, etc.

Both CP and CAB are more dimensionally stable and have better low temperature properties than CA. These materials are useful for signs and displays as there is little electrostatic dust attraction.

9 POLYCARBONATE

1 Common name: Polycarbonate.

2 Abbreviation: PC.

3 Systematic chemical name: Poly(oxy-1,4-phenylene dimethylmethylene-1,4-phenylene-oxy-carbonyl).

4 Some suppliers:	*5 Trade names or trade marks:*
Anic	Sinvet
Bayer	Makrolon
Ciba-Geigy	Verafil
General Electric	Lexan
Mobay	Merlon
Mitsubishi	Novarex

6 Material properties: Polycarbonates are a type of long-chain polyester in which aromatic groups are linked by the carbonate ester group (OCOO). The material is strong, stiff, tough and transparent; it maintains its properties over a wide temperature range and is commonly classed as an engineering plastic. The material is also virtually self-extinguishing and has good electrical insulation characteristics but is not recommended for use in the presence of an electric arc. Special care is required during processing and the material can have limited resistance to notches, chemicals and ultra-violet light (see Section 30). It is fairly expensive and susceptible to crazing when strained and this last point mars an otherwise excellent resistance to creep. PC is used in blends with ABS as such materials are readily coloured, strong, have good light-fastness, high heat distortion temperatures and are relatively easy to mould.

7 Ease of flow: Poor. This material has a high melt viscosity.

8 Shrinkage: An amorphous material which shrinks about 0·7 %, i.e. about 0·005 in/in. Shrinkage appears to be uniform, i.e. the same both with and across the direction of flow.

9 Resistant to the following: Dilute acids, oil, petrol, high molecular weight alcohols, detergents, aliphatic hydrocarbons, oxidising agents and trichloroethylene. Good resistance, for a thermoplastic, to ionising radiation.

10 Not resistant to: Caustic soda, amines, ammonia, strong acids, certain chlorinated hydrocarbons (e.g. chloroform and methylene chloride), dioxan, cyclohexanone, pyridine and hot phenols. Attacked by water at temperatures greater than 60 °C. Frozen-in strain may be assessed by noting resistance to carbon tetrachloride.

Swollen by acetone, benzene and carbon tetrachloride; aromatic hydrocarbons, esters and ketones may cause stress-cracking. Heavy loads, applied to PC gears, can result in unacceptable wear resistance or high friction levels.

11 Material detection or identification: With a density of 1·2 g cm^{-3} this material sinks in water. When placed in a flame it burns (but not very easily) with a yellow flame and the moulding chars and blisters; the moulding will not continue to burn if the flame is removed and a faint odour of phenol is produced. Soluble in methylene chloride and ethylene dichloride. Polycarbonates have a very high carbon content and give characteristic infra-red absorption spectra. Softens at temperatures above 150 °C but serviceable up to 135 °C.

12 Colouring: Due to the high processing temperatures (and therefore degradation problems) the choice of colours may be restricted compared to other thermoplastics, e.g. some organic colourants can be adversely affected by the high temperatures. Dry colouring is used but ensure that the colourant is dry. Compounded material is preferred by many moulders as the use of this eliminates such problems.

13 Materials and component handling: Any trace of water will cause problems, such as frothing, to occur. Therefore rigorous predrying is essential if components with superior properties (e.g. impact strength) are to be obtained. Water content must be reduced or maintained below 0·02 % by heating the sealed (unopened) tins at 115 °C for 3 h—maintain this temperature by using a heated hopper (120 °C) on the machine. Wetness in the granules may be tested for by heating at 270 °C for 1 min on a covered glass slide (TVI test). After cooling bubbles will be seen in damp material and the number of bubbles may be related to the water content.

14 Mould and gate considerations: Follow polystyrene (PS) practice.

Pinpoint gates may be used for small articles. In general, gates should be of a reasonable depth, e.g. 70% of the maximum wall thickness and with a minimum size of 1 mm. Flash, ring and diaphragm gates are also used; use short, full round runners if possible. Hot-runner moulds are successfully used with this material.

15 Flow path: wall thickness ratio: At 1 mm wall thickness the maximum flow path: wall thickness ratio for this material is lower than that found for UPVC or PMMA—this is partly due to the high melt viscosity but is also due to the high set-up temperature.

16 Projected area: Usually 3–5 tsi (45–75 MN m^{-2}) clamping pressure is used.

17 Cylinder equipment: The inside barrel surface can be damaged or etched when PC is processed. Hence replaceable barrel liners are sometimes employed as these can be changed relatively easily. Open nozzles may be employed but the screw is usually fitted with a back-flow valve. If drooling occurs with an open nozzle do not lower the nozzle temperature. If drooling is excessive fit a shut-off nozzle, e.g. a needle shut-off nozzle.

18 Screw cushion: About 4 mm.

19 Shot capacity: Up to 85% of the cylinders rated capacity may be used—as little as 20% may be used if necessary.

20 Melt temperature: Accurate temperature control is required: 260–310 °C (500–600 °F). During purging a 'good' melt will not show bubbles (moisture) or a silver lining (overheating). Periodically the injection unit should be withdrawn so that the melt can be inspected, this is because the bubbles may not show up in the moulding and 'bad' material will not make good mouldings.

21 Barrel residence time: This depends upon the temperature employed and the polymer grade—some colourants may be degraded very quickly at the high melt temperatures necessary for this material.

22 Temperature settings: See Table 9. The use of high mould temperatures does not necessarily lengthen the cycle unduly as this material has a high glass transition temperature and therefore a high set-up temperature.

23 Injection speed—mould filling speed: As high as possible consistent with other considerations, e.g. the pressure available, surface finish required, etc. When the flow is upset (e.g. by changes in cross-section) the surface appearance may be marred, therefore programmable injection speed is an advantage.

24 Injection pressure: The machine should be capable of giving the following: First stage: up to 2000 bars; 200 MN m^{-2}; 29 000 psi. Second

TABLE 9

Zone No.	*Location*	*Temperatures °C* *From*	*To*	*Temperatures °F* *From*	*To*
1	Barrel rear (near the hopper)	250	260	482	500
2	Barrel middle	270	280	518	536
3	Barrel middle	280	300	536	572
4	Barrel front	280	300	536	572
5	Nozzle	290	300	555	572
	Mould	85	120	185	248

stage (dwell or follow-up pressure): up to 1200 bars; 120 MN m^{-2}; 17 400 psi.

25 Screw rotational speed (rpm): Because of the high melt viscosity high screw speeds are not recommended. If possible a screw motor which can develop high torque should be used. Use a speed of about 45 rpm, e.g. on a 40 mm (1·57 in) screw.

26 Back pressure: Up to 300 bars; 30 MN m^{-2}; 4350 psi. Degradation (due to oxidation) may occur if back pressure is not used.

27 Shutting down: PC has a reputation for strong adhesion to metal and if the injection barrel is allowed to cool with PC inside then metal may be pulled out of the barrel wall. Therefore if there is a break in production (e.g. an overnight stop) empty the barrel and keep the heat on, e.g. at 170 °C. At the end of a run purge out with a very stiff flow PE. Do not, repeat not, put PVC, POM or celluloses into a cylinder set at PC temperatures. For thorough cleaning, it will be necessary to remove the screw from the barrel so that mechanical cleaning can be employed.

28 Reprocessing: Up to 20 % of regrind can be added provided that it is dry; if more is added then the mechanical properties will suffer. Drying may take 20 h at 115 °C. Do not re-use material which was processed when wet or damp.

29 Finishing: Large sprue gates may be removed by milling (e.g. end-milling) and side gates may be removed by routing. For less critical applications, cropping may be employed. Take care to avoid scratching or galling—PC can even be scratched with PMMA.

Components may be decorated by silk screening, spraying and hot foiling. Mouldings may be joined by ultrasonic welding or cementing, e.g. using methylene chloride or ethylene chloride—it may be better to make a cement by dissolving 8 % of PC in one of these solvents. Mouldings can also be joined with epoxide resins and with hot-melt adhesives, e.g. based on polyamides.

30 Other comments: Shrinkage is reduced if glass fibre is added to the material—with 30% glass fibre the shrinkage could be only 0·002 in/in but the density could be 1·43 g cm^{-3}. When correctly stabilised the material has good outdoor weathering properties. (Weather resistance can be obtained by applying solutions containing ultra-violet absorbers.)

31 Typical components: Because of their high heat distortion temperature (HDT), low moisture absorption, low creep and high strength these materials are widely used in electronic and electrical applications. The electrical properties show little dependence on current frequency and are not substantially changed after mouldings have been immersed in water for long periods or after being heated to high temperatures, e.g. 130 °C. Their clarity, impact resistance and good weathering resistance (when suitably stabilised) makes them suitable for glass replacements where breakage is a problem, e.g. due to vandalism. Therefore the material is used in outdoor lighting applications, safety helmets, filters, car lamp housings, goggles, lenses, food mixer parts, tableware, housings for computers, guard panels, terminal boxes, etc. Polycarbonates based on trimethyl bisphenol A (as opposed to bisphenol A—used for conventional PC) have Vicat softening points 50 °C higher, e.g. 195 °C. They are not so resistant to impact, however, and they have a poorer colour.

10 POLYETHERSULPHONE

1 Common name: Polyethersulphone or polysulphone or polyarylsulphone.

2 Abbreviation: PES.

3 Systematic chemical name: Not commonly used as it is very complex. For example, a suggested chemical name for Victrex is poly(oxy-1,4-phenylene-sulphonyl-1,4-phenylene).

4 Some suppliers:	*5 Trade names or trade marks:*
ICI	Victrex
Union Carbide	Udel
3 Ms	Astrel

These materials differ chemically and have, for example, differing high temperature properties; the upper use temperature increases in the order Udel; Victrex; Astrel.

6 Material properties: These materials all contain aromatic groups (*p*-

phenylene), sulphone groups (SO_2) and ether groups (—O—); other groups (diphenyl) may also be incorporated in the polymer chain. Any plastic which contains *p*-phenylene groups (e.g. PC) has a high heat distortion temperature, is rigid at room temperature and requires a high processing temperature. PES materials are tough and have a wide temperature range of use (e.g. from −100 to +150 °C). Because of their chemical structure the materials are resistant to chemical changes (e.g. oxidation) on heating; electrical insulation properties are good but the tracking resistance is not outstanding. The materials have good creep resistance, are transparent, have self-extinguishing characteristics and are tough. Like PC they are notch sensitive; however, they have lower impact strengths than PC, are more expensive but possess higher heat distortion temperatures and better creep resistance.

7 Ease of flow: These materials have a high melt viscosity. In an effort to reduce the viscosity during processing, be careful that overheating does not occur as the colour of the material will change, i.e. it will darken.

8 Shrinkage: As these are amorphous materials shrinkage is low, e.g. approximately 0·6 % for the unfilled material. The presence of fillers, e.g. glass fibre, reduces the shrinkage but increases the density, e.g. from 1·37 to 1·6 g cm^{-3} (when 30 % glass fibre is used). The unfilled material may give shrinkages of 0·007 in/in.

9 Resistant to the following: Unaffected by most inorganic reagents. Aqueous solutions do not generally attack PES although there may be some slight water absorption. Has good resistance to alkalis and to dilute acids. Aliphatic hydrocarbons, alcohols, benzene, petroleum spirit, oils and fats do not usually attack this material.

10 Not resistant to: Concentrated oxidising mineral acids. PES is soluble in highly polar organic solvents and in a number of chlorinated and aromatic hydrocarbons. A large number of organic materials will therefore attack this material, e.g. chloroform, cresols, acetone, cyclohexanone, ethyl acetate, methylethylketone, etc. If this material is to be used outside then it should be light stabilised, e.g. with carbon black.

11 Material detection or identification: This material has a high density (e.g. 1·37 g cm^{-3}) and mouldings therefore sink when placed in water. Being an amorphous material it can be obtained as transparent mouldings, e.g. pale-straw colour. The mouldings are tough and have a very high softening point, e.g. a Vicat softening point of greater than 220 °C. When placed in a flame the mouldings do not burn very easily but when they burn they do so with a white flame and give off an odour of sulphur. When the mouldings are struck with a hard object they can give off a metallic sound or ring.

12 Colouring: The natural colour is a transparent amber shade and so a wide colour range is possible. It can be dyed or pigmented; dyes give transparent colours and pigments give opaque colours. Dyes can be incorporated, in some grades, by the moulder but in general the addition of pigments is not recommended as agglomerates may form. Such inclusions could act as stress-raisers and cause failure during impact.

13 Material and component handling: The material contains absorbed water and should be dried at 150 °C for at least 3 h before any attempt is made to mould. Use a heated hopper if possible. A component will absorb water—the rate of absorption and the amount of absorption will depend on the humidity, moulding, shape, etc. Newly moulded specimens will increase in dimensions by about 0·2 % when exposed to air or water. Stability is obtained rapidly (e.g. within two weeks) at an equilibrium water content of, about, 2 %.

14 Mould and gate considerations: Because of the rigid chemical structure of these materials high set-up temperatures are obtained. Large amounts of frozen-in strain can therefore be easily incorporated into the mouldings unless care is taken. The problem is made more difficult because large runners and gates need to be used (necessary because of the high melt viscosity). The use of such large runners and gates can easily result in overpacking. Reduce such undesirable effects by keeping dwell time and pressure to the minimum, using high mould temperatures, and annealing the mouldings if necessary (e.g. by heating in glycerine at 160 °C). To improve the strength of the moulded part avoid sharp corners in the mould and put a generous radius in the gate entry region. Keep pressure losses as small as possible by using short, generously sized runners and use full-round sections where possible. Moulds are commonly heated by circulating hot oil and/or electrically powered band or cartridge heaters.

15 Flow path: wall thickness ratio: Because these materials have high viscosities and high set-up temperatures the ratio of flow path: wall thickness is low, e.g. at 1 mm wall thickness it may be of the order of 60:1.

16 Projected area: These materials can have a very high viscosity and a high set-up temperature. To get such grades to flow into relatively thin sections and/or to have a good surface finish, high pressures will be required, therefore high clamping pressures may be needed, e.g. up to 10 tsi (150 $MN\,m^{-2}$). More easy flow grades can be handled using clamping pressures recommended for PC.

17 Cylinder equipment: Because these materials have high melt viscosity it is not necessary to use shut-off nozzles. An open nozzle has the added attraction that it is more streamlined and does not provide a site where stagnation can occur. Because of the high melt temperatures necessary it is very important to ensure that there are no stagnation points in the system.

18 Screw cushion: A small screw cushion is normally employed, e.g. 4 mm.

19 Shot capacity: Try to match the shot size to that of the injection unit. Do not run small shots on a barrel with a large capacity as degradation may occur.

20 Melt temperature—as measured in the nozzle or by an air-shot technique: The melt temperature required depends on the particular material and the mould employed. Set the machine controls so that a thermally homogeneous melt is produced with a temperature of between 330–400 °C (626–752 °F). Those materials that can withstand the highest end-use temperatures usually require the highest melt and mould temperature, e.g. 400 °C may be required for Astrel on some jobs.

21 Barrel residence time: Because of the high temperatures employed, barrel residence times are comparatively short. If the material is left to cook in the barrel, degradation will occur; reduce the barrel temperature settings if a stoppage occurs. Periodic purging may also be required as with time a layer of black, degraded polymer may be formed and this could give black streaks in the moulding. PC and PP have been used where extensive purging is required.

22 Temperature settings: Please note that it is the melt temperature which is important; those in Table 10 are only suggested, initial settings. The temperature of the hydraulic oil and of the material in the hopper should not vary excessively. Accurate temperature control is essential when processing this material.

23 Injection speed—mould filling speed: As high as possible consistent with good quality mouldings—high speeds may cause surface markings in the gate area and may even cause degradation (through shear).

TABLE 10

Zone No.	*Location*	*Temperatures °C*		*Temperatures °F*	
		From	*To*	*From*	*To*
1	Barrel rear (near the hopper)	320	370	608	698
2	Barrel middle	330	390	626	734
3	Barrel middle	350	410	662	770
4	Barrel front	360	430	680	806
5	Nozzle	360	420	680	788
	Mould	130	200	248	356

24 Injection pressure: The machine should be capable of giving the following: First stage: up to 2000 bars; 200 MN m^{-2}; 29 000 psi. High pressure may be required in order to overcome high viscosity. Special high pressure machines have been built to process these materials. Second stage (dwell or follow-up pressure): up to 1200 bars; 120 Mn m^{-2}; 17 400 psi. Be careful not to overpack the mould.

25 Screw rotational speed (rpm): Use a relatively low screw speed, e.g. less than 60 rpm, and use a screw motor that is capable of giving high torque. This material's high melt viscosity can result in high shear being developed—as this will cause degradation keep the speed low.

26 Back pressure: 400 bars; 40 MN m^{-2}; 5800 psi. Working against a small back pressure can help to eliminate air, however, too high a back pressure may cause degradation. Adjust the back pressure so that when sprue break occurs the material will just weep from the nozzle—then reduce slightly.

27 Shutting down: Turn off the heat and purge until clear; if restarting with PES maintain the temperatures at 200 °C until start-up. If changing to another material, before turning down the temperatures, purge clear and then introduce a stiff-flow (low MFI) polyolefin. Purge this through and then drop the temperature settings gradually until (about) 260 °C is reached. Purge out the polyolefin and shut down.

28 Reprocessing: Up to 30 % regrind may be blended in with virgin material. As moisture does not cause chemical degradation, scrap mouldings (made from wet material) may be ground, dried and used again. Because of the material's excellent thermal stability regrind can be used several times before the physical properties change.

29 Finishing: Finishing processes include machining, welding (thermal and ultrasonic), solvent welding (accomplished with methylene chloride or dimethylformamide), painting, etching, metallising, etc. Thin films of metal may be deposited by vacuum metallising or electroplating. Metal adhesion is good and copper coated components are of interest in printed circuitry.

Because of the toughness of this material, gates are removed (by cropping) when the components are still hot.

30 Other comments: An idea of the amount of frozen-in stress can be obtained by dipping mouldings in ethyl acetate; high levels of frozen-in strain will cause cracks to develop very quickly, e.g. 2 s.

31 Typical components: Because these materials are expensive they are not large tonnage materials. They are used when a thermoplastic is required which has good long-term thermal stability, dimensional stability, toughness and resistance to burning. These properties together with their good electrical insulation properties have promoted their use in the microwave cooking field, e.g. for grilles, dishes, etc. These materials have extended the range of use of thermoplastics as, for certain applications, they may replace thermosets and metals, e.g. in cameras.

PES is used in dishwasher and automotive components, the electrical and electronics industry (e.g. for printed circuit boards, television components, etc.); hair dryer parts, oven, projector and fan heater components are also moulded. That is, the material is used when other cheaper thermoplastics (such as PC) are not suitable.

Within this family of materials there is a significant difference in the maximum upper end-use temperature of approximately 100 °C (the highest upper end-use temperature is approximately 250 °C). Some electrical components are required to perform satisfactorily after extensive oven ageing tests, e.g. 10 000 h at 200 °C.

11 POLYPHENYLENE OXIDE (MODIFIED)

1 Common name: Polyphenylene oxide (modified).

2 Abbreviation: PPO or modified PPO.

3 Systematic chemical name: Modified poly(oxy-1,4-phenylene) has been suggested.

4 Some suppliers:	*5 Trade names or trade marks:*
General Electric	Noryl
Engineering Polymers	Noryl

6 Material properties: The term 'Noryl' covers a wide range of materials as these materials are mixtures of PPO and other polymers, e.g. PPO and PS or PPO and TPS; they are easier to process and cost less than PPO itself. In general, modified PPO is a tough, stiff material with a wide temperature range of use (e.g. from −40 to 130 °C); at elevated temperatures it maintains its good load bearing characteristics. Because of this and their low water absorption, mouldings have good dimensional stability. The material also has excellent dielectric properties, a low coefficient of expansion and may be rated as self-extinguishing and nondripping, it is an opaque material with good resistance to hydrolysis.

7 Ease of flow: Modified PPO has a fairly high viscosity but it is lower than PPO; the viscosity is greater than PC but lower than ABS.

8 Shrinkage: This amorphous material exhibits low shrinkage, e.g. 0·007 in/in which can be considerably reduced by the addition of glass fibre (e.g. down to 0·002 in/in when 30 % glass is used). The shrinkage observed varies only slightly with changes in injection pressure, melt temperature, mould temperature, wall thickness and the flow direction.

9 Resistant to the following: Outstanding resistance to aqueous environments—only absorbs about 0·2% moisture from boiling water at equilibrium. Not attacked by dilute acids, alcohols, detergents, dilute alkalis.

Can withstand exposure to elevated temperatures; the heat distortion temperature ranges from 90 to 150 °C (dependent on grade).

10 Not resistant to: Halogenated and aromatic hydrocarbons, ketones and low molecular weight esters. Unless adequately stabilised, ultra-violet light resistance can be poor.

11 Material detection or identification: Has a fairly low density (1·06 $g\,cm^{-3}$) which can be reduced (e.g. to 0·9 $g\,cm^{-3}$) by the use of blowing agents. As the natural material is not very attractive it is usually seen as coloured opaque mouldings. Mouldings are tough and relatively difficult to cut. When burnt a yellow, sooty flame results and there is an odour of phenol.

12 Colouring: In-house colouring is not normally performed as the use of masterbatches, dry colours, etc. may lead to a serious loss of properties, delamination, etc. It is normally purchased already compounded and a wide colour range in opaque colours is available. Black or grey colours will have the best resistance to ultra-violet light.

13 Material and component handling: Will absorb 0·07% water in 24 h at room temperature (similar to PBT). Normally predrying is not necessary, if drying is necessary use 100 °C and dry for up to 2 h (conditions depend on grade). Do not exceed 6 h drying.

14 Mould and gate considerations: Tab or fan gates are preferred as these eliminate jetting, gate size should be as large as possible so that fast mould filling is obtained. Sprues and runners should also be relatively large, e.g. the small end of the sprue should be greater than 3 mm diameter and the sprue should be well tapered. Generous tapers, rounded corners, etc. help to ease ejection—problems may be caused through the low shrinkages experienced with this class of material. Where long flows are envisaged consideration should be given to hot runners (insulated runners are not recommended). On hot-runner moulds good temperature control is essential, use an adequate number of cartridge heaters, and in order to get long heater life under-rate the heaters, i.e. use 350 $W\,kg^{-1}$ of steel. Avoid material stagnation points, e.g. by rounding the ends of flow channels and avoiding the use of heaters in the melt stream.

15 Flow path: wall thickness ratio: Very dependent on the mixture employed. It can be tailored to suit the application within reasonable limits. It can, for example, cover the range quoted for ABS.

16 Projected area: Where venting (and therefore burning) is a problem reduce the clamping pressure to the lowest value possible so as to allow the escape of trapped volatiles. Usually 2–3 tsi (30–45 $MN\,m^{-2}$) is sufficient.

17 Cylinder equipment: Open nozzles with an orifice diameter of 3 mm ($\frac{1}{8}$ in) or greater should be used. Drooling can be prevented by good temperature control and/or by decompression; shut-off nozzles are not recommended. General purpose screws may be used providing that the screw length is adequate, e.g. 17:1 and that the compression ratio is not excessive, e.g. below 3·5:1.

18 Screw cushion: About 5 mm.

19 Shot capacity: 40–80% of PS shot capacity. Avoid small shots on large equipment as such severe under-utilisation will cause stagnation and therefore degradation.

20 Melt temperature: 250–290 °C. Higher temperatures than this have been used (e.g. 310 °C) with unreinforced resins when the shot size and barrel capacity have been closely matched on fast-running jobs. Some heavily filled grades are run at 300 °C in order to improve flow.

21 Barrel residence time: At 280 °C the barrel residence time is approximately 10 min. If a stoppage of more than 10 min is anticipated then reduce the temperature settings, e.g. so that melt is about 180 °C (356 °F).

22 Temperature settings: See Table 11.

23 Injection speed—moulding filling speed: As high as possible consistent with surface finish, venting requirements, etc. Improved fusion of the material around the cores is achieved by the use of high speeds at high mould temperatures.

24 Injection pressure: The machine should be capable of giving the

TABLE 11

Zone No.	*Location*	*Temperatures °C* *From*	*To*	*Temperatures °F* *From*	*To*
1	Barrel rear (near the hopper)	190	210	374	410
2	Barrel middle	210	240	410	464
3	Barrel middle	230	270	446	518
4	Barrel front	250	290	482	554
5	Nozzle	240	275	464	527
	Mould	80	110	176	230

following: First stage: up to 1500 bars; 150 MN m^{-2}; 21 400 psi. Second stage (dwell or follow-up pressure): up to 1000 bars; 100 MN m^{-2}; 14 100 psi.

25 Screw rotational speed (rpm): Adjust speed to suit the moulding cycle, a high torque motor should be fitted and screw speed shall be kept low, e.g. below 100 rpm.

26 Back pressure: 35 bars; 3·5 MN m^{-2}; 500 psi.

27 Shutting down: Turn off heat and empty the barrel. PS and PMMA are effective purging materials for PPO.

28 Reprocessing: Up to 100 % regrind can be used; however, it is best to re-use the material as it is formed in blends with virgin material, such blends may contain up to 25 % regrind. Do not use degraded or contaminated material.

29 Finishing: Because of the ductile nature of this class of materials, trimming can be performed without shattering. Mouldings can be decorated by a wide variety of techniques, e.g. hot stamping, hot foiling, painting, printing, etc.

30 Other comments: The stiffness and creep resistance of this material is considerably enhanced by the addition of glass fibre. Flame retardant grades are available which are halogen-free.

31 Typical components: By incorporation of polystyrene into polyphenylene oxide the heat distortion temperature is lowered but it is still high enough for many engineering applications. The presence of the PS makes processing easier. The material's good resistance to detergents, and its ability to be moulded to close tolerances, has helped it to become established for washing machine, dishwasher and pump components. It has a useful combination of properties, e.g. heat resistance, impact strength and flame resistance. Most important markets are in the electrical and automotive industries, e.g. for the backplates of TV sets and for car dashboards. This material is more expensive than ABS but cheaper than PC. Structured foam components are a major market for Noryl, e.g. for machine housings. The material is reported to be under active development as a material of construction for motorcycle engines, e.g. blocks and crank cases.

12 LOW DENSITY POLYETHYLENE

1 Common name: Low density polyethylene or low density polythene.

2 Abbreviation: LDPE.

3 *Systematic chemical name:* Low density poly(methylene).

4 *Some suppliers:*	5 *Trade names or trade marks:*
Anic SpA	Eraclene
Anic SpA	Riblene
Ato Chimie	Lacqtene
BASF	Lupolen LD
CDF Chimie	Lutrène
Chemie Linz	Duplen LDPE
Dow	Dow Polyathylene
DSM	Stamylan LD
Du Pont	Alathon
Hoescht	Hostalen LD
ICI	Alkathene
Montedison	Fertene
Shell	Carlona
Union Carbide	—
USI Europe	Petrothene

6 *Material properties:* LDPE is a crystalline thermoplastic and as such it is not available in transparent grades. Its natural colour is a milky white and it has a soft wax-like feel. Because of chain branching the crystallinity level is low—both long-chain and short-chain branches are present. The material has a density less than water ($0{\cdot}92\,g\,cm^{-3}$), is tough but only has moderate tensile strength. The chemical resistance and electrical insulation properties (over a wide range of temperatures) are excellent. Polyethylene may be divided into three density ranges: low density range from 0·910 to $0{\cdot}925\,g\,cm^{-3}$, medium density range from 0·926 to 0·940 and high density range from 0·940 to $0{\cdot}965\,g\,cm^{-3}$. 'Low' is also known as Type I and 'medium' is known at Type II. HDPE is sometimes known as Type III. Density is related to crystallinity; crystallinity levels are related to the number of short-chain branches and their distribution. Linear low density polyethylene (LLDPE) contains little or no long-chain branching. This affects its flow behaviour and its thermal properties. For example, it has a higher maximum service temperature than LDPE but is lower than HDPE.

7 *Ease of flow:* LDPE is an easy flow material. Ease of flow is rated by the melt flow index (MFI); the lower the number, the stiffer is the flow. However, low MFI materials have better stress-crack resistance, solvent resistance and exhibit higher impact strength (due to higher molecular weight). At 200 °C a material with an MFI of 2 will have a higher viscosity than PS; one with an MFI of 0·2 would have a lower viscosity.

8 Shrinkage: When the density is 0·910–0·925 g cm^{-3} the shrinkage is approximately 0·020–0·050 in/in (Type I). When the density is 0·926–0·940 g cm^{-3} (Type II) the shrinkage is about 0·015–0·04 in/in—Type III or HDPE can have very high shrinkage. The shrinkage for LDPE is about 2%.

9 Resistant to the following: No solvents at room temperature. Not chemically attacked by nonoxidising acids, alkalis and many aqueous solutions. Low water absorption even after long immersion times; addition of carbon black (to improve weathering) will increase water absorption. Absorption of other liquids (e.g. acetone and benzene) will be greater for LDPE than for HDPE.

10 Not resistant to: Ultra-violet light, oxygen at elevated temperatures and strong nitric acid. Will dissolve in benzene at 60 °C. Aromatic and chlorinated hydrocarbons will cause swelling. Certain materials, e.g. detergents, may cause stress cracking of grades with a MFI greater than 0·4. HDPE is also more prone to stress cracking.

11 Material detection or identification: Floats in water and cuts easily with a knife. Insoluble in most common solvents but soluble in hot benzene or toluene. Ignites easily when placed in a flame and burns rapidly. Burning drops usually fall and there is an odour of burning wax or paraffin; the flame is blue but tipped with yellow. The melting point is approximately 105 °C. It can be distinguished from HDPE and PP by density and melting point determinations. LDPE is an inert material which does not dissolve in common chemicals.

12 Colouring: Dry colouring is used for noncritical applications but masterbatching is the preferred system where accurate and consistent colour matching is required. Compounded material is more expensive and is not so widely used as good colouring can be achieved with masterbatches. Liquid colour is under active development, e.g. to overcome problems such as poor feeding.

13 Material and component handling: Predrying is not usually necessary unless hygroscopic additives are present. When necessary, 3 h at 65 °C is usually sufficient.

14 Mould and gate considerations: As the MFI increases the smaller the feed system of the mould needs to be. However, the runner sizes should not be reduced too much (e.g. below 6 mm or $\frac{1}{4}$ in) as it is necessary to supply a reasonable amount of dwell pressure in order to compensate for shrinkage. A wide range of gate types is used. There is now considerable interest in the use of hot runners and insulated hot tip runners; such systems reduce scrap production and components are ejected from the mould already finished.

15 Flow path:wall thickness ratio: This material is a very easy flow material. At 1 mm wall thickness, the maximum flow path:wall thickness ratio would be higher (for a material with a high MFI) than PP or PS, e.g. LDPE, 200:1; PS, 150:1. Very easy flow grades may reach 300:1 on a high injection speed machine.

16 Projected area: The clamping pressure required is of the order of 1–2 tsi (15–30 MN m^{-2}).

17 Cylinder equipment: To prevent material weeping from the nozzle it is usual to fit a valve in the nozzle (e.g. spring loaded); to minimise material flowing back down the screw a valve is also fitted to the screw. Decompression (suck-back) is used as an alternative to nozzle valves, e.g. in conjunction with hot-runner moulds.

18 Screw cushion: From 2 to 6 mm.

19 Shot capacity: Because of the material's good heat stability as little as 10% of the cylinders rated capacity may be used; the maximum possible will be about 95%.

20 Melt temperature—as measured in the nozzle or by an air-shot technique: 180–280 °C.

Low (Type I)	180 → 280 °C (356 → 536 °F)
Medium (Type II)	165 → 220 °C (330 → 428 °F)
High (Type III)	205 → 260 °C (400 → 500 °F)

21 Barrel residence time: If the material is overheated oxidation occurs and there is a serious loss of electrical insulation ability; it is therefore necessary to incorporate antioxidants, e.g. di-β-naphthyl-*p*-phenylenediamine, so as to minimise change. It is good practice not to allow the material to cook in the barrel.

22 Temperature settings: Please note that it is the melt temperature which is important; those in Table 12 are only suggested, initial settings. The temperature of the hydraulic oil and of the material in the hopper should not vary excessively.

23 Injection speed—mould filling speed: To get fast injection speeds, high pressures are not required.

24 Injection pressure: The machine should be capable of giving the following: First Stage: up to 1300 bars; 130 MN m^{-2}; 18 800 psi. Second stage (dwell or follow-up pressure): up to 800 bars; 80 MN m^{-2}; 11 600 psi. Careful control over both dwell time and pressure is needed in order to avoid overpacking of the mould and yet still be able to compensate for shrinkage.

TABLE 12

Zone No.	Location	Temperatures °C From	Temperatures °C To	Temperatures °F From	Temperatures °F To
1	Barrel rear (near the hopper)	120	200	248	392
2	Barrel middle	160	230	320	446
3	Barrel middle	180	260	356	500
4	Barrel front	200	280	392	536
5	Nozzle	210	270	410	518
	Mould	10	60	50	140

25 Screw rotational speed (rpm): Can be very high. High screw speeds can, however, result in melt temperature variations, therefore adjust the speed to suit the cooling time.

26 Back pressure: Up to 200 bars; 20 MN m^{-2}; 2900 psi.

27 Shutting down: No need to purge with another material.

28 Reprocessing: Up to 100 % regrind may be used; ensure that the scrap is kept clean and dry. If the regrind is mixed with virgin polymer then the material can be re-used more than other plastics. Check that the additives in one batch are not interacting with those from another lot.

29 Finishing: Difficult to deflash because of the flexibility of the material. Hand scalpels, cropping, etc. may be used to remove the feed system. Better to design the mould so that degating is achieved during mould opening. Decoration of mouldings is not very common. The surface can be made more receptive to inks or adhesives by a corona discharge or by ozone.

30 Other comments: Close tolerances are difficult to mould on LDPE and they are difficult to hold in service as the material has a high coefficient of thermal expansion.

31 Typical components: Because of this material's ease of moulding and its relative low cost it has become established as a general purpose moulding material, e.g. for toys, caps and lids for containers, bowls, beakers, pipe couplings, pots, bins, etc. Transmission of gases and vapours decreases as crystallinity is increased; HDPE is better than LDPE in this respect. Resistance to creep and component stiffness (modulus) also increase as crystallinity increases. As the MFI increases, in general, the tensile strength decreases. Exposure to high energy radiation causes crosslinking and thus a change in properties, e.g. heat resistance.

LLDPE is used in place of LDPE when improved hot or cold resistance is required, e.g. for household appliances.

13 HIGH DENSITY POLYETHYLENE

1 Common name: High density polyethylene.

2 Abbreviation: HDPE.

3 Systematic chemical name: High density poly(methylene).

4 Some suppliers:	*5 Trade names or trade marks:*
Anic SpA	Eraclene HD
BASF	Lupolen HD
BP	Rigidex
Chemische Werke Hüls	Vestolen A
Chemplex	Chemplex
Dow	Dow Polyethylene
DSM	Stamylan HD
Hoescht	Hostalen
Montedison	Moplen RO
Solvay	Eltex
Union Carbide	—
Wacker Chemie	Wacker Polyäthylen

6 Material properties: In the presence of stereospecific catalysts, and under conditions of low temperature and pressure, ethylene will polymerise to give a substantially linear polymer. Because of its regularity this material has a higher level of crystallinity than LDPE. This increase gives plastics which have a higher density, rigidity, tensile strength, hardness, heat distortion temperature, chemical resistance, viscosity and resistance to permeability; however the impact strength is lower. For some applications the weathering resistance is adequate but is improved by compounding with carbon black. Compared to PP homopolymer, HDPE has better resistance in low temperature impact and to oxidation. PP has a higher softening point, better resistance to flexing, a higher hardness, a higher tensile strength and elongation and will also cycle faster. Moulding gloss can be similar for PP and HDPE.

7 Ease of flow: This material is a relatively easy flow material but compared to LDPE the high density materials are usually offered as lower MFI materials in the injection moulding grades.

8 Shrinkage: Because of the linear nature of the molecule high levels of crystallinity can be obtained; high shrinkage values can therefore be obtained and these can be markedly different in two directions, i.e. across the flow and with the flow. Distortion can therefore be a problem; shrinkage values of up to 0·05 in/in are possible for thick sectioned mouldings. However, shrinkage values from 0·015 to 0·040 in/in are more usual.

9 Resistant to the following: No solvents at room temperature. Not chemically attacked by non-oxidising acids, alkalis and many aqueous solutions. Some organic materials (e.g. white spirit, carbon tetrachloride) will cause swelling; in this respect HDPE is better than PP. Theoretically, HDPE is more resistant to oxidation than LDPE; in practice they are similar (this could be due to catalyst residues).

10 Not resistant to: Ultra-violet light, oxygen at high temperatures and oxidising acids. Resists aromatic and chlorinated hydrocarbons more than LDPE. Stress-cracking can be a problem; that is cracking caused by a stress (internal or external) in the presence of an active liquid, e.g. detergent or metal soap. PP is best in this respect.

11 Material detection or identification: Similar to that described for LDPE. The high density material is more chemically resistant (it dissolves in benzene at 85 °C; LDPE at 60 °C), more opaque and has, of course, a higher density and rigidity. It is slightly more difficult to cut than LDPE but the cut also has smooth edges. HDPE has a higher melting or softening point than LDPE but lower than PP; all three materials will float in water but can be separated, or identified, by the use of melting point and density determinations (e.g. using alcohol/water mixtures).

12 Colouring: A wide range of pelleted, coloured compounds is available from the manufacturers although masterbatching is the main technique used at the present time (the masterbatches may be based on waxes, HDPE or LDPE). Liquid colour is used successfully; care should be taken as too much may cause screw slip. The use of certain colourants may alter crystallinity levels (and therefore shrinkage and/or distortion) no matter which way they are added, e.g. phthalocyanine pigments as used for blues and greens.

13 Material and component handling: Predrying is not normally necessary but if required it may be performed at 85 °C for approximately 3 h; as with many black materials (based on carbon black) predrying may be required for this colour. Some large components (e.g. crates) may need to be held on cooling jigs after moulding for a few minutes (e.g. 10) in order to reduce warping and to maintain sizes—this is necessary as crates must fit together so that they can be stacked.

14 Mould and gate considerations: Mouldings may exhibit warping caused through differential shrinkage. Warping may be minimised by differential cooling, i.e. maximum cooling in the gate region, or by proper grade selection. A grade with a narrow molecular weight distribution is required and the affect of additives on crystallinity (see Section 13 above) should not be forgotten. Flow path lengths should also be evened out as much as possible. For large components this may mean using multiple gates, e.g. crates commonly have four gates (one in each corner).

Hot-runner moulds are widely used and the material is good for insulated runners because of the thermal properties (high specific heat, low thermal conductivity, etc.).

15 Flow path: wall thickness ratio: The ratio of maximum flow path to wall thickness is usually less for HDPE than for LDPE. For example at 1 mm wall thickness HDPE may be 170:1 whereas LDPE could be 200:1. The material is similar to PP in this respect.

16 Projected area: Similar to PP, i.e. 2 tsi (30 MN m^{-2}) is usually sufficient.

17 Cylinder equipment: Barrels are usually equipped with a sealing type nozzle and the screw usually has a back-flow valve. Decompression may prove advantageous when easy flow grades (high MFI grades) are being moulded.

18 Screw cushion: From 2 to 6 mm.

19 Shot capacity: 80–85 % of the cylinders rated capacity (on PS) may be used.

20 Melt temperature—as measured in the nozzle or by an air-shot technique: 205–260 °C (400–500 °F). Usually thin walled components use the higher temperature; crate moulding may be done at 230 °C.

21 Barrel residence time: A fairly tolerant material, i.e. one is not worried about thermal decomposition provided that reasonable care is taken. It is good practice not to leave any material to 'cook' in the barrel if moulding is interrupted; turn down the temperature settings if there is a break in production. A more thermally stable material than PP.

22 Temperature settings: Please note that it is the melt temperature

TABLE 13

Zone No.	*Location*	*Temperatures °C*		*Temperatures °F*	
		From	*To*	*From*	*To*
1	Barrel rear (near the hopper)	160	200	320	392
2	Barrel middle	170	230	338	446
3	Barrel middle	200	260	392	500
4	Barrel front	220	280	428	536
5	Nozzle	210	270	410	516
	Mould	5	50	41	122

which is important; those in Table 13 are only suggested, initial settings. The temperature of the hydraulic oil and of the material in the hopper should not vary excessively.

Higher melt temperatures usually result in mouldings having higher gloss but, of course, cycle times are correspondingly longer.

23 Injection speed—mould filling speed: As high as possible consistent with other requirements, e.g. surface gloss.

24 Injection pressure: The machine should be capable of giving the following: First stage: up to 1800 bars; 180 $MN\,m^{-2}$; 26 100 psi. Second stage (dwell or follow-up pressure): up to 1500 bars; 150 $MN\,m^{-2}$; 21 700 psi. Avoid overpacking as this will result in mouldings which will be brittle in the gate region.

25 Screw rotational speed (rpm): Run at as low a screw speed as possible so that uniform plasticisation is obtained; this material requires a large heat input.

26 Back pressure: Up to 200 bars; 20 $MN\,m^{-2}$; 3000 psi.

27 Shutting down: Purging with another material is not necessary when the machine is being shut down.

28 Reprocessing: The material is commonly regranulated and blended in (e.g. 25 to 30 %) with the virgin polymer—100 % scrap can be run if required, provided that the material is kept clean and dry. HDPE is more tolerant than PP in this respect.

29 Finishing: HDPE is more difficult to electroplate than PP (little electroplating is done as markets do not demand such a finish). It can be decorated by vacuum metallisation, hot foiling and printing (after the surface has been made receptive, e.g. by a flame treatment). Hot embossing of mouldings (with a name or legend) is common for items such as crates. Flash removal is easy (easier than for PP); sprues and runners are easily removed by cutting or cropping (e.g. with a knife or side cutters).

30 Typical components: This material is strong and stiff, even at low temperatures. For these reasons it is used to produce components which must have a reasonable impact strength at low temperatures, e.g. milk-bottle crates, containers used in refrigeration, fish boxes, shipping pails for paint, adhesives, etc. It is also used for containers where its rigidity and improved resistance to gas permeability is an asset compared to LDPE, e.g. food storage boxes. Widely used for housewares (buckets, bowls, etc.) but because of its higher density it is at a disadvantage compared to PP. At the same price per tonne, PP would have a 10 % price advantage (assuming of course that the moulding sections are the same). Because of this material's resilience, and its better resistance to stress whitening compared to PP, it is widely used to make snap-on caps, e.g. for aerosols. In storage or in use a change in colour, caused by flexing, is a disadvantage for such an application.

14 POLYPROPYLENE

1 Common name: Polypropylene.

2 Abbreviation: PP.

3 Systematic chemical name: Poly(propylene).

4 Some suppliers:	*5 Trade names or trade marks:*
Amoco Chemicals	Amoco PP
Anic SpA	Kastilene
Ato Chimie	Lacqtene P
BASF	Novolen
Chemische Werke Hüls	Vestolen P
Chemie Linz	Duplen PP
Eastman Chemical	Tenite
Hercules Chemicals	Profax
Hoescht	Hostalen PP
ICI	Propathene
Montedison	Moplen
Shell	Shell PP
Soltex	Fortilene
Solvay	Eltex P

6 Material properties: PP usually refers to both polypropylene and to copolymers in which propylene is the major constituent. Such materials are stiffer than polyethylene and have a lower density (0·9 g cm^{-3}). It is naturally translucent with a milky white colour; many different colours are possible. These materials have a high softening temperature, are stiff and strong and have very good stress-crack resistance. Electrical insulation is good and a high gloss, scratch-resistant surface is possible. By copolymerisation (e.g. with ethylene and/or by rubber modification) materials with improved resistance to low temperature embrittlement are produced; rubber modification improves the impact strength for both homopolymers and copolymers but reduces the Vicat softening point.

7 Ease of flow: A relatively easy flow material but higher barrel and mould temperatures are needed compared to polyethylene. PP has a higher viscosity than LDPE but a lower viscosity than PMMA. In general, high viscosity grades give lower cavitation or void problems.

8 Shrinkage: A crystalline material which shrinks by approximately 2 %; it shrinks less than polyethylene (PE) but more than polystyrene. As thick sections cool more slowly than thin sections, shrinkage in thick sections is greater. Shrinkage is approximately 0·018 in/in and is nearly uniform in all directions; in this respect it is better than HDPE.

9 Resistant to the following: Acid (dilute and concentrated), alkalis, alcohols, detergents, salt solutions, fruit juices, oils, aromatic hydrocarbons, chlorinated hydrocarbons; also resistant to materials that cause environmental stress-cracking of PE. Aromatic hydrocarbons (e.g. xylene) and chlorinated hydrocarbons (e.g. chloroform) will cause swelling at room temperatures; also swollen by ethers, esters and aqueous oxidising agents.

10 Not resistant to: Outside weathering (unless stabilised), concentrated oxidising acids, e.g. strong nitric or acidic potassium dichromate. Dissolved by aromatic and chlorinated hydrocarbons at elevated temperatures, e.g. 85 °C. High temperatures and contact with copper will cause serious degradation very quickly.

11 Material detection or identification: With a density of 0·9 g cm^{-3} this material floats in water and cuts easily with a sharp knife. At room temperature it is insoluble in all solvents but will dissolve in boiling toluene. When placed in a flame it burns readily with a yellow tipped, blue flame and burning drips are produced together with a candle-like smell; on removal from the flame it continues to burn. Little smoke is produced. Components feel hard and waxy; they are virtually unbreakable unless cut. Has a higher softening point and a lower density than HDPE. PP will float in isophorene but sink in a 80:20 mixture (by volume) of alcohol and water—this comment refers to unfilled grades.

12 Colouring: A wide range of colours is possible; the accepted method is to use compounded material. Dry colouring is possible but dispersion may be difficult and streaking may occur; if possible use PP powder when dry colouring. Masterbatches are also used and in this case the granule size of the masterbatch should be similar to that of the base polymer.

13 Material and component handling: Predrying is not normally required. A great deal of PP is used to manufacture crates and with such products creep may occur due to the components being overstacked; the product may become distorted or the stack may collapse if the creep is excessive.

14 Mould and gate considerations: Gate into the thickest part of the moulding if possible and keep the wall thickness as uniform as possible so that distortion is minimised. To avoid sink marks use large full round sprues (also full round runners if possible) and keep the size of ribs small, e.g. 50 % of the adjacent wall thickness. Insulated-runner and hot-runner moulds are widely used; pinpoint gates are useful with some components but if high injection speeds are used be careful that jetting does not occur. Tab gates are widely used.

Differential cooling of the mould (so as to achieve uniform cooling of the product) can help to minimise warping. Voiding may be a problem with thick sections of homopolymer.

15 Flow path: wall thickness ratio: High ratios of flow path: wall thickness are possible, e.g. at 1 mm wall thickness 175:1 may be possible. Very easy flow grades are available for use on fast cycling machines, in which the ratio may reach 350:1.

16 Projected area: Usually 2 tsi (30 MN m^{-2}) of clamping pressure is sufficient; high MFI materials (easy flow grades) may be moulded at 1·5 tsi (22 MN m^{-2}) and low MFI (stiff flow grades) may require 4 tsi (60 MN m^{-2}).

17 Cylinder equipment: Nozzle usually equipped with a sealing type nozzle and the screw usually has a back-flow valve. Decompression may prove advantageous when high MFI materials (easy flow) are being moulded.

18 Screw cushion: About 4 mm.

19 Shot capacity: Up to approximately 85 % of the cylinder's rated capacity in polystyrene may be used.

20 Melt temperature—as measured in the nozzle or by an air-shot technique: Below 230 °C the melt viscosity is relatively high; process above this temperature, i.e. 230–275 °C (446–525 °F). Do not exceed 275 °C as the viscosity changes rapidly and oxidation may easily occur.

21 Barrel residence time: Do not leave the machine standing for long periods at high temperatures otherwise degradation will occur. PP is more susceptible to oxidation than polyethylene as it possesses many tertiary carbon atoms.

22 Temperature settings: See Table 14.

23 Injection speed—mould filling speed: High speed needed, therefore set as high as possible consistent with other requirements, e.g. surface gloss and cavitation (in thick mouldings).

24 Injection pressure: The machine should be capable of giving the following: First stage: up to 1800 bars; 180 MN m^{-2}; 26 100 psi. Second stage (dwell or follow-up pressure): up to 1500 bars; 150 MN m^{-2}; 21 700 psi. High dwell pressure may need to be applied for long times in order to avoid voids or excessive shrinkage; however, avoid overpacking.

TABLE 14

Zone No.	Location	Temperatures °C From	Temperatures °C To	Temperatures °F From	Temperatures °F To
1	Barrel rear (near the hopper)	170	210	338	410
2	Barrel middle	210	230	410	446
3	Barrel middle	220	250	428	482
4	Barrel front	230	260	446	500
5	Nozzle	240	270	464	518
	Mould	5	80	41	176

25 Screw rotational speed (rpm): High screw speeds are possible; it is best to set the speed so that plasticisation suits the cooling cycle, i.e. as slow as possible.

26 Back pressure: Up to 200 bars; 20 MN m^{-2}; 3000 psi.

27 Shutting down: No need to purge with another material when shutting down.

28 Reprocessing: A small amount of regrind (e.g. 20 %) can be blended into the virgin polymer.

29 Finishing: When suitably treated (e.g. etching) components may be electroplated or vacuum metallised. Hot foiling, printing and painting are also possible. Removal of thin flash may be difficult with a knife as a rough edge may be produced and the product may be scarred; such flash has been removed with a flame. Automatic degating is possible if submarine gates have been used. Large gates may be removed by hand cropping or by machine cropping.

30 Other comments: Because of this material's resilience and flexibility some undercuts may be stripped from the mould; snap-fits are possible. Integral hinges can be moulded into a component and the hinge may have virtually indefinite flex-life.

31 Typical components: Housings with moulded integral hinges, luggage, housewares (buckets, bowls, etc.), toys, interior parts for cars, picnic ware, washing machine drums, pump components, etc. Rubber modified PP (e.g. which contains 30 % rubber) is widely used for car bumpers and rubber modified (10 % rubber) copolymers are used for bottle crates.

Components of high rigidity and/or high heat distortion temperature can be obtained with talc or glass filler. However, the addition of both these fillers raises the density, e.g. talc-filled polymers may have densities up to

$1{\cdot}3\,g\,cm^{-3}$ and glass-filled grades may reach $1{\cdot}2\,g\,cm^{-3}$ (depends on filler content).

Components made from PP can be sterilised (e.g. by steam or radiation) and so the material is used to make hospital equipment, e.g. disposable syringes. Moulding will suffer creep if subjected to high stress and such conditions should therefore be avoided.

15 POLYBUTYLENE TEREPHTHALATE

1 Common name: Polybutylene terephthalate or terephthalate.

2 Abbreviation: PBT also known as PTMT and/or PBTD.

3 Systematic chemical name: Poly(oxytetramethylene-oxyterephthalate).

4 Some suppliers:	*5 Trade names or trade marks*
Akzo Plastics	Arnite
Amcel	Kelanex
Ato Chimie	Orgater
BASF	Ultradur
Bayer	Pocan B
Celanese Plastics Co	Celanex
Chemische Werke Hüls	Vestodur
Ciba-Geigy	Tenatine
Dynamit Nobel	PTMT
Eastman Chemical	Tenite
GAF Corp	Gafite
General Electric	Valox
Montedison	Pibitor

6 Material properties: The reaction between acids and alcohols (to produce esters) is commonly used to make high polymers. If terephthalic acid is reacted with 1,4-butane-diol then a polyester is produced—polybutylene terephthalate (PBT). This is also known as polytetramethylene terephthalate (PTMT). It is a partially crystalline, thermoplastic polyester and is classed as an engineering plastic. The material has a fairly high heat distortion temperature, is stiff, tough and hard. This material also has good abrasion resistance, resists failure by fatigue and creep, and displays good electrical properties (over a wide range of temperatures and humidities). The combustion behaviour is similar to that of polyamides. Compared to other materials, which also contain *p*-phenylene groups (e.g. PC), PBT has a lower heat distortion temperature and low water absorption. The key features of this material are good resistance to chemical stress-cracking coupled with excellent dimensional stability. Rapid cycles are possible because of the fast set-up rate. Mouldings have very low coefficients of friction and a hard surface.

7 Ease of flow: This is an easy flow material with melt viscosities similar to those of the polyamides, e.g. PA6. Purging with PBT is difficult because of this low melt viscosity; with some grades (very low melt viscosity) stripping and cleaning of the screw and barrel may be required before PBT can be run.

8 Shrinkage: Shrinkage is dependent on mould temperature and part thickness—usually 0·017–0·023 in/in. Anisotropic shrinkage can cause warping—shrinkage may be greater (e.g. by a factor of 2) transverse to the flow direction. If this is a problem use a mineral-filled grade. If components are to be used at elevated temperatures (e.g. above 60 °C) then postmoulding shrinkage may occur. To minimise this either anneal (2 h at 150 °C) or mould at high mould temperatures, e.g. 130 °C.

9 Resistant to the following: Alcohols, ethers, the high molecular weight esters and ketones, aliphatic hydrocarbons, dilute aqueous solutions (salts, acids and bases), oils, detergents, chlorinated hydrocarbons, petrol, brake fluid, etc.

10 Not resistant to: Strong acids and bases, low molecular weight esters and ketones, aromatic hydrocarbons and phenols. Complex phenols, e.g. *o*-chlorophenol, will dissolve PBT and are used for intrinsic viscosity measurements. The material has relatively poor resistance to hot water and long-term contact should be avoided.

11 Material detection or identification: With a density of 1·31 $g\,cm^{-3}$ this material sinks in water and has a high, sharp melting point, e.g. 225 °C; above this temperature the viscosity of the material is low. It is attacked by sodium hydroxide and ethylene dichloride. Mouldings have a hard, scratch resistant surface with a very low coefficient of friction; it is not available in clear grades.

12 Colouring: A range of compounded colours is available ranging from cream to black; as properties are dependent on the colouring system it is important to evaluate a product using mouldings made from the intended formulation. Masterbatches are used for small mouldings but not for large (may get colour streaking).

13 Material and component handling: Can be moulded without predrying; if drying is necessary this may be done at 120 °C, e.g. for 3 h. Only open the container when it has reached room temperature; reseal immediately and only take out what can be used in a short time, e.g. in less than 4 h.

14 Mould and gate considerations: To counteract shrinkage high holding (or dwell) pressures are required and as high injection speeds are also recommended gate sizes are usually large, e.g. pinpoint gates should be larger than 1·5 mm diameter. Gate size and location should be chosen so that mould filling proceeds smoothly and uniformly, i.e. no jetting. Use flash gates for box-shaped parts. Good mould venting is required and the use of glass-filled grades may necessitate the use of tungsten-carbide facings in fast-flow areas, e.g. the gate. Use a 3° taper on the sprue and try (because of the shrinkage) to taper all cores adequately; with some moulds a large number of ejector pins may be required.

15 Flow path: wall thickness ratio: Despite the easy flow nature of this material very high ratios of maximum flow path: wall thickness are not obtained; this is because of this material's high set-up temperature. At 1 mm wall thickness this ratio could be 125:1. To improve the flow behaviour raise the mould temperature or injection speed; do not use high melt temperatures.

16 Projected area: Because of the low melt viscosity, and the high injection speeds employed, high clamping pressures are recommended, e.g. 3–$4\frac{1}{2}$ tsi (45 MN m^{-2}–67 MN m^{-2}). Glass-filled grades require similar pressure to the unfilled grades.

17 Cylinder equipment: The screw should be equipped with a back-flow valve. High compression ratio screws or those with short compression zones (e.g. a nylon screw) should not be used. Use conventional 3 zone screws with an L:D ratio of about 20:1 and with a compression ratio of less than 2:1. The channel depth in the feed zone should not be too deep, e.g. for a 40 mm screw it should be about 4 mm.

18 Screw cushion: About 4 mm.

19 Shot capacity: This material is associated with fast cycling and so the maximum shot capacity is approximately 70% of that for PS.

20 Melt temperature: 260–275 °C (500–527 °F). Do not exceed 280 °C (e.g. in an effort to get better flow properties) as severe thermal degradation may occur.

21 Barrel residence time: Do not allow the material to stand in the barrel for more than 20–30 min. If there is a hold up in production reduce the temperatures (maximum 200 °C) until the problem is solved, then, and only then, may the machine be reset. If a flame resistant grade is being run (and a problem occurs) purge the barrel and turn off the heaters until you are ready to start producing mouldings again. If thermal degradation of any PBT melt occurs then the cylinder and screw should be stripped and cleaned (remove the solidified melt after cooling).

TABLE 15

Zone No.	Location	Temperatures °C From	Temperatures °C To	Temperatures °F From	Temperatures °F To
1	Barrel rear (near the hopper)	240	250	464	482
2	Barrel middle	250	260	482	500
3	Barrel middle	250	270	482	518
4	Barrel front	260	275	500	527
5	Nozzle	255	270	491	518
	Mould	70	140	158	284

22 Temperature settings: See Table 15. If the screw channel depth in the feed zone is large then an ascending temperature gradient is recommended (low at the feed zone, high at the barrel front). Reduced power consumption may result with a descending temperature gradient in the case of screws with shallow depth, feed sections.

23 Injection speed—mould filling speed: Slow injection speeds may result in a poor surface finish, therefore inject as fast as possible. This will also help to give good knitting of converging melt fronts. Programmed injection speed setting (e.g. with slow initial cavity filling) may result in a bad surface finish around the gate.

24 Injection pressure: The machine should be capable of giving the following: First stage: up to 2000 bars; 200 MN m^{-2}; 29 000 psi. Second stage (dwell or follow-up pressure): up to 1200 bars; 120 MN m^{-2}; 17 400 psi.

25 Screw rotational speed (rpm): High screw speeds may cause thermal degradation; adjust the speed to suit the cooling cycle, e.g. up to 100 rpm.

26 Back pressure: 400 bars; 40 MN m^{-2}; 5800 psi.

27 Shutting down: Because of the risk of thermal degradation, reduce the temperatures as soon as practicable and purge the barrel thoroughly; if necessary use a low melt index PE. Most changes from PBT to other thermoplastics can be accomplished without cleaning the cylinder.

28 Reprocessing: Scrap material should be dried unless it is re-used immediately.

29 Finishing: Can be decorated by hot foiling using high temperatures (e.g. in the region of 175 °C). Because of this material's good chemical resistance, coating and solvent bonding may prove difficult; friction and ultrasonic welding can work well. Because of this material's high heat distortion temperature, machining (drilling, turning, etc.) presents few problems.

30 Other comments: Glass fibre filled grades have a higher density but reduced mould shrinkage and improved stiffness. For example, with 15% glass the density is 1·42 g cm^{-3} and the shrinkage is (about) 0·006 in/in; with 30% glass the density is 1·52 g cm^{-3} and the shrinkage may be 0·004 in/in. However, mould shrinkage is very much dependent on the flow direction for glass-filled grades; for such grades postmoulding shrinkage may be eliminated by using mould temperatures of 90 °C.

31 Typical components: PBT is used in pump housings, impellers, etc., that is where advantage can be taken of this material's strength, chemical and heat resistance. In the electrical and electronics industries this material is used in connectors and fuse cases; the mouldings are strong and stiff and the electrical insulation properties are virtually independent of temperature, humidity and frequency.

Some exterior parts for cars have been made of PBT as the material can withstand the solvents and temperatures encountered in painting operations. Bicycle wheels have also been made from these materials as the resultant product is lighter than the metal wheel. Appliance parts and housings (e.g. parts for toasters, hair dryer casings) are other growth areas as a wide colour range is possible, the mouldings have a good surface finish, low flammability ratings and good electrical insulation characteristics.

Gloss has been improved by blending small amounts of polyethylene terephthalate (PET) in with PBT. Addition of elastomers is also being used to improve gloss and impact resistance, e.g. for mineral-filled grades.

PET is stiffer, tougher and glossier than PBT but PBT has better electrical properties, is easier to process and cycles faster.

16 NYLON

1 Common name: Nylon.

2 Abbreviations: PA6 or PA66.

3 Systematic chemical name: PA6 is poly(imino-1-oxohexamethylene). PA66 is poly[imino(1,6-dioxohexamethylene)iminohexamethylene].

4 Some suppliers:	*5 Trade names or trade marks:*
PA6	
Aicar	Naycar
Akzo Plastics	Akulon
Ato Chimie	Orgamide
BASF	Ultramid B
Bayer	Duretham B

4 Some suppliers:—contd.	*5 Trade names or trade marks: —contd.*
PA6	
Bergmann	Bergamid B
BIP	Beetle
Degusa Wolfgang	Degalan 6
Emser Werke	Grilon
Frisetta GMBH	Frisetta Nylon B
ICI	Maranyl
Rhone Poulenc	Technyl
Snia Viscosa	Sniamid
PA66	
Aicar	Naycar
Akzo Plastics	Akulon
BASF	Ultramid A
Bergmann	Bergamid A
Celanese Plastics Co.	Celanese nylon 66
Ciba-Geigy	Verafil
Du Pont	Zytel
Dynamit Nobel	Trogamid
Emser Werke	Grilon T
Fabelta	Fabelnyl
Frisetta GMBH	Frisetta nylon A
ICI	Maranyl
Rhone Poulenc	Technyl
Snia Viscosa	Sniamid

6 Material properties: Nylon is a generic term used to describe thermoplastics which contain the amide group —CONH—; as the polymers contain many such groups they are also known as 'polyamides'. Such materials are tough, resist fatigue, impact and solvents, have low friction and absorb some water (reversibly). Many different types are possible, e.g. 6; 7; 9; 11; 12 and 6 6; 6 10; 6 12, etc. Where one number is used it indicates the number of carbon atoms in the raw material (e.g. a lactam); where two numbers are used (e.g. 6 6) then the first number indicates the number of carbon atoms in the parent diamine and the second indicates the number of carbon atoms in the parent acid. Nylon 66 and nylon 6 are the best known members of the group and as the amide groups are separated by sections based on methylene groups, they are also known as aliphatic polyamides.

Such materials compete against each other and with the acetals; nylons have superior abrasion resistance and toughness than the acetals. Nylon 66 is probably the most widely used nylon moulding material. It is slightly stronger and stiffer than nylon 6; it has a higher heat distortion temperature, absorbs less water and has a higher flexural strength. The material is made by reacting hexamethylene diamine and adipic acid.

Nylon 6 is made from caprolactam and in many respects it is similar to type 66. It is marginally lighter in colour and with a slightly lower density ($1{\cdot}13\,g\,cm^{-3}$ as compared to $1{\cdot}15\,g\,cm^{-3}$). The material has a lower melting point and can therefore be processed at a lower temperature; the shaping temperature range is also wider and processing can therefore be considered easier. Nylon 6 is slightly cheaper than type 66 (in the UK), more flexible, has a higher impact strength and slightly better low temperature properties. In general it has better solvent, grease and detergent resistance than PA66 but its resistance to dilute mineral acids is poorer.

7 *Ease of flow:* Both these materials are very easy flowing; type 6 does not set up as sharply as PA66 and consequently is useful as a glass fibre filled grade where the ability to apply pressure, so as to avoid voiding, is an advantage.

8 *Shrinkage:* Polyamides are crystalline materials and so high shrinkage values may be recorded. For example, PA66 may yield values of 0·010 to 0·020 in/in; type 6 may give values of 0·010 to 0·015 in/in. The amount of shrinkage is dependent on, for example, part thickness because as part thickness increases so does the opportunity for increased crystallinity. Nylon 66 exhibits high after-moulding shrinkage, e.g. for up to 2 years after moulding.

9 *Resistant to the following:* In general these materials are resistant to alkalis, dilute acids (but not nitric acid), detergents, fuels, greases, oils, chlorinated hydrocarbons, aromatic hydrocarbons, ketones and esters. Some of these materials may cause polyamide mouldings to swell, e.g. chlorinated hydrocarbons.

10 *Not resistant to:* Strong acids, bleaching agents (e.g. hydrogen peroxide). Sunlight can cause embrittlement unless the polymer is stabilised (e.g. with carbon black). Soluble in phenol, cresol, formic acid and glacial acetic acid. Water and, for example, some alcohols may cause swelling. Water acts as a plasticiser and makes the material softer and tougher but worsens the electrical insulation properties.

11 *Material detection or identification:* The density of type 6 is slightly lower than type 66 (e.g. 1·13 compared to $1{\cdot}15\,g\,cm^{-3}$) therefore both will sink in water. Both are opaque, stiff, hard, tough materials which are easily cut so as to yield cuts with smooth edges. Difficult to ignite but once alight

they are not self-extinguishing. The polymer melts to a free flowing liquid (which can be pulled into strings) which drips and carries the flame with it as it falls. The flame burns blue, has a yellow edge and emits a smell of burning hair. The easiest way to distinguish 6 from 66 is to measure the melting point; 66 will melt sharply but at a higher temperature, e.g. 265 °C. Type 66 is also more resistant to chloroform, benzyl alcohol and trichloroethylene. When type 66 is heated in boiling N_1N-dimethylformamide it does not dissolve—PA66 does. Nylon 6 is also soluble in 4N hydrochloric acid whereas 66 is insoluble.

12 Colouring: The fully compounded material is preferred; it is found that the addition of pigment can dramatically affect properties (e.g. impact strength) and this can cause problems if trials are performed with natural polymer. A limited amount of colour masterbatch is used and liquid colour is being actively developed. Many of the grades now being moulded are fully colour compounded materials which are also lubricated and nucleated so as to get good colour, ease of flow and fast cycling.

13 Material and component handling: These materials absorb water (type 6 more than 66, e.g. 2·5 % as opposed to 1·5 %) and must therefore be stored in a clean, dry space. A first-in first-out system should be adopted and the moisture-proof containers should only be opened when required and after they have reached the workshop temperature. If the moisture content is greater than about 0·2 % problems will occur. Dry if possible in a vacuum drying oven at 80 °C for up to 18 h; the absence of air will help to prevent oxidation.

14 Mould and gate considerations: It is difficult to predict the shrinkage of nylon. Correct the production moulds after trial mouldings (made from the production compound) have been produced using production conditions. For large mouldings a short, direct sprue gate is frequently used with a taper angle of 3°. For smaller mouldings, pinpoint gates may be used—for section thicknesses up to 4 mm use a gate with a diameter of about 1·5 mm. Such small gates may be used as tunnel gates so that automatic degating is obtained. Ring and diaphragm gates are used on circular components whereas film or flash gates are used for parts of large surface areas which have relatively thin sections—such gates give the best filling pattern. Hot runner moulds may be used with nylons provided that (i) good sealing is obtained, (ii) accurate temperature control is possible and (iii) the melt does not come into contact with copper (this metal can cause degradation). Successful moulding is being achieved with insulated hot-tip systems, i.e. as an alternative to conventional hot runner.

15 Flow path: wall thickness ratio: Because of the high set-up temperature of these materials the ratio of maximum flow path:wall thickness is not high as would be expected. It is better than acetals (POM) but worse than ABS. At 1 mm wall thickness it is approximately 100:1.

16 Projected area: The clamping pressures required for these materials are relatively high. Type 6 would probably need slightly lower pressures but the biggest difference is between unfilled grades and glass-filled grades. Unfilled grades would need between 3 and 5 tsi (45 to 75 MN m^{-2}) and filled materials may need from 4 to 10 tsi (60 to 150 MN m^{-2}). Materials may be required to flow into thin sections or the mouldings may need to have a good surface finish. Both types require high injection pressures and therefore high clamping pressures.

17 Cylinder equipment: Both these materials have a relatively sharp melting point and their melt viscosity is low. The screw is therefore fitted with a back-flow valve and the cylinder is fitted with a nozzle whose temperature can be accurately controlled. To prevent material drooling from the nozzle a shut-off nozzle (thermal seal, hydraulic seal, spring loaded needle, etc.) is usually used; drooling can also be prevented by using decompression. To prevent excess air being drawn into the cylinder (i) arrange for decompression to occur before sprue break, (ii) decompress relatively slowly and (iii) use decompression in conjunction with a thermal seal nozzle. These measures should help to prevent oxidation. The thermal seal nozzle must be designed to eliminate stagnation points and the back-flow valve must seat well (to prevent leakage).

18 Screw cushion: Between 2 and 6 mm.

19 Shot capacity: If the machine is fitted with a general purpose screw then the shot capacity will be limited, e.g. 80 % of the maximum for PS. To avoid material stagnation do not use less than 20 %. The 'standard' screw may give unacceptable melting (under-plasticised or over-plasticised melt) and in order to overcome such problems screws designed for nylon are used. If the machine has been fitted with a screw designed to suit nylon (i.e. a nylon screw) then the maximum rated capacity may be utilised—such a screw may be essential for some products (e.g. pharmaceuticals) where very high quality mouldings are required.

20 Melt temperature: For PA6 the melt temperature is usually in the region of 220 °C but can, for some grades, be 270 °C (428 to 518 °F). For nylon 66 the melt temperature is usually in the region of 270 °C; because of the sharp melting point and because of the risk of oxidation, the processing temperature range is much smaller for 66, e.g. 265–300 °C (509–572 °F). Temperature control must therefore be precise.

21 Barrel residence time: Nylon 66 will tend to degrade slightly more rapidly than PA6 at its (high) processing temperature. If there are any serious hold-ups in production turn the barrel temperature down (e.g. to 170 °C) while the problem is being solved. If a grade containing a flame retardant is being processed take extra care as some grades burn up relatively quickly. Purge the barrel and turn off the heaters in the event of a serious stoppage.

22 Temperature settings: See Table 16. All temperatures must be capable of being precisely set and controlled, e.g. PID control using a deep thermocouple. If the machine is equipped with a thermal seal nozzle then the temperature of this is set slightly lower than the melting point of the polymer. Accurate mould temperature control is essential if variations in crystallinity are to be avoided.

TABLE 16

Zone No.	*Location*	*Temperatures °C*				*Temperatures °F*			
		PA66		*PA6*		*PA66*		*PA6*	
		From	*To*	*From*	*To*	*From*	*To*	*From*	*To*
1	Barrel rear (near the hopper)	260	280	210	250	500	536	410	482
2	Barrel middle	270	290	215	260	518	554	419	500
3	Barrel middle	270	300	220	270	518	572	426	518
4	Barrel front	270	300	225	270	518	572	435	518
5	Nozzle	265	280	220	270	509	536	428	518
	Mould	60	100	60	90	140	212	140	194

23 Injection speed—mould filling speed: Usually fast for thin sections and slower for thicker sections.

24 Injection pressure: The machine should be capable of giving the following: First stage: up to 1500 bars; 150 MN m^{-2}; 21 400 psi. Second stage (dwell or follow-up pressure): up to 300 bars; 30 MN m^{-2}; 4300 psi. The screw forward time (SFT) is very important in nylon moulding and may be longer than the cooling time.

25 Screw rotational speed (rpm): To minimise temperature variations this should be set to suit the cooling cycle; it should not exceed 100 rpm. With some colours, high screw speeds can cause colour changes.

26 Back pressure: 150 bars; 15 MN m^{-2}; 2140 psi. Excess back pressure may cause degradation due to large amounts of shear heat being generated.

27 Shutting down: Purging with a different material is not normally necessary—when restarting moulding, allow the machine to heat soak for a few minutes after the set temperature has been reached. This should ensure that the remaining material in the barrel has melted and the risk of screw breakage will be minimised.

28 Reprocessing: Nylon is noted for its toughness; however it is only tough when it has absorbed a certain amount of water. When it comes from the mould it will be dry and will therefore be more brittle. Regranulate immediately if possible. Use up to 30% of regrind but do not regrind contaminated material or damaged material (e.g. overheated); predry regrind if required.

29 Finishing: Degating may be by drilling, milling, cropping, etc.; some edge gates may be broken by tumbling and the gate angle (and length) determine the finish obtained. Normal machinery procedures, e.g. drilling, tapping, etc., are used. Mouldings can be finished by silk screening, hot stamping, etc.

30 Other comments: If mouldings of improved stability are required then they should be annealed by heating (e.g. by immersion in a nonoxidising oil) at a temperature 20 °C higher than the maximum service temperature, e.g. 20 min at 150 °C for PA6; 170 °C for PA66. The mechanical properties of polyamides are markedly dependent on cooling; slow cooling (using a hot mould) increases crystallinity and this gives harder mouldings with higher stiffness and tensile strength.

31 Typical components: Polyamides have a useful combination of properties which makes them attractive for many applications. 'Under the bonnet' applications (fuel pipes, fuel filter components, etc.) take advantage of nylon's resistance to hydrocarbons and heat. Gears, cams, levers, door furniture, etc. utilise the material's toughness, abrasion resistance, low friction and fatigue resistance. Other applications are appliance housings, fans, impellers, door handles, safety belt components, bathroom fittings, etc.

Glass-reinforced polyamides are used where the components must have a higher heat distortion temperature, greater rigidity and dimensional stability than that offered by the unreinforced material. A typical application that takes good advantage of this material's properties, is a power tool outer case or housing; a novel application is a bicycle frame.

These materials need higher injection speeds, and barrel and mould temperatures, than the unfilled grades. Cycles may be shorter as the parts can be ejected at higher temperatures; it is sometimes discoloration which limits ejection temperature. Mould shrinkage is reduced but density is increased by glass addition. For example, for PA66 with 20% glass fibre, the shrinkage is approximately 0·005 in/in and the density is 1·28 $g\,cm^{-3}$;

with 50% glass fibre the shrinkage is 0·003 in/in and the density is 1·51 g cm^{-3}. To avoid problems associated with glass fibre (warping, etc.) glass spheres are now being used.

The end uses of 6 and 66 are often duplicated; the choice between the two being governed by material availability, personal preference, price and processability.

17 ACETALS

1 Common name: Acetals, polyoxymethylene, polyformaldehyde.
2 Abbreviation: POM.
3 Systematic chemical name: The homopolymer is poly(oxymethylene).

4 Some suppliers:	*5 Trade names or trade marks:*
Amcel	Kematal
BASF	Ultraform
Celanese	Celcon
Du pont	Delrin
Hoescht	Hostaform

6 Material properties: Acetals are polymers based on formaldehyde and both homopolymers and copolymers are available. Both categories of material are crystalline and they have roughly similar properties, i.e. they are stiff materials, with good fatigue endurance, resistance to creep, a low coefficient of friction and an attractive appearance. Of the two types the homopolymer has the highest tensile strength, flexural strength, resistance to fatigue and hardness. The copolymer is more thermally stable, easier to process, resists degradation by hot water better, has superior alkali resistance and higher elongation.

This type of material is commonly referred to as an engineering material and as such they compete with the nylons. The nylons have better resistance to abrasion and to impact but acetals have better fatigue endurance, creep and water resistance and they are stiffer.

7 Ease of flow: Acetals flow relatively easily but because of the high temperature at which they set-up, a relatively low flow path:wall thickness ratio may be obtained.

8 Shrinkage: Being crystalline the shrinkage is high, e.g. about 2% (approximately 0·020 in/in). After-shrinkage is usually complete within 48 h of moulding.

9 Resistant to the following: Most organic reagents. Acetals are unaffected by washing in solvents such as acetone, ethyl alcohol and petrol at room temperature. Resistant to weak acids and alkalis. May be swollen by long immersions in certain solvents, e.g. phenol, acetone, toluene, etc.

10 Not resistant to: Strong organic acids and oxidising agents, decomposed by dilute mineral acids; strong alkalis which attack the homopolymer do not attack the copolymer so readily. Hypochlorite solutions will also attack mouldings.

11 Material detection or identification: With a density of 1·41 g cm^{-3} these materials sink readily in water and being crystalline they cannot be obtained in clear grades. Natural grades give hard, white mouldings which burn with a pale blue flame, giving off a pungent odour of formaldehyde but emitting little smoke. Fairly difficult to cut but the cut has smooth edges.

12 Colouring: In-house colouring is best done by means of masterbatches, e.g. at 3% concentration. Dry colouring is not so successful for colour matched jobs. Liquid colour is being developed but care should be taken to ensure that problems such as slippage, plate-out, etc. do not develop.

13 Material and component handling: These materials are less hygroscopic than the nylons but they still should be stored in a dry space. Predrying is not usually necessary but when it is, dry for about 2–3 h at 110 °C. Acetal mouldings are generally small and are stored collectively.

14 Mould and gate considerations: To assist in rapid mould filling, and to minimise any risk of degradation caused through air compression, venting is often used when acetals are moulded. As the material has a high hot strength the ejector pins can be relatively small; mouldings can also be sprung off quite large undercuts. For small mouldings pinpoint gates may be used but in general gate thickness should be half the moulding's wall thickness and the gate land length should be kept low, e.g. half the gate thickness.

15 Flow path: wall thickness ratio: Copolymers are easier flowing than homopolymers. At 1 mm wall thickness, the maximum flow path: wall thickness ratio for POM would be about 100:1 (PA 110:1, PMMA 95:1).

16 Projected area: For easy flow materials the clamping pressure required is approximately 3–4 tsi of projected area (45–60 MN m^{-2}) for general purpose material it is 4–5 tsi (60–75 MN m^{-2}).

17 Cylinder equipment: If an open nozzle is inadequate then use a nozzle valve which contains a spring loaded needle—in the event of degradation this type of nozzle will act as a safety valve. Use a ring type check valve on the screw.

18 Screw cushion: From 2 to 6 mm.

TABLE 17

Zone No.	Location	Temperature °C From	Temperature °C To	Temperature °F From	Temperature °F To
1	Barrel rear (near the hopper)	150	170	302	338
2	Barrel middle	160	180	320	356
3	Barrel middle	180	200	338	392
4	Barrel front	190	215	374	419
5	Nozzle	195	215	383	419
	Mould	40	120	104	248

19 Shot capacity: From 15 to 75% of the cylinder's rated capacity may be used.

20 Melt temperature: 195–215, e.g. 210 °C (e.g. 410 °F). Ensure that the nozzle does not become blocked because of, for example, the temperature falling.

21 Barrel residence time: As copolymers are more thermally stable than the homopolymer, longer residence times are possible for the copolymer grades. Do not allow these materials to stagnate in the barrel. If a stoppage occurs purge out every 10 to 15 min. Acetals can decompose violently.

22 Temperature settings: See Table 17.

23 Injection speed—mould filling speed: A rough or porous surface may result if the mould filling speed (or material temperature) is too low.

24 Injection pressure: The machine should be capable of giving the following: First stage: <1500 bars; 150 MN m^{-2}; 21 400 psi. Second stage (dwell or follow-up pressure): <750 bars; 75 MN m^{-2}; 10 700 psi.

25 Screw rotational speed (rpm): Keep the screw speed as low as possible so as to minimise the danger of degrading the material (by shear heating) and to give better temperature uniformity.

26 Back pressure: <200 bars; 20 MN m^{-2}; 2300 psi.

27 Shutting down: Ten minutes before the end of the run turn off the power to the barrel heaters, withdraw the carriage away from the mould and then purge the barrel empty and leave the screw in the fully forward position. If you are changing from a halogen-containing polymer (e.g. PVC or a plastic containing a flame retardant) to acetal then purge with another material (e.g. HDPE) first. If a change is being made to a high temperature plastic, e.g. PC, also purge with PE before introducing the next polymer into the barrel. To minimise fume production drop purgings into cold water.

28 Reprocessing: Providing that the material has been kept clean and dry up to 100 % regrind can be used—usually in lower grade mouldings. In premium grade mouldings the amount of regrind is kept low, e.g. <25 %. Lubricants may help the processing of reground material.

29 Finishing: Typical finishing operations performed on acetal components include (i) machining, i.e. milling, drilling (threading no problem); (ii) sprue removal (using side cutters or scalpels); and (iii) decoration, e.g. by chrome plating and hot foiling. Components may be joined by hot-plate (300 °C) welding.

30 Other comments: If glass fibre (e.g. 20 %) is added to acetals then materials with higher stiffness, strength and creep resistance result. Overheating can result in the production of formaldehyde. As this is a gas then high pressures could be developed within a moulding machine (high enough to be dangerous) by the decomposing material being trapped between a sealed nozzle and the valve on the screw. Therefore do not use nozzles which seal more tightly as the pressure increases (some types of mechanical-seal nozzle do this).

31 Typical components: Acetals are fairly expensive materials which may be classed as engineering materials and they are used, for example, as metal replacements. This is because of their desirable properties, e.g. little change of impact strength with temperature, resistance to fatigue, etc. Typical parts are gears, bearings, moulded sprockets and chains, cams, fan-blades, carburettor bodies, pump impellers, aerosol components, etc.

POMs are said to be free from biological attack but as they are susceptible to ultra-violet radiation they should not be exposed to ultra-violet unless they are protected.

18 POLYVINYLIDENE FLUORIDE

1 Common name: Polyvinylidene fluoride.

2 Abbreviation: PVDF or PVF_2.

3 Systematic chemical name: Not known but poly(1,1-difluoroethylene) has been suggested.

4 Some suppliers:	*5 Trade names or trade marks:*
Dynamit Nobel	Dyflor
Kay-Fries	Dyflor
Kureha	KF
Laporte	Solef
Penwalt	Kynar
Solvay	Solef
Ugine Kuhlmann	Foraflon

6 Material properties: There are several fluoropolymers that can be processed by injection moulding and the group includes fluorinated ethylene-propylene (FEP), ethylene tetrafluoroethylene (ETFE), ethylene-chlorotrifluoroethylene (ECTFE), chlorotrifluoroethylene (CTFE), perfluoroalkoxy tetrafluoroethylene (PFA) and polyvinylidene fluoride (PVDF). In general these materials have excellent chemical and heat resistance, they are good electrical insulators and have high impact strengths. PVDF has the highest dieletric constant heat deflection temperature, flexural strength and modulus; its yield strength and creep resistance is high for a fluorinated thermoplastic. The abrasion resistance of this material is similar to that of polyamides. As the thermal and light stability is very good no heat stabilisers or UV stabilisers are necessary and this means that the material can be nontoxic.

By keeping the molecular weight relatively low a PVDF material can be produced which is highly crystalline—this material is used for moulding as products are stiff and strong. PVDF is relatively easy to process for a fluoroplastic.

7 Ease of flow: Behaves similarly to PP; the melt is more viscous than a PA melt. Flows easily at 225–245 °C. The MFI depends on the grade but at 230 °C with a 5 kg load it could be between 1 and 18.

8 Shrinkage: As this is a highly crystalline material the shrinkage is high, e.g. approximately 2–3 %. Shrinkage increases slightly as mould temperature increases. Shrinkage may be reduced to 1·5 % by using reinforced grades, i.e. from 0·030 in/in to 0·020 in/in.

9 Resistant to the following: PVDF has excellent resistance to a wide range of chemicals, e.g. halogens, salt solutions, inorganic acids and bases, aliphatic and aromatic hydrocarbons, carboxylic acids and acid chlorides, mercaptanes and chlorinated hydrocarbons. Resists degradation by ultraviolet light, α and β radiation.

10 Not resistant to: Hot concentrated sulphuric acid, strongly basic amines, hot concentrated bases and alkali metals. Swollen by polar solvents such as acetone and ethylacetate. Dissolves with difficulty in solvents such as dimethylformamide, dimethylsulphoxide and tetramethylurea. Some stress-cracking in strong alkalis.

11 Material detection or identification: The natural colour of this material is a translucent milky white and as it is a crystalline material it is not possible to produce transparent mouldings. It has a high density (1·78 g cm^{-3}) and will therefore sink readily in water. The Vicat softening point is approximately 130 °C and the crystalline melting point (as measured by differential thermal analysis (DTA)) is approximately 170 °C.

Relatively easy to cut and/or machine. The material has good fire resistance and will not propagate a flame. Pungent fumes of hydrofluoric acid are liberated at (about) 320 °C.

12 Colouring: Some pigments (e.g. titanium dioxide) and some other inorganic materials (boron oxide) may cause rapid decomposition of PVDF. Before adding any compounding ingredients check with the supplier. Do not add glass fillers, e.g. glass fibre, as such fillers may cause rapid decomposition. Before processing PVDF ensure that the machine does not contain anything which could cause decomposition, e.g. a glass-reinforced polymer. Use the PVDF in ready compounded form or only use masterbatches or dry colours specified by the supplier.

13 Material and component handling: Due to the low moisture absorbency of this material (e.g. 0·05 %) predrying is not normally necessary. If the material becomes contaminated with water, e.g. due to condensation, dry at 80 °C for 2–4 h. Internal strains may be removed from moulded components by annealing, e.g. 2 h at 160 °C in an oil bath.

14 Mould and gate considerations: The use of light-alloy moulds is not normally recommended as the PVDF sticks to this type of metal. Moulds designed for PP have been successfully used with this material. As shrinkage is very high, moulding around inserts can create high stress levels—it may be best to use a filled grade in such applications. Ensure that the mould is well vented as otherwise the air trapped may cause the PVDF to decompose in some cases. Hot-runner moulding is difficult.

15 Flow path: wall thickness ratio: Depends on grade and type—can range from 25:1 up to 300:1.

16 Projected area: 2 tsi (30 MN m^{-2}) is usually sufficient as PVDF is commonly moulded into thick sectioned components.

17 Cylinder equipment: Ensure that the screw and barrel assembly is scrupulously clean before introducing PVDF into the system—some contaminants can reduce the thermal stability dramatically and as hydrofluoric acid may be emitted there is an obvious hazard. Do not clean equipment by heating in an oven or a flame unless there is very good ventilation. Strip and clean the machine before moulding. Highly crystalline material can be used without a shut-off nozzle.

18 Screw cushion: As small as possible.

19 Shot capacity: Under-rate the machine, e.g. by 25 %. The barrel size should be such that at most two to three times the shot weight is contained in the heating cylinder.

20 Melt temperature: 200–260 °C. Higher temperatures than this, e.g. 300 °C (572 °F) may lead to discoloration. Low crystalline material may require 380 °C and the barrel must be constructed of acid-resistant material as decomposition can occur more quickly at this temperature.

21 Barrel residence time: Temporary stoppages (e.g. up to 30 min) can be tolerated without reducing barrel temperatures. If a longer stoppage is envisaged reduce the temperatures (e.g. to 175 °C) until processing can recommence.

22 Temperature settings: See Table 18.

TABLE 18

Zone No.	*Location*	*Temperatures °C*		*Temperatures °F*	
		From	*To*	*From*	*To*
1	Barrel rear (near the hopper)	160	210	320	420
2	Barrel middle	170	220	338	428
3	Barrel middle	180	240	356	464
4	Barrel front	190	260	374	500
5	Nozzle	180	250	356	482
	Mould	30	120	86	248

23 Injection speed—mould filling speed: PVDF components may be of heavy cross-section and in such cases slow filling speeds should be employed. High speeds can cause a local temperature rise which may result in discoloration.

24 Injection pressure: The machine should be capable of giving the following: First stage: up to 1500 bars; 150 MN m^{-2}; 21 400 psi. Second stage (dwell or follow-up pressure): 750 bars; 75 MN m^{-2}; 10 700 psi.

25 Screw rotational speed (rpm): If the screw is a tight fit in the barrel then at high speeds discoloration or even degradation, may occur. Keep the speed as low as is practicable, e.g. below a peripheral speed of 7·5 m min^{-1} (for a 30 mm screw this would be below 80 rpm).

26 Back pressure: 300 bars; 50 MN m^{-2}; 4500 psi.

27 Shutting down: If PVDF is to be re-run then purge the barrel before switching off. If another material is to be run purge clean and then use high viscosity PE or PMMA. This material is only miscible with polyamides and acrylics. Purge with virgin polymer only; do not use purging compounds.

28 Reprocessing: Material can be reprocessed in many cases without obvious impairment of properties. If dust is generated during the regrinding operation then smoking should not be permitted as the PVDF dust falling on to the cigarette end can generate fluorinated compounds. Keep grinders, etc. scrupulously clean.

29 Finishing: PVDF components can be welded together, e.g. by hot plate (250 °C), ultrasonic and friction welding. If components require machining then use the techniques and equipment recommended for PA.

30 Other comments: Grades with improved mechanical properties (e.g. stiffness and wear resistance) can be obtained if the material is compounded with fibrous fillers. Asbestos, and carbon fibre have been used. Glass-filled compounds have been known to cause degradation of PVDF at processing temperatures. Antistatic grades are also available.

31 Typical components: The chemical process industry uses this material for valves, pumps, bearings, etc.; this is because of its outstanding chemical resistance, relative ease of processing, high strength, rigidity and abrasion resistance. Because it has a high, sharp, crystalline melting point this material maintains a great deal of its chemical resistance even at a high, continuous-use temperature of 130 °C. It can be made to have exceptional piezo-electric properties—this explains current interest in uses such as transducers, microphones, loud-speakers, etc. Price is similar to that of PTFE but because processing is easier, finished articles tend to be cheaper.

19 POLYPHENYLENE SULPHIDE

1 Common name: Polyphenylene sulphide.

2 Abbreviation: PPS.

3 Systematic chemical name: Poly(thio-1,4-phenylene).

4 Some suppliers:	*5 Trade names or trade marks:*
Phillips	Ryton

6 Material properties: Crystalline, thermoplastics material based on benzene groups linked with sulphur atoms. Usually sold already compounded with glass fibres (e.g. 40 %) or with mixtures of glass fibres and mineral fillers. As a result of these additions and because of its chemical structure, it is classed as an engineering polymer. It retains its strength at high temperatures and/or in aggressive environments. Some grades can have high impact strengths but most have low elongations at break. The materials are excellent electrical insulators which are unaffected by humid environments.

7 Ease of flow: Materials are rated as easy flow. At processing temperatures it flows better than PC and PPO.

8 Shrinkage: Low mould shrinkages are possible, e.g. 0·2 %. Shrinkage may not be uniform as PPS is commonly filled with a fibrous filler (i.e. glass)—thicker sections give higher shrinkages than thin sections. Exposure to temperatures above the glass transition temperature (e.g. 90 °C) will cause additional shrinkage, particularly if low mould temperatures (e.g. below 130 °C) were used during production.

9 Resistant to the following: Can withstand high temperatures without deformation. Combines chemical resistance with hydrolytic and dimensional stability at high temperatures. Absorbs only small amounts of water, e.g. 0·05 %.

10 Not resistant to: Some chemicals at high temperatures (e.g. 90 °C) have some slight affect, e.g. aromatic and chlorinated solvents. Strong oxidising acids, e.g. concentrated sulphuric acid, will attack the material.

11 Material detection or identification: With a density greater than $1\,g\,cm^{-3}$ (e.g. $1{\cdot}5$–$2{\cdot}1\,g\,cm^{-3}$) mouldings sink quickly in water. High density materials contain mineral fillers in addition to glass fibres. Mouldings are stiff and hard and do not cut easily with a knife. They have outstanding inherent nonflammability and can therefore only be burnt with difficulty. To set fire to a moulding a temperature of 540 °C must be reached—it does not drip even at this temperature. PPS chars and blisters with a smell of sulphur when heated in a strong flame.

12 Colouring: The base colour is dark brown and therefore the colour range is comparatively limited. Mouldings made from this material must be capable of performing under arduous conditions (e.g. high temperatures) and these conditions of service also restrict the colours available. Sold already compounded in pellet form.

13 Material and component handling: Grades which contain mineral fillers will require drying before moulding. This is best done in a dehumidifying drier although convection oven heating is also used; drying temperatures may reach 175 °C. Try 150 °C for 6 h initially. Components which are required to have the best dimensional stability should be annealed at 225 °C for up to 4 h after being moulded in a hot mould (e.g. 140 °C). Components moulded at low mould temperatures may only have heat distortion temperatures of 150 °C (before annealing).

14 Mould and gate considerations: Because of the wear problems associated with heavily filled polymers the tool steel should be selected carefully; carbon steels which contain high chromium levels (also molybdenum and vanadium) should be used. Mould surfaces should be highly polished and hardened—this will give a good finish to the moulding and longer life to the mould. Protective plating (e.g. hard chrome) is also useful for reducing erosion. Moulds may be heated electrically (e.g. to 140 °C) and in such cases cartridge heaters are commonly employed—allow $0{\cdot}5\,kW\,kg^{-1}$ of mould weight. The feed system should be designed to eliminate welds and to minimise warpage caused through fibre orientation. Diaphragm and flash gates are thus useful gate types. Because PPS does not suffer from overpacking problems to the same extent as other thermoplastics, it is not essential to use balanced runner systems; hot-runner systems have been successfully used. Because of the low mould shrinkages the sprue, cavities and cores should be adequately tapered.

15 Flow path: wall thickness ratio: Markedly dependent on grade, melt and mould temperatures, the flow path:wall thickness ratio can be of the order of 150:1 (at a wall thickness of 1 mm). If a mould can be filled with another material then PPS can be made to fill that cavity despite the high filler loadings employed.

16 Projected area: If insufficient clamping pressure is available flashing will occur and/or the surface finish of the moulding will be poor. For thick sectioned components (e.g. greater than 3 mm) allow 2 tsi (30 MN m^{-2}); for thinner sectional mouldings 3 tsi (45 MN m^{-2}) may be required.

17 Cylinder equipment: Nozzles should be equipped with shut-off valves so as to prevent material drooling. Open nozzles (with accurate temperature control) have, however, been used. Screws should be fitted with back-flow valves and adequate allowance should be made for the maintenance of cylinder equipment (due to the abrasive nature of the melt).

18 Screw cushion: 4–6 mm.

19 Shot capacity: Use from 50 to 85% of the machines rated shot capacity.

20 Melt temperature—as measured in the nozzle or by an air-shot technique: 300–360 °C. For a particular grade increasing the melt temperature will dramatically alter (i.e. reduce) the melt viscosity. The affect of alterations in melt temperature is mainly on viscosity as such changes do not dramatically alter strength properties. Do not exceed 370 °C as irritating gases may be produced.

21 Barrel residence time: At the high processing temperatures employed it is necessary to keep the material moving through the barrel. However, at an average moulding temperature of 310 °C short stoppages can be tolerated.

22 Temperature settings: Please note that it is the melt temperature which is important; those in Table 19 are only suggested, initial settings. The temperature of the hydraulic oil and of the material in the hopper should not vary excessively. The mould temperature range may be 30–90 °C but not 90–120 °C. Over this second temperature range minor differences in cooling can cause major differences in crystallinity (and therefore properties).

TABLE 19

Zone No.	Location	Temperatures °C From	To	Temperatures °F From	To
1	Barrel rear (near the hopper)	290	300	554	572
2	Barrel middle	300	310	572	590
3	Barrel middle	310	320	590	608
4	Barrel front	310	360	590	680
5	Nozzle	305	320	599	608
	Mould	130	160	266	320

23 Injection speed—mould filling speed: Very high injection speeds can cause burning of the moulding—if burning occurs improve venting and/or reduce speed.

24 Injection pressure: The machine should be capable of giving the following: First stage: up to 1500 bars; 150 MN m^{-2}; 21 800 psi. Second stage (dwell or follow-up pressure): up to 1000 bars; 100 MN m^{-2}; 14 500 psi.

25 Screw rotational speed (rpm): As low as possible (e.g. below 100 rpm) so as to minimise fibre degradation.

26 Back pressure: Up to 50 bars; 5 MN m^{-2}; 700 psi. Little or no back pressure is needed. Back pressure should be as low as possible consistent with good shot weight reproducibility.

27 Shutting down: For overnight stops empty the barrel, leave the screw forward and turn off the heaters. To clean out PPS use an extrusion grade of HDPE.

28 Reprocessing: Keep material that is to be reground in sealed containers as this will minimise moisture pick-up and contamination. Assume that regrind must be dried and keep level as low as possible, e.g. <25%. Many polymers will degrade at the melt temperatures required for PPS; purge before and after a PPS moulding run with a stiff flow (low MFI) HDPE.

29 Finishing: Gates readily removed, especially if a deliberate notch is designed into the gate area—this causes the feed to break off easily. Can be hot stamped provided that the blocking machine can reach and hold the required high temperatures.

30 Other comments: Where necessary mouldings with improved dimensional stability at elevated temperatures (e.g. 135–200 °C) are obtained by using high mould temperatures (e.g. 135 °C) and postmould annealing.

31 Typical components: The ability of this material to withstand high temperatures without deformation has allowed this material to be used in connectors, terminal blocks, sockets, coil formers, bobbins and relay components. Arc resistant grades are available which are used to mould relay bases, motor housings, lamp holders and switch components. Used as a metal replacement in the automotive industry, e.g. carburettor parts, ignition plates, lamp sockets and flow control valves (for heating systems). Because of the increasing use of quartz–halogen lamps, components may reach temperatures above 200 °C—PPS replaces ceramics in such applications.

20 RUBBER REINFORCED POLYPROPYLENE

1 Common name: Rubber reinforced polypropylene, or elastomer modified thermoplastic or olefin thermoplastic elastomer.

2 Abbreviation: EMT or PP/EP/(D)M or OTE.

3 Systematic chemical name: Blend of poly(propylene) and poly-(methylene-*co*-propylene).

4 Some suppliers:	*5 Trade names or trade marks:*
PP—see recommendation for PP	—
EP(D)M—see recommendation for elastomers	—
Compounded material	
DSM	Keltan
Dynamit Nobel	Dynaform
Esso	Vistaflex PP
Hoescht	Hostalen
ICI	Propathene OTE
Montedison	Moplen SP
Uniroyal	Uniroyal TPR

6 Material properties: When ethylene is copolymerised with propylene (e.g. the ratio of ethylene to propylene is 70:30) an amorphous rubbery material results called EPM. If a third monomer is used then the product is called EPDM—an abbreviation that covers both of them is EP(D)M. Both are compatible with PP and if the ratio of rubber is less than 50 % the blend is called an elastomer modified thermoplastic (EMT). Rubber contents from 10 to 40 % are commonly used for injection moulding compounds. By using different grades of the base polymers and/or by altering the ratios used a wide range of properties can be offered. All of these blends can be processed on conventional thermoplastics equipment and vulcanisation is not required. The blends have good resistance to impact (even at low temperatures), can still have high Vicat softening points (e.g. >100 °C), good ageing and weathering resistance (when protected), a neutral colour (a wide colour range is therefore possible), good electrical insulation properties and a low density (0·86–0·9 g cm^{-3} when unfilled).

7 Ease of flow: Broadly speaking it is similar to that found for PP; blends based on copolymers give easier flowing grades.

8 Shrinkage: The addition of the EP(D)M will reduce shrinkage and minimise voiding compared with the base polymer; the shrinkage is reduced by about 0·2%. Because the crystallinity is reduced the difference in shrinkage between the longitudinal and transverse directions is decreased.

9 Resistant to the following: Oils, fats, alcohols and glycols. Because of the amorphous nature of the rubber, the blends have good resistance to impact at low temperatures. Commonly assessed by testing at −40 °C.

10 Not resistant to: Relatively limited resistance to petrol and other hydrocarbon solvents; as the rubber content increases the solvent resistance decreases. However, as the general chemical resistance of EP(D)M is good and as the chemical resistance of PP is excellent, the chemical resistance of the blends is very good.

11 Material detection or identification: When unfilled these materials float in water as they have a density of 0·9 g cm^{-3}. Transparent mouldings are not possible and such mouldings can be cut fairly easily with a knife. If the materials are not well mixed together then on bending or flexing the mouldings may easily break. The blends burn readily and have no distinctive odour. As with most rubber-like materials, infra-red analysis provides a convenient method of identification. As the rubber concentration in the blend increases then the hardness, modulus and softening point decreases. Such blends can be easily scratched or marked, e.g. with a thumbnail.

12 Colouring: A wide range of colours is possible using masterbatches or dry colours. Masterbatches may be based on PE or PP; if possible use a masterbatch which has a higher MFI than the EMT as this will give better dispersion. Compounded material gives the best colours.

For outdoor use, use grades which contain carbon black (e.g. 2%). Light colours may be stabilised using sterically hindered amines.

13 Material and component handling: If compounded blends are being moulded then these may be treated as though they were PP, e.g. predrying is not necessary. If blends are being prepared in-house then ensure that separation does not occur after tumble mixing the materials together. This problem may be eliminated by metering the PP and the EP(D)M into the throat of the machine—sometimes referred to as direct blend injection moulding (DBIM). Blends are sometimes mixed in internal mixers and in such a case the mix temperature should reach 200 °C. Continuous compounding is normally preferred; the particle size of the rubber (in the compounded material) should be about 0·5 μm.

14 Mould and gate considerations: These materials are softer than PP and greater care should be taken over ejecting the mouldings, e.g. a large number of large area ejectors may be needed. Softer grades may need to have a large taper on the sprue, e.g. 5°. As the follow-up pressure may need to be high (and applied for a long time so that cavitation is avoided), the gate and runner system should be designed to minimise pressure losses, e.g. short, round runners should be used. The gates and runners may need to be larger than those used for PP so that adequate follow-up pressure can be applied; do not overpack. Ensure that the mould is adequately vented as mould filling speeds can be high—see Section 23 below.

15 Flow path: wall thickness ratio: Depends on the types of polymers selected, ratios used, etc.—high ratios of flow path: wall thickness are possible. As the rubber content increases with some grades, the ease of flow also increases (the rubber can function as a flow promoter).

16 Projected area considerations: Similar to PP, e.g. approximately 2 tsi (30 MN m^{-2}).

17 Cylinder equipment: When tumble mixes, pellets or fully compounded materials are being moulded then a screw with a back-flow valve is required; the nozzle should be valved or decompression may be used. When pellets of PP and EP(D)M are being fed to the machine, or tumble mixes, special mixing screws may be used to get optimum dispersions; such screws have regions of high shear and/or mixing pins along their length. Devolatilising units should function as good mixers for these materials.

18 Screw cushion: About 4 mm.

19 Shot capacity: Up to 90 % of the material's rated capacity (in PS) may be used.

20 Melt temperature—as measured in the nozzle or by an air-shot technique: 220–270 °C (428–518 °F). Higher temperatures can give larger flow lengths and better weld strengths (if multiple-gated moulds are being used). The use of low temperatures may, however, improve dispersion, e.g. if high screw speeds and a high back pressure is used. As too high a melt temperature can cause degradation it is best to work at approximately 250 °C.

TABLE 20

Zone No.	*Location*	*Temperatures °C* *From*	*To*	*Temperatures °F* *From*	*To*
1	Barrel rear (near the hopper)	170	210	338	410
2	Barrel middle	210	230	410	446
3	Barrell middle	220	250	428	482
4	Barrel front	230	260	446	500
5	Nozzle	240	270	464	518
	Mould	20	60	68	140

21 Barrel residence time: These materials are relatively stable—it is still good practice not to let the machine stand for long periods at processing temperatures.

22 Temperature settings: Please note that it is the melt temperature which is important; those in Table 20 are only suggested, initial settings. The temperature of the hydraulic oil and of the material in the hopper should not vary excessively.

23 Injection speed—mould filling speed: As high as possible consistent with other requirements. When multiple gated moulds are being used it is not necessary to use very fast filling speeds in order to get high weld strength; employ a steady filling speed through large gates and do not overpack.

24 Injection pressure: The machine should be capable of giving the following: First stage: up to 1800 bars; 180 MN m^{-2}; 26 100 psi. Second stage (dwell or follow-up pressure): up to 1500 bars; 150 MN m^{-2}; 21 700 psi.

25 Screw rotational speed (rpm): High screw speeds are possible but not necessarily desirable—in general adjust the screw speed to suit the cooling cycle, i.e. as slow as possible.

26 Back pressure: Up to 200 bars; 20 MN m^{-2}; 3000 psi.

27 Shutting down: No need to purge with another material when shutting down.

28 Reprocessing: When reprocessing be careful as slight colour changes can occur if the material has been overheated. As exposure to shear at high temperatures affects dispersion then reprocessed material may have different properties.

29 Finishing: Because these materials are softer than PP, components moulded from such blends are easier to finish (e.g. desprue) than PP mouldings. As automotive components may need to be finished by painting it is best to avoid the use of silicone mould releases.

30 Other comments: Considerable interest is being expressed in extending the range of use of these materials by adding fillers and fibres, e.g. talc and glass fibre. As EP(D)M is noted for its ability to accept large quantities of fillers and oils, there is considerable scope for experiment.

31 Typical components: The big advantage of these materials is versatility, i.e. a wide range of grades is possible. As rubber content increases there is an increase in impact strength but a decrease in modulus, hardness, dimensional stability on heating and Vicat softening point; chemical resistance is also decreased as rubber concentration increases. The automotive industry now makes extensive use of such blends for use as car bumpers, protective strips, interior panels, dashboards, radiator grilles, etc. These materials can withstand abuse and when they break they fail in a ductile manner, i.e. they are safer for the occupants. There is now a tendency to use PP copolymers (rather than homopolymers) as by doing so the concentration of 'expensive' EP(D)M may be reduced, e.g. by 10%.

In general, products moulded from melt-compounded material exhibit higher Vicat softening points, higher elongations at break and better low temperature properties than blends based on tumble mixtures of PP and EP(D)M.

If optimum results are expected from any blend of PP and EP(D)M then it is best to melt-compound the blend and then mould it under those conditions suggested for ABS mouldings which are to be electroplated, i.e. relatively high melt and mould temperatures, slow filling speeds and with minimum overpacking. Such conditions are particularly suitable for components of heavy wall thickness.

21 THERMOPLASTIC POLYURETHANE

1 Common name: Thermoplastic polyurethane; thermoplastic urethane elastomer.

2 Abbreviation: TPU standing for thermoplastic urethanes would seem reasonable.

3 Systematic chemical name: Not known.

4 Some suppliers:	*5 Trade names or trade marks:*
Elastogran (BASF)	Elastollan and/or Caprolan
Bayer	Desmopan
Davathane	Davathane
Goodrich	Estane
Uniroyal	Roylar
Upjohn	Pellethane

6 Material properties: Thermoplastic polyurethanes are thermoplastics materials which exhibit rubbery properties. They are obtained by reacting a hydroxyl-terminated polyester or polyether with a diisocyanate so as to give linear polymers which contain the urethane group —NHCOO. Such polymers are noted for their resistance to wear and abrasion; they have high strength, resilience and flexibility. The materials are suitable for use over a wide temperature range, e.g. −40 to 140 °C, and can be transparent. High compressive strength and low compression set are other desirable attributes. Both polyester TPU and polyether TPU are available in a number of grades, the hardness of which can be varied over a wide range.

However, at the same hardness as polyester TPU, polyether TPU will have better low temperature flexibility, hydrolytic stability, fungus resistance and the highest resilience. It is not as good as polyester TPU for abrasion and oil resistance nor as tough; at the same hardness polyether TPU has the lowest specific gravity.

7 Ease of flow: TPU can be very easy flowing. Grades are available which can be processed on low-pressure machines developed for moulding shoe soles on to uppers. However, most TPUs are of higher molecular weight and as such are relatively viscous materials. Do not add lubricants in an effort to reduce viscosity as degradation may result.

8 Shrinkage: Depends on the grade employed and the thickness; in the range 0·5–2%. The material absorbs water and this causes dimensional changes; up to 3% water may be absorbed as the mouldings come into equilibrium with their surroundings. Harder grades exhibit less shrinkage than softer grades.

9 Resistant to the following: Both classes of material have good resistance to oil, petrol, grease and aromatic hydrocarbons and excellent resistance to aliphatic hydrocarbons. They have good stress-crack resistance.

10 Not resistant to: Acids, alkalis, alcohols, ketones, chlorinated hydrocarbons and synthetic lubricants. As TPU materials contain aromatic diisocyanates, discoloration will occur after a short-term exposure to weathering unless they are adequately protected, e.g. by carbon black incorporation. Polyester TPUs have the best resistance to oils and weathering and the best physical properties.

Polyether TPUs are more expensive but have the best resistance to hydrolysis (although neither material should be used in hot, wet conditions as degradation will occur). Polyether TPUs based on polytetramethylene ether have the best resistance to hydrolysis; amongst the polyester materials those based on polycaprolactone (also known as caproesters) are probably the best.

11 Material detection or identification: As the specific gravity is greater than 1 (e.g. 1·11–1·22) these materials sink readily in water. Natural colour varies from milky white to clear depending on the source but is usually translucent. Being thermoplastic these materials soften on heating, e.g. above 140 °C. The materials will burn if placed in a flame but the flame is not self-supporting. Attacked by a wide variety of reagents, e.g. dilute alkalis. Mouldings are strong, resilient and have good cut resistance. High elongations are possible, e.g. 500 %.

12 Colouring: A range of standard colours is available and manufacturers will, for large lots, colourmatch to requirements. The natural state may range from colourless to amber and so a wide colour range is possible which may be achieved by using masterbatches. May be supplied already preblended with masterbatches, i.e. pellets of masterbatch distributed in the pellets of TPU. It is best to use a TPU-based masterbatch, e.g. at 2 % concentration. Universal masterbatches are not so good.

13 Material and component handling: These materials are hygroscopic and are therefore supplied in foil-lined sacks so as to minimise moisture absorption. Adopt strict control over storage so that material is in store for the minimum time. Open sacks only when required and only remove what is strictly necessary from each sack—reseal sacks immediately and if possible use a heated hopper on the moulding machine. If more than 0·1 % moisture is absorbed then the material must be dried, e.g. 3 h at 80 °C in an air-circulating oven or 1 h in a vacuum oven.

14 Mould and gate considerations: Because of the flexibility of these materials the sprue should be kept short and should have a generous taper, e.g. 5°—this will assist ejection. To minimise frictional heat build-up, runners and gates should be as short and as generous as possible. As shrinkage is affected by processing conditions, conditioning, etc. the finalisation of mould dimensions should only be undertaken after trial mouldings have been produced and conditioned. Cast moulds (e.g. based on aluminium) have been used for short-run jobs. Undercuts are possible with flexible grades. To stop shrinkage variations it is essential that the temperature of the mould is held very accurately and adequate venting shall be provided.

15 Flow path:wall thickness ratio: High ratios are possible; these are obtained by using low molecular weight grades and/or flow promoters. In general, however, the higher molecular weight grades, as used for most moulding jobs, will not give high flow ratios—it is worse than plasticised PVC in this respect.

16 Projected area: This can vary from 0·5 to 2·5 tsi (5 to 34 $MN\,m^{-2}$).

Shoe soling may only need low clamping pressures because of the easy flow nature of the materials and the wall thickness of the components. The harder grades will require clamping pressures similar to those employed for HDPE, e.g. 2 tsi. Precision engineering components may require even higher pressures.

17 Cylinder equipment: A conventional 'thermoplastics' screw should be employed—to minimise thermal degradation risks do not use very high compression ratio screws, greater than, for example, 3:1. Use a back-flow valve on the screw; do not use a valve on the nozzle if it can be avoided as over-shearing can occur. If this over-shearing becomes excessive then the temperature rise produced may cause foaming or frothing.

18 Screw cushion: Approximately 3 mm.

19 Shot capacity: Try and avoid under-using the machine as this can lead to long residence times with its attendant risk of degradation.

20 Melt temperature—as measured in the nozzle or by an air-shot technique: 180–225 °C dependent on grade (356–437 °F). The more flexible grades require lower temperatures than the harder grades. Temperature control must be precise—if the temperature is too high then frothing and blistering will be seen when an air shot is taken. However, high melt temperatures can give mouldings with the best transparency provided that overheating has not occurred. If this occurs then a frothy, low viscosity melt will result.

21 Barrel residence time: Should be kept as short as possible so as to minimise the risk of 'yellowing'. As residence time increases the amount of regrind which can be used decreases as thermal history is cumulative. Do not allow to stand in the barrel for longer than 30 min.

22 Temperature settings: Please note that it is the melt temperature which is important; those in Table 21 are only suggested, initial settings.

TABLE 21

Zone No.	*Location*	*Temperatures °C*		*Temperatures °F*	
		From	*To*	*From*	*To*
1	Barrel rear (near the hopper)	150	180	302	356
2	Barrel middle	160	190	320	374
3	Barrel middle	170	200	338	392
4	Barrel front	180	210	356	410
5	Nozzle	190	220	374	428
	Mould	15	65	59	149

The temperature of the hydraulic oil and of the material in the hopper should not vary excessively.

23 Injection speed—mould filling speed: Fast injection speeds can cause degradation if the shear rate is too high as excessively high melt temperatures may result.

24 Injection pressure: The machine should be capable of giving the following: First stage: 1200 bars; 120 MN m^{-2}; 17 200 psi. Second stage (dwell or follow-up pressure): 750 bars; 75 MN m^{-2}; 10 700 psi.

25 Screw rotational speed (rpm): To minimise any risk of shear degradation keep the screw speed low, e.g. below 100 rpm. The motor should be capable of developing high torque at low speeds.

26 Back pressure: 15 bars; 1·5 MN m^{-2}; 215 psi. High back pressures can cause excessive shear and so back pressure should be kept as low as possible.

27 Shutting down: No need to purge with another material when shutting down. Turn off heat, close hopper slide and purge the barrel clean. If there is material left in the dryer reduce the temperature to 60 °C for overnight stops.

28 Reprocessing: Up to 50 % regrind has been used but a more realistic figure is probably around 20–25 %. Regrind immediately after moulding and then dry, e.g. 3 h at 80 °C. If the material is 'wet' then bubbles may appear in the melt and/or the material properties will deteriorate even at low (e.g. 0·1 %) moisture content.

29 Finishing: Desprueing is relatively easy because these materials are relatively soft. For some outdoor applications the components are decorated after removal from the mould as this protects them against weathering. Decoration may be by hot foiling, painting etc.

30 Other comments: Do not use components until they have been conditioned or annealed. For example, by being heated at 100 °C for 24 h or by being stored at room temperature for one month. Room temperature storage is not as good as high temperature conditioning as this permits stress relaxation and allows crystalline regions to stabilise—annealed components have a higher maximum service temperature, better creep resistance and lower compression set. Annealing is sometimes referred to as postcuring; it is not always necessary. If the colour of the components is not uniform this may be masked by adding a small amount of colour.

31 Typical components: Compared to traditional rubbers these materials are expensive but as they can be handled like thermoplastics, fast production on standard machines is possible. A very wide range of grades is available so that, for example, the hardness can range from 75 Shore A to 70

Shore D. The 'A' scale is used for rubbers and the D for plastics (the higher the number the harder a material will be). Linear TPU can bridge the hardness gap between rubbers and plastics as PE has a hardness of approximately 65 Shore D; PP of 75; and PS of 100. These thermoplastics elastomers are used where a flexible, durable material with good oil resistance, is required. They are used in the form of grommets, seals, washers, wheels, rollers, etc. and in industries such as automotive, mechanical handling and mining. The relatively poor resistance of some grades to synthetic lubricants (phosphates and diesters) should not be forgotten. On outdoor exposure, fading (particularly of pale colours) may occur—use ultra-violet stabilised grades for outdoor use, e.g. for animal ear tags. Fibre-modified grades are under active development, e.g. for car components.

22 SBS BLOCK COPOLYMERS

1 Common name: SBS block polymers, teleblock copolymers.

2 Abbreviation: SBS.

3 Systematic chemical name: One suggested name for a linear block is poly(1-phenylethylene-*b*-1-butenylene-*b*-1-phenylethylene).

4 Some suppliers:	*5 Trade names or trade marks:*
Anic	Europrene
Phillips	Solprene
Polyfleet Ltd	Polyfleet TR
Shell	Cariflex TR and/or Kraton

6 Material properties: SBS polymers are thermoplastic rubbers; they are block copolymers in which the styrene and butadiene monomers have been reacted so as to give long segments or blocks of polybutadiene and polystyrene, e.g. two end-blocks of PS connected to a central BR block. Although the segments are chemically connected they are incompatible and this results in the formation of PS domains, or clusters, in a continuous rubbery phase. These domains act as crosslinks and/or reinforcing fillers, so that the rubbery polybutadiene phase exhibits high elasticity just like a conventional, vulcanised elastomer.

On heating, the polystyrene domains are destroyed so that shaping in a mould is possible, i.e. the material behaves as a thermoplastic at 'high' temperatures and as a rubber at 'low' temperatures. Various arrangements of the segments are possible, e.g. linear polymers (Shell materials) and radial or branched molecules (Phillips). To achieve rubbery properties the styrene content is kept below 40%. SBS can be used without additives but additives are commonly added to reduce cost, improve processability and to achieve specific properties (see Section 12 below).

7 *Ease of flow:* Not so fluid as plasticised PVC—more like a stiff grade of LDPE. At the same molecular weight, radial polymers will exhibit a lower viscosity. For a given viscosity the use of a radial block polymer will result in compounds with higher hot strength and greater flex strength. However, because of the molecular weights used it is commonly found that the flow of branched compounds is poorer.

8 *Shrinkage:* These are amorphous thermoplastics which exhibit low shrinkage, e.g. below 1·0%. High filler loadings will reduce shrinkage and improve hot strength. Shrinkages range from 0·4 to 1% (depending on composition).

9 *Resistant to the following:* Water, alcohols, weak acids, weak bases. Addition of up to 15% EVA helps to prevent ozone attack. Benzophenone-type stabilisers improve ultra-violet resistance.

10 *Not resistant to:* Hydrocarbons, esters and ketones; such materials dissolve or cause excessive swelling. This is useful for adhesive purposes but poor solvent resistance limits the application of SBS; solvent resistance may be improved by adding a small amount of another rubber, e.g. nitrile.

11 *Material detection or identification:* The density ranges from 0·9 to 1·2 g cm^{-3} dependent on composition, i.e. mouldings may or may not sink in water. Components can be light coloured and cut easily with a knife. When burnt there is a smell of styrene and also a rubbery odour. Mouldings may be capable of large extensions even at low temperatures, but they will exhibit a very low softening point, e.g. a Vicat softening point of 70 °C. Attacked by a wide range of organic solvents. Good solvents include cyclohexane, toluene, methyl ethyl ketone and diethyl ether. Uncompounded material is translucent; however mouldings are usually based on compounds.

12 *Colouring:* Usually supplied to moulders already compounded and therefore coloured. Compounds based on these materials consist of SBS polymer, another polymer (e.g. PS), fillers (e.g. calcium carbonate), plasticisers (e.g. naphthenic oils), stabilisers (e.g. phenolic antioxidant) and a colouring system.

The mix should be melt-compounded, e.g. in a Banbury mixer at 140 °C. Dry blends can be prepared in high speed or ribbon mixers—do not exceed 65 °C. Complete the fluxing inside the injection moulding machine using, if necessary, a mixing screw.

13 *Material and component handling:* Normally these materials do not

require any special storage procedures. Some additives, e.g. carbon black, may make the compositions hygroscopic. Foil-lined sacks may be used to contain such grades. If it is necessary to dry SBS compound then the drying times can become quite long as high temperatures cannot be used.

Raw polymer is available in a white, irregular crumb form and also as oil-extended material. Fully compounded material is available from specialist compounders (e.g. Polyfleet Ltd.) in the form of granules.

14 Mould and gate considerations: SBS compounds can be used in moulds designed for other plastics, e.g. TPU or plasticised PVC—due allowance should, of course, be made for the different shrinkages of these materials. Runners should preferably be round (e.g. 10 mm) and as short as possible; if possible use a round gate with a minimum diameter of 0·8 mm. The gate should be as large as possible so as to reduce the amount of frictional heat generated, e.g. at least half the maximum part thickness. If the gates are small then filling will be slow and voids, streaking, etc. may result. Undercuts can often be removed from the mould simply by jumping or stripping the projection over the obstruction. Because of the low shrinkage, and the high coefficient of friction, ejection may be difficult—use the maximum amount of taper possible and round all corners and edges. Avoid overpacking; the use of stripper plates and/or air ejection may prove advantageous.

When very soft compounds are being moulded then a conventional sprue-puller may be inadequate—use a barbed probe. This projects into the sprue and it, in turn, is cleared by means of a sleeve-type ejector.

As compounds are often thick, an adequate supply of mould coolant should be available.

15 Flow path: wall thickness ratio: Information not available; the ratio would depend on grade, shear rate, etc. Would be expected to be of the order of 150:1.

Footwear is commonly produced from these materials on special machines in which the mould filling pressure is generated by screw rotation (injextruders)—this is possible as the materials are easy flowing and the wall thicknesses of the components are large. On such low shear rate machines the temperature control should be precise. This is because at low shear rates the affect of changes in melt temperature on viscosity can be pronounced—more so than at high shear rates.

16 Projected area: On conventional in-line screw machines this is of the order of 2 tsi (30 MN m^{-2}). Adequate venting is essential with these materials—one way of improving venting is to adjust the locking force so that flashing is just prevented.

17 Cylinder equipment: No special requirements—use a back-flow valve on the screw and decompression (if available). This will prevent drooling and avoid the need for complex valves.

18 Screw cushion: About 3–6 mm.

19 Shot capacity: Because of the thermal stability of this class of material, machines may be severely under-rated. However, try to match the machine to the job.

20 Melt temperature—as measured in the nozzle or by an air-shot technique: 150–200 °C (302–392 °F). Delamination, excessive orientation, etc., may occur if the temperature is too low, whilst degradation (crosslinking) will occur if the temperature is too high.

21 Barrel residence time: Materials are normally very stable—if barrel residence times can be kept below 15 min, a melt temperature of 220 °C may be used with some grades.

22 Temperature settings: Please note that it is the melt temperature which is important; those in Table 22 are only suggested, initial settings. The temperature of the hydraulic oil and of the material in the hopper should not vary excessively.

TABLE 22

Zone No.	*Location*	*Temperatures °C*		*Temperatures °F*	
		From	*To*	*From*	*To*
1	Barrel rear (near the hopper)	120	150	248	302
2	Barrel middle	140	160	284	320
3	Barrel middle	150	180	302	356
4	Barrel front	160	200	320	392
5	Nozzle	160	200	320	392
	Mould	15	65	59	149

23 Injection speed—mould filling speed: Resistance to flow is very dependent on shear rate; use as high an injection rate as possible.

24 Injection pressure: The machine should be capable of giving the following: First stage: <1300 bars; 130 MN m^{-2}; 18 800 psi. Second stage (dwell or follow-up pressure): <800 bars; 80 MN m^{-2}; 11 600 psi.

25 Screw rotational speed (rpm): High screw speeds are possible but are not desirable as large quantities of frictional heat can be generated; run the screw as slowly as possible, e.g. 100 rpm.

26 Back pressure: <200 bars; 20 MN m^{-2}; 2900 psi.

27 Shutting down: Because of the good thermal stability of this class of materials no special precautions are necessary—this material is a good purging compound for PVC.

28 Reprocessing: As this material behaves as a thermoplastic then reprocessing is possible. Up to 100 % regrind has been used, it is best to use regrind as a percentage of virgin material.

29 Finishing: Many components can be separated from their runner system by a simple twisting action. Knives, scalpels and secateurs may also be used for trimming.

30 Other comments: The range of use of this material may be extended by blending with more temperature-resistant plastics (e.g. PP) and reinforcing fillers, e.g. fine particle size silicas. The use of a saturated rubber segment (i.e. instead of polybutadiene) will improve weathering resistance.

31 Typical components: SBS thermoplastic elastomers exhibit reasonable strength and elasticity at room temperatures—they can be compounded to give a wide range of properties or grades. In general these materials exhibit properties which are similar to those associated with vulcanised rubbers—they exhibit good grip, resilience, high elongation at break, low set values and good low temperature properties. A wide range of hardnesses is available, e.g. from 40 to 90 Shore A. At room temperature the stress–strain properties can be similar to those found for polyisoprene rubber, i.e. high elongation at break for a low level of stress application. (SBS has been used to make elastic bands.)

Most applications of melt processed SBS are in the footwear industry, e.g. unit soles, heel units, top pieces and casual shoes and it seems to be replacing PVC and vulcanised rubber. When SBS is moulded excessive orientation should be avoided as such orientation makes the material feel stiffer in the flow direction.

Also used for swim-fins, ski-boots, suction cups, closures and bathroom mats. As the temperature is raised then elasticity is lost—do not use above 55 °C.

23 THERMOPLASTIC POLYETHERESTER

1 Common name: Thermoplastic polyetherester or thermoplastic polyester rubber.

2 Abbreviation: Not commonly used; an ASTM suggestion is YBPO.

3 Systematic chemical name: Not known. Materials are copolymers of butylene terephthalate and polytetramethylene glycol terephthalate.

4 Some suppliers:
Akzo
Du Pont

5 Trade names or trade marks:
Arnitel
Hytrel

6 Material properties: These materials are a class of thermoplastic elastomers which are available in a range of hardnesses, e.g. from 40 to 70 Shore D. Each molecular chain of these block polymers consists of segments, some of which can crystallise together. These 'hard' segment crystalline regions link the chains together by thermally reversible network structures. The other (soft) segments are joined to the 'hard' segments by ester linkages and it is these soft segments which are responsible for elastomeric behaviour. The structure consists of a crystalline network superimposed on an amorphous network. As the crystal structures have a high melting point these materials resist relatively high temperatures.

7 Ease of flow: They perform like conventional crosslinked elastomers over a wide temperature range but soften and flow like more conventional thermoplastics at elevated temperatures. The MFI ($g\,10\,min^{-1}$) depends on the grade, e.g. a soft (40 Shore D grade) material could have a MFI of 5 at 190 °C; a 72 Shore D material (hard) could have a MFI of 7·5 (at 230 °C).

8 Shrinkage: Part shrinkage can be minimised by using low mould temperatures. Soft grades can be compressed by excessive dwell pressures so that 'negative' shrinkage may result on ejection. Generally, shrinkages range from 0·004 to 0·016 in/in.

9 Resistant to the following: These materials can withstand very low temperatures without becoming brittle (e.g. −55 °C) and are resistant to most oils and solvents. Harder compounds are more resistant than soft. Serviceable in dilute acids and bases, hydrocarbons, alcohols, ketones and esters. Resistant to hydrolysis. Hydrolysis resistance may be improved by blending with polycarbodiimide masterbatch, e.g. at 2%.

10 Not resistant to: Attack by hot, concentrated acids and bases. Soluble in phenols, glycols, cresols and some chlorinated solvents, e.g. chloroform.

11 Material detection or identification: In the natural state these materials are opaque, creamy white solids and mouldings can feel soft and rubbery. They will sink when placed in water as the density is greater than one (e.g. 1·17–1·25 $g\,cm^{-3}$). The harder the grade, the higher is the density and the crystalline content. The mouldings do not feel as 'waxy' as polyethylene as their frictional characteristics are approximately half-way between those of PE and a traditional NR vulcanisate (e.g. for the softest grades). Mouldings can be cut easily with a knife and high elongations are possible, e.g. greater than 400 %. Good chemical resistance but the material will burn if placed in a flame. The flame is blue, edged with yellow and the fumes smell sweet but are irritating.

12 Colouring: Colour concentrates based on thermoplastic polyester (aromatic) should be used, as efficient dispersion is essential for good weathering. Other polymers, used as pigment bases, can impair properties.

13 Material and component handling: Materials are hygroscopic and should be stored accordingly. A first-in, first-out policy should be adopted and the moisture-proof sacks should only be opened when the material has reached workshop temperature. Reseal immediately after opening and if moisture is picked up, dry for 1 h at 150 °C (moisture content must be kept below 0·1 %). Do not blend with acidic materials (e.g. clays) as these will catalyse degradation.

14 Mould and gate considerations: Moulds should be vented at the parting line—use broad shallow vents. Ejectors should be large so that component damage does not result; taper the side-walls as much as possible and if necessary use a mould release (if the components are to be painted use a nonsilicone). Use full, round runners and keep them as short as possible; most gate types can be used. Hot-runner moulds have been used but the temperature control must be precise to avoid freezing and/or degradation. Such degradation may cause high and therefore dangerous gas pressures. Decomposition may also be caused by long residence times due to stagnation or resistance to flow in the feed system.

15 Flow path: wall thickness ratio: The material is capable of easy flow and should give high flow path: wall thickness ratios, e.g. similar to the flow for PE of equivalent MFI.

16 Projected area: As components are often moulded into precision components, adequate clamping pressure must be available, e.g. 3 tsi (45 MN m^{-2}).

17 Cylinder equipment: Special equipment is not required—use a screw with a gradual transition zone (similar to those used for PE). The screw should be equipped with a nonreturn valve. Do not use a screw with a short decompression zone as excessive frictional heating results which can cause degradation. The cylinder should be fitted with a valve or decompression should be used. Hopper heaters should minimise the risk of moisture pick-up.

18 Screw cushion: 4 mm.

19 Shot capacity: Under-rate the machine by approximately 30 % if possible; this will give good plasticisation with acceptable residence time in the barrel.

20 Melt temperature: These materials have a crystalline melting point of from 170 to 220 °C dependent on the grade selected (e.g. 170 °C for a 40 Shore D grade; 220 °C for a 72 Shore D grade). Melt temperatures range from 175 to 250 °C. Thick-walled components can be moulded at temperatures close to the crystalline melting point.

21 Barrel residence time: Materials have good stability but if overheated (e.g. to 300 °C) then tetrahydrofuran (THF) will be produced. As this is toxic it is good practice not to allow this type of material to 'cook' in the barrel and the moulding areas should be adequately ventilated. Turn off heat if the stoppage is liable to be longer than 10 min and then purge clean.

22 Temperature settings: See Table 23.

TABLE 23

Zone No.	*Location*	*Soft grade Temperatures °C From*	*To*	*Hard grade Temperatures °F From*	*To*
1	Barrel rear (near the hopper)	140	155	180	200
2	Barrel middle	145	165	190	210
3	Barrel middle	160	180	200	230
4	Barrel front	170	190	210	240
5	Nozzle	160	180	200	230
	Mould	10	70	10	70

23 Injection speed—mould filling speed: As high as possible. Avoid jetting as this will cause unsightly marks and/or a rough surface finish.

24 Injection pressure: The machine should be capable of giving the following: First stage: <1300 bars; 130 MN m^{-2}; 18 000 psi. Second stage (dwell or follow-up pressure): <800 bars; 80 MN m^{-2}; 11 600 psi. Be careful that overpacking does not occur, particularly for very soft grades.

25 Screw rotational speed (rpm): High rotational speeds can be used but in general the screw speed should be kept as low as possible so that temperature inhomogeneities do not result, e.g. below 100 rpm.

26 Back pressure: <200 bars; 20 MN m^{-2}; 2900 psi.

27 Shutting down: It should not be necessary to purge the machine for overnight stops—reduce the heat whilst running the barrel dry. Purging may be accomplished with LDPE or PMMA, e.g. use a low MFI (0·25) grade of LDPE.

28 Reprocessing: May be reclaimed on equipment designed for PE. Blend with virgin polymer but keep ratio of regrind low (max. 10 %) so as to maintain quality.

29 Finishing: Degating is relatively easy as these materials are quite soft. Painting, foiling, stamping, etc. are employed.

30 Other comments: When used in outdoor weathering applications the materials must be protected, e.g. by incorporation of an ultra-violet masterbatch or a black masterbatch (available from resin suppliers).

31 Typical components: These materials have good low temperature flexibility, abrasion resistance and flex-fatigue resistance. Some grades can withstand prolonged exposure to temperatures exceeding 120 °C. They have been used for segmented tracks of snow vehicles, seals, gaskets, gears, sportswear (e.g. ski-boot soles) and diaphragms.

24 FIBRE REINFORCED THERMOPLASTICS

1 Common name: Fibre reinforced thermoplastics.

2 Abbreviation: FRTP.

3 Systematic chemical name: Not commonly used; materials are usually referred to as glass-reinforced nylon, etc. Major materials reinforced are the engineering plastics such as nylon, acetals, polycarbonates and polymers containing *p*-phenylene groups, e.g. PBT.

4 Some suppliers: As for the base polymer.

5 Trade names or trade marks: Similar to those employed for the base polymer in many cases; other letters or numbers may be added for identification.

6 Material properties: Depends on the base polymer and on the type and concentration of the fibrous filler. Glass is very popular and is used over the range 5–50 % (glass content by weight), 20–40 % being most common, as below 20 % there may be little reinforcement whereas above 40 % processing is difficult. Because of the affect of processing, final fibre length is of the order of 0·5 mm (0·02 in) and the effect on strength (or reinforcement) is not as great as it is for thermosets (where the fibre length is greater). This means that although the final mouldings may be a lot stiffer than the unreinforced material they may not necessarily be stronger.

Materials are stiffer than the base polymer, have lower coefficients of thermal expansion, higher heat distortion temperatures and higher densities. FRTPs are available as precompounded materials and in masterbatch form; some in-house compounding of polymer and fibrous filler is done although it is not easy to do well. Coupling agents can improve properties, with glass fibre reinforcement, but unfortunately there is no universal coupling agent, i.e. one suitable for all polymers. FRTPs may also contain inorganic mineral fillers.

7 Ease of flow: In general the viscosity of FRTPs is higher than that of the base polymer. However, as some glass-filled melts are markedly pseudoplastic (e.g. nylon 66) the viscosity at high filling speeds will be lower than that found at low filling speeds.

8 Shrinkage: Compared to the unfilled polymer the shrinkage is much reduced but there may be a problem caused by fibre orientation, i.e. warping. Shrinkage reduction depends on fibre concentration (like the other properties affected). For example, the shrinkage of PBT is approximately 0·02 in/in—the addition of 15% glass reduces it to 0·006 in/in and the addition of 30% reduces it to 0·003 in/in (approximately).

9 Resistant to the following: The stress-cracking tendency of materials such as PS, SAN and PC is reduced; the creep resistance of FRTPs is better than that of the parent polymer. Low temperature impact resistance is sometimes better than that of the parent plastics material.

10 Not resistant to: Those substances which attack the base polymer. The interface between resin and glass can be attacked by alkalis. Use correct moulding conditions so that there is no glass at the surface which will help to 'wick' alkali into the moulding.

11 Material detection or identification: The presence of inorganic fibres (such as glass and asbestos) may be shown by burning off the base thermoplastic. By weighing a known sample (before and after burning) the fibre concentration may be calculated. As the glass content increases surface gloss decreases but stiffness and heat distortion temperature (HDT) increases.

12 Colouring: The presence of the fibre can result in a colour difference between the reinforced and the unreinforced material even though they have the same colourants. A matt finish is usual on FRTP mouldings and this can affect colour.

13 Material and component handling: Ensure that the moisture level is low as the steam produced can cause moulding problems and/or wear. It can be worse for plated or case-hardened metal surfaces as these may be brittle and relatively thin. Most of the plastics used are prone to water absorption and due care should be taken in storage to ensure that water pick-up does not occur.

14 Mould and gate considerations: A major problem with FRTPs is warpage—this is caused through orientation (fibre and polymer) during mould filling and/or mould packing. This can cause particular problems with flat mouldings and although it can be minimised by good part design and care in processing, it may be best to use fillers of low aspect ratio, i.e. of a shorter L:D ratio. Such fillers may stiffen the material but may not strengthen it.

Hot-runner moulds and hot-tip insulated runner moulds can be used, e.g. with acetals—good tool design, accurate temperature control and repeatable machine settings (e.g. shot volume) are important. Because of the reduced shrinkage, tapers may need to be increased in order to get satisfactory ejection. The mould should be designed so that parts can be easily replaced when wear occurs. Some moulders object to FRTPs because of the mould and machine wear that such materials may cause. Use a high chrome, high carbon tool steel for the mould.

15 Flow path: wall thickness ratio: This depends on fibre content, fibre size, size distribution, base polymer, etc. As the fibre concentration increases, the flow length may decrease. However, provided the mould can be filled quickly, e.g. by increasing injection pressure, mould or melt temperature, thin sections can be moulded. Fast filling is necessary because these materials set up more quickly than the base polymer—the filler acts as a nucleating agent. Nylon with 40 % glass content can still possess 70 % of the unmodified material's flow properties.

16 Projected area: FRTPs are often used in 'engineering' applications where close dimensional tolerances are required; high clamping pressures may be required, e.g. 8 tsi (120 MN m^{-2}).

17 Cylinder equipment: Similar to that required for the base polymer. A screw with a long feed section is to be preferred as this will ensure that the fibre is encapsulated by the melt before it reaches the compression zone—this zone should be short. Feed zone length, compression zone length and metering zone length should be 10, 4 and 6 screw diameters, respectively, and the compression ratio should be 3:1. Hardened screws and barrels are preferred as these have higher wear resistance, achieved, for example by ion implantation of nitrogen, case hardening, etc. If the moulding temperatures are high (e.g. for nylon 66) and a long run is envisaged, the use of a bimetallic barrel and a special alloy screw should be considered.

As FRTPs may also contain mineral fillers, which could fuse at high temperatures, it is safer to mechanically clean the screw rather than burn it clean.

18 Screw cushion: About 4 mm. If unable to maintain the screw cushion then repair or replace worn valve components immediately as otherwise severe wear may occur.

19 Shot capacity: Usually below that specified for PS. This is because, for example, some of the materials are crystalline and most have much higher processing temperatures than PS; both factors help to reduce shot capacity.

20 Melt temperature: Slightly higher than for the base polymer (e.g. by 15 °C) if this can be tolerated.

21 Barrel residence time: The materials often used are usually moulded at high temperatures and they also have relatively poor thermal stabilities. Do not allow the material to stand and cook in the barrel in the event of a stoppage; reduce the set temperatures and purge out (if in doubt).

22 Temperature settings: To stop fibre attrition the barrel and nozzle temperatures should be 10–30 °C higher (if possible) than those used for the base polymer. Feed zone temperatures should be kept high so as to minimise the risk of conveying cold material into the compression zone (compression of solid/melt mixtures increases abrasion). High mould temperatures improve gloss and maximise flow. At melt temperatures greater than 250 °C corrosion and wear (with glass fibre) may be more severe. See the temperatures recommended for the base polymer.

23 Injection speed—mould filling speed: Fast injection speeds may minimise weld line faults and reduce orientation as fibre-filled melts can be markedly shear thinning; fast speeds also help to stop the material freezing prematurely in the gates. Large runners and good venting are necessary; the mould should be hard, tough and chromium plated so as to withstand abrasion.

24 Injection pressure: First stage: to get high speeds the injection pressures may need to be higher than those used for the base polymer, e.g. up to 40 % higher. Second stage: (dwell or follow-up pressure): see recommendations for the base polymer.

25 Screw rotational speed (rpm): Keep low, as high screw speeds cause fibre degradation which in turn means more fibre ends and thus greater wear. In general, high screw speeds can also mean melt temperature inhomogeneities.

26 Back pressure: Keep as low as possible as high back pressures may cause excessive wear, e.g. on the screw root and on the conveying face of the screw flight. If uncompounded material and is being handled then relatively high back pressures may be needed to get adequate dispersion.

27 Shutting down: See recommendations given for the base polymer.

28 Reprocessing: Regrind concentration should be kept low, e.g. below 25%, as reprocessing reduces filler size which can (i) increase wear on the screw and barrel and (ii) reduce product properties. Regranulated scrap may also require drying.

29 Finishing: Degating may be a problem because of the strength of the material and also because the fibres cause tool blunting and wear. Gates may be removed by a machining operation (routing, milling, etc.) as such processes give cleaner parts than cropping which may produce more strand ends.

30 Other comments: FRTPs are usually based on expensive plastics and good housekeeping (scrap recovery, material storage, etc.) is therefore essential if costs are to be kept down. Impact strength at low temperatures is usually improved by fibre addition; this benefit may not be retained at higher temperatures.

31 Typical components: FRTP components are used in the following major areas:

Equipment and appliances (about 40%): Hairdryer casings, power tool casings, etc.

Automotive (about 30%): Energy absorbing bumpers, under the bonnet parts (e.g. petrol filters, brake fluid tanks, radiator tanks, etc.).

Electrical (about 20%): Coil formers for relays, transformer housings, motor housings, etc.

Miscellaneous (about 10%): Cigarette lighters, humidifiers, bearings, gears, etc.

In 1978 approximately 48 000 tonnes of glass-reinforced nylon was used in Europe (this material is the major FRTP). Glass is the major reinforcement as it is relatively cheap, dimensionally stable, resistant to chemical attack and does not burn. However, it is difficult to both stiffen and toughen the plastics material when glass (or other 'conventional' fillers) are used.

What is needed is an FRTP which is both stiffer and tougher than the parent material; fillers based on carbon fibre or aromatic polyamides are being investigated.

For heat resistance, asbestos offers useful properties. For example, a van heater unit is now moulded from asbestos-reinforced PP as the mouldings are light in weight, corrosion resistant, have good strength at low temperatures and can withstand 110 °C for long periods.

At the present time polyethylene terephthalate (PETP) reinforced with 45% glass fibre (Rynite 545) is said to be the most rigid thermoplastic available and because of its stiffness it is being used to replace die-cast metals.

25 STRUCTURAL FOAM

1 Common name: Structural foam (single component).

2 Abbreviation: SF.

3 Systematic chemical name: Not commonly used; materials are usually referred to as, for example, foamed PS or structural foam PS.

4 Some suppliers:	*5 Trade names or trade marks:*
Polymers	See recommendation for appropriate polymer
Blowing agents	
Boots–Fisons	Genitron
Fairmount	Azocel

Compounded materials are also available from specialist compounders.

6 Material properties: A cellular plastic may be described as a plastic the apparent density of which has been decreased substantially by the presence of numerous gas cells distributed throughout its mass. If the density of the outside layer (or skin) is much greater than that of the material in the centre of the component then the term 'structural foam' is employed. If a plastics melt, containing gas pockets under pressure, is injected into a closed, cool mould then the high density skin is formed as the gas at the surface escapes and the structure collapses. The skin has a 'swirl finish' in such a case.

A major limitation of many plastics is their low stiffness or modulus; by increasing part thickness (through expansion) then for a given weight of plastic, component stiffness will be increased. This is because stiffness is proportional to thickness. Expansion is commonly achieved by using azodicarbonamide—where this is unacceptable (due to colour, odour, etc.) then alternative azo compounds are used or systems based on bicarbonate are available.

7 Ease of flow: Melt should be more easy flowing because of internal gas pressure.

8 Shrinkage: For structural foam work-toughened polystyrene is commonly used, e.g. easy flow grades of medium toughness. Shrinkage is therefore low (similar to the base polymer) and is generally found to be uniform. Shrinking, even in thick sections, can be eliminated. In PP, shrinkage is of the order of 2 %, although this can be reduced considerably by using fillers. A rule of thumb is to allow 0·006 in/in for all materials unless otherwise advised.

9 Resistant to the following: Similar to the base polymer but because of the structure, SF components may be more quickly attacked by solvents. Because of the lower stress levels in the mouldings, components can exhibit better stress-crack resistance and a reduced tendency to warp.

10 Not resistant to: Those chemicals which attack the parent polymer. The impact strength of SF components is often lower than that of a component moulded from the same weight of (unfoamed) material.

11 Material detection or identification: When only one plastics material has been used (sandwich mouldings can have two) the surface commonly has a 'swirled' appearance which may feel slightly rough if a fingernail is drawn across the component surface. This can be masked, e.g. by sanding or painting. The component may or may not float in water; this depends on the base polymer, degree of expansion, etc. Components usually cut fairly easily to show the lower density centre. Components are commonly decorated by printing (woodgrains) and are thicker than 5 mm (0·2 in). Clear mouldings are not possible. For PS, densities of 0·7 g cm^{-3} are used and the skin thickness may be 1 mm (0·04 in).

12 Colouring: The material may be purchased already compounded (i.e. with blowing agents, colouring systems) already incorporated. By using a multicolour system 'woodgrain' effects are possible (e.g. teak, oak, etc.) without veneering or printing; such mouldings are used, for example, in the furniture industry to produce handles.

13 Material and component handling: The polymer may be tumbled together with the solid blowing agent, e.g. 1 % for 20 min at 30 rpm. To stop the blowing agent powder separating from the polymer granules, liquid paraffin or butyl stearate may be used to coat the granules. Multi-component pigment blends are available (matched to the shade required) and their use simplifies the colouring operation. More uniform cell structures are obtained if compounded material (containing blowing agent, colouring system, etc.) is used. Excess blowing agent (e.g. greater than 2 %) can cause extended cycle times and large voids to be produced in the moulding.

14 Mould and gate considerations: As fast mould filling is necessary in SF moulding, the feed system should be designed accordingly, i.e. to minimise pressure losses. Single point and multiple gate systems are used—multiple gating reduces flow path length and gives a more uniform density distribution. Circular and fan gates are used and gating is usually into the thinnest wall section. Good mould venting is essential if uniform density is to be achieved—it is better to maintain a slight pressure in the cavity during mould filling so as to minimise gas loss from the melt. Good cooling is essential in SF moulding because of part thickness. Moulds need not be constructed from steel (aluminium is widely used) but if beryllium copper is used it should be chrome plated as otherwise ammonia (from the blowing agent) will dull the surface. Mould surfaces are not highly polished; they are commonly grained so that the mouldings simulate wood. Hot-runner tools can be used but shut-off valves must be fitted to maintain gas pressures. To minimise density differences caused by gravity, a vertical clamping machine should be used if possible. To minimise water penetration via the trimmed sprue, it is common to gate into the inside of a component. Mouldings are then retained on the fixed mould half and are ejected by pins actuated by chains (connected to the moving platen).

15 Flow path: wall thickness ratio: For other materials this ratio has been given for 1 mm part thickness. Such thin sections are not possible with SF; part thickness must be greater than 3 mm to allow the structure to develop. Because of this, and because of the internal gas pressure, the melt flows fairly easily, however, a non-uniform structure may develop (through gas loss) if the flow path is too long. Flow path:wall thickness ratios of 100:1 can be obtained. As the flow path length increases, so does the density; to minimise this, and to get faster filling, multiple gates are common.

16 Projected area: One of the big advantages of SF moulding is that the clamp pressure required per unit of projected area is low for thick-sectioned mouldings >6 mm, e.g. $\frac{1}{3}$ tsi (5 MN m^{-2}). Cast, aluminium moulds may be used and the production of large components by injection moulding is commercially attractive because of the low clamping pressures required. Special SF machines are made and these have enormous platens for their size. The clamp pressure required depends upon the material and the mould used; in the case of thick-walled parts with short flow lengths the clamp pressure (compared to the parent polymer) may be reduced by a factor of 10; a factor of five is common, however. Large mouldings can be made with low clamping forces, e.g. machine housings (mould size about 600 × 300 mm) have been produced with only 10 tonnes lock.

17 Cylinder equipment: It is necessary to stop the gas from escaping from the cylinder (whilst the previous shot is cooling in the mould). The screw must have an efficient back-flow valve, the barrel must also be equipped with a valve, and it is advisable to use a screw delay timer.

18 Screw cushion: As the mould is not filled with melt (unlike the sandwich moulding process) the screw must bottom; the screw cushion is therefore zero.

19 Shot capacity: On an 'ordinary' machine (i.e. one not specifically designed for SF work) it is the platen size which would limit the shot size, not the ability of the plasticising unit to produce uniformly plasticised melt. A 'special' machine should be capable of dealing with the material for which it was designed.

20 Melt temperature: The blowing agent should decompose, as specified, at the polymers' melt processing temperature—the melt temperature should therefore be uniform and reproducible. For polystyrene materials this will be from 220 to 240 °C when azodicarbonamide is used; higher temperatures (within the specified range) usually mean lower densities. In general, melt temperatures are similar to those used for the base polymer.

21 Barrel residence time: This should be as short as possible so as to prevent gas loss. When the polymer plus blowing agent is fed into the barrel then, providing the temperature exceeds the decomposition temperature of the blowing agent, a gas is formed. Because of the thickness of the mouldings cooling cycles can be quite long and if the polymer melt was kept at the front of the barrel then the gas would be lost. To help minimise this a screw delay timer is used. This timer comes into operation when injection is complete and delays screw rotation until as late a time as possible.

TABLE 24

Zone No.	*Location*	*Temperatures °C* *From*	*To*	*Temperatures °F* *From*	*To*
1	Barrel rear (near the hopper)	130	140	266	284
2	Barrel middle	150	160	302	320
3	Barrel middle	170	190	338	374
4	Barrel front	200	220	392	428
5	Nozzle	220	240	428	464

22 Temperature settings: Adjust the feed zone temperatures so that decomposition of the blowing agent does not occur in the feed zone. Temperatures given in Table 24 are suggested temperatures for polystyrene plastics. Mould temperatures are similar to those used for the base polymer. High mould temperatures mean better surface finish.

23 Injection speed—mould filling speed: This must be high (speeds of up to 5000 $cm^3 s^{-1}$ have been used) so that the mould fills before the gas can escape—if it does escape then expansion will obviously be hindered or a non-uniform structure will result. If a woodgrain is being moulded then slowing down the injection speed may improve the 'grain' effect.

24 Injection pressure: The machine should be capable of giving the following: First stage: up to 1500 bars; 150 $MN m^{-2}$; 21 400 psi. Second stage (dwell or follow-up pressure): follow-up pressure as such is not necessary—the screw remains forward under 'screw contact pressure' until the gate has sealed.

25 Screw rotational speed (rpm): Provided the melt temperature uniformity can be kept within acceptable limits this should be as high as possible—this will give better dispersion.

26 Back pressure: Up to 150 bars; 15 $MN m^{-2}$; 2140 psi. Back pressure is used to homogenise the melt; after the screw has finished plasticising it must not be forced back by the gas pressure in the barrel as premature foaming will result. If necessary, delay plasticisation until the last possible moment by using 'screw delay time.'

27 Shutting down: Purge the barrel clean before shutting down; purge out with PS or TPS.

28 Reprocessing: Mouldings can be regranulated and re-used but ensure that colours and/or colour distribution is not affected.

29 Finishing: In many respects SF mouldings are handled like wooden components; they can be drilled, sawn, glued, etc. Appearance may be improved by sanding and painting. Rough surfaces are also filled before painting. Sprue scars may be filled before painting to stop moisture penetration. Sprue removal is made easier by keeping the angle between the sprue and the mould sharp, i.e. not radiused. This allows the sprue to be snapped off—even for large diameters. Surfaces may be veneered with woodgrain films or printed with the appropriate pattern. Parts are also decorated by hot foiling and hot stamping.

30 Other comments: Mechanical properties are usually obtained from the testing of moulded test specimens. Owing to differences in pore size and pore size distribution there can be a significant difference between results obtained by testing specimens and those obtained by testing the component.

31 Typical components: Styrene polymers are widely used for SF work because they are relatively cheap, give smoother finishes on mouldings than PE and have low heat contents (e.g. compared to the PEs)—this means that they can be cooled more rapidly. Because of their structure and thickness, SF mouldings cool slowly. In the case of thick-walled parts, eject from the mould and complete the cooling in a vat or bath.

Moulded housings for computers, business machines, hi-fi appliances and TV sets have been made from SF. To reduce fire risks PPO has been used in such applications. Because SF looks, feels and sounds like wood it has been widely used to produce loud-speaker cabinets, music centre parts, door panels, etc. SF mouldings are also used to produce parts for the car industry and the transport industry (pallets, boxes, etc.).

Other structural foam processes are possible, e.g. integral skin counter pressure, sandwich moulding, etc.; the process described which uses conventional equipment is, however, the most popular. The other processes can eliminate some of the defects found in SF mouldings, e.g. by injecting the gas filled melt into a hermetically sealed mould, filled with nitrogen gas, a surface free from swirl effects can be obtained. This process (gas counter pressure process) is expensive but can give a good surface finish. Sandwich moulding allows two different materials to be used for the skin and the core.

26 AMINOPLASTICS

1 Common name: Aminoplastics or aminos.

2 Abbreviation: UF and MF.

3 Systematic chemical name: Not known; the following have been suggested: Urea-methanal (UF); melamine-methanal (MF).

4 Some suppliers:	*5 Trade names or trade marks:*
MF	
Bakelite GMBH	Bakelite
Ciba-Geigy	Melopas
Dynamit Nobel	Ultrapas
Monsanto	Resimene
Montedison	Melbrite
Perstorp	Isomin
Raschig	Resipas
Resert GMBH	Resart
Sud-West-Chemie	Supraplast
N V Vynckier	Vyncolite
UF	
Allied	Plaskon
BIP	Beetle
James Ferguson	Nestorite
Montedison	Gabrite
Sterling	Sternite
Synres Nederland	Symesines

6 Material properties: Aminoplastic moulding powders are combinations of resins (either urea-formaldehyde or melamine-formaldehyde) with various reinforcing fillers (e.g. woodflour, alpha-cellulose, glass fibre, etc.) and with pigments, lubricants, hardeners, etc. They are thermosetting plastics which are available in fine powder or granular form and a wide colour range is possible. Both materials have good electrical properties (e.g. arc resistance), have high surface hardness, good gloss, an attractive appearance and do not burn easily. In general, however, they have low impact strength and their dimensions change as the humidity changes. The mouldings are hard and stiff and have good heat resistance; cellulose-filled grades have good antistatic properties. MF mouldings can withstand up to 120 °C continuous exposures, UF will degrade at 80 °C if continuously exposed.

Thermosetting plastics (PF:UF, etc.) differ from thermoplastics in that in general they have superior creep resistance, resistance to distortion at elevated temperatures and have hard abrasion resistant surfaces. They are not, however, transparent and do not have the resistance to impact associated with some thermoplastics. In general the materials are brittle and notch sensitive but their behaviour is very dependent on the filler used.

7 Ease of flow: A range of grades are available; the material must have good flow retention so that it does not precure in the runners.

8 Shrinkage: Approximately up to 1 % (e.g. 0·7 %) for UFs and MFs; on heating at 100 °C for 2 days after-shrinkage will occur, e.g. 0·4 % for UF and 0·7 % for MF.

9 Resistant to the following: Solvents (e.g. alcohols, ketones, aromatic and chlorinated hydrocarbons), detergents, greases and oils. MFs have better resistance to water (hot or cold), acids and alkalis, staining (by tea or coffee) and dry heat; they also have superior electrical insulation properties.

10 Not resistant to: Strong acids as these will cause decomposition; weak acids may attack UFs but MFs are usually unaffected. Strong alkalis will attack both materials but both are unaffected by weak alkalis. Sunlight may cause a colour change with some grades.

11 Material detection or identification: Mouldings are opaque and can be any colour from white to black; such mouldings have a stiff hard 'solid' feel and cannot be cut easily. Both burn with difficulty on heating; on burning, swelling and cracking may occur and the edge of the burnt portion will turn white. The flame will have a pale yellow colour and the flame edge will be a light blue-green. The UF will give off a strong odour of formaldehyde and also a musty odour; MF will produce a fish-like odour. MFs resist scuffing with a finger nail better than UFs. Both have densities greater than 1, e.g.

$1{\cdot}5\,g\,cm^{-3}$ (dependent on filler) and the mouldings will sink in water. UF mouldings are relatively easily stained on immersion in hot dye solutions; MFs will resist such immersions even in acidic dye solutions.

12 Colouring: Translucent colours are possible but most commonly mouldings are seen as opaque components; colours range from white to black and a wide colour range is available. The colouring is carried out at the compounding stage.

13 Material and component handling: As the material is slowly curing during storage, strict control over material storage and turnover should be employed. If the powder contains a large fines content, or if fine powder is being employed, then hopper bridging may be a problem. It is best to avoid such materials—agitators or auger feeders may, however, be of assistance.

14 Mould and gate considerations: During the hardening or setting process a large amount of volatile matter (e.g. water) is produced. Adequate venting is therefore essential using wide, shallow slots around the periphery of the mould. The technique known as breathing may help; just before the mould fills, injection is stopped and the mould is slightly parted—after mould closing, injection is then completed. Vacuum evacuation of the mould cavities is now also employed as this technique permits fast cavity filling and eliminates the need for 'breathing'.

Pressure losses are reduced by using gates and runners of circular cross-section. Runners should be kept as short as possible and sharp bends should be avoided. The most commonly used gates are the tab gate and the flash gate—although submarine and pinpoint are gaining in popularity as they minimise finishing and assist in automating the process. Although gate wear is not so severe as it is for PFs it is still advisable to carry the gates on removable pads. Each cavity should be surrounded by a land which is approximately 9 mm ($\frac{3}{8}$ in) wide as this will help avoid flashing; do not locate holes or slots close to the cavity as this may make flash difficult to remove. During ejection use an air blast to remove loose flash. Warm sprue and warm runner moulding are also used with these materials in an effort to reduce material usage.

15 Flow path: wall thickness ratio: Not usually quoted; these materials have inherently poor flow characteristics. As the materials are forced through the barrel/mould assembly they gain heat and this causes the viscosity to change dramatically; they are worse than PFs in this respect.

16 Projected area: Should be greater than 2·5 tsi ($37\,MN\,m^{-2}$). Easy flow grades may require 5 tsi ($75\,MN\,m^{-2}$) if flashing is to be avoided. On the clamp side a 'mould-halt' facility is useful for insert loading.

17 Cylinder equipment: A deep flighted screw may be used in an effort to minimise the heat build-up caused by shear. It is now more general to use a relatively shallow flighted screw as this gives better homogeneity; the heat generated must be capable of being removed by the barrel.

The screw may be temperature controlled by fluid circulation so as to improve plasticisation and to reduce overheating. The screw may also be capable of being moved in both directions (by hydraulic piston)—this is of assistance when hardened material must be removed from the barrel. To streamline the flow, back-flow valves are not fitted to the screw, this means that during injection, screw rotation may occur and this should be prevented. To minimise excessive shear heating the screw may have a zero compression ratio (or even a negative one). The pitch of the screw may be 0·75 of the screw diameter; the L:D ratio is about 15:1. The shape of the screw tip should match the internal shape of the nozzle; the nozzle may need to be temperature controlled and the machine must be equipped with sprue break.

18 Screw cushion: As small as possible, e.g. 2 mm. During mould packing (and before curing occurs) the material is lost from the cavity in the form of flash due to viscosity reduction. This means that often the cushion is not very effective.

19 Shot capacity: The full theoretical shot capacity will not be available as material leaks back down the screw during injection; approximately 80% of the theoretical shot capacity will be available.

20 Melt temperature: Usually in the region of 100–120 °C. Cure times are usually in the range of 10–90 s with such temperatures; adjust cure time until there are no blisters in the component. Blisters may be caused by under-cure or over-cure.

21 Barrel residence time: Should be kept to a minimum so as to prevent premature curing. The material should be capable of accommodating short stoppages in production without precuring.

22 Temperature settings: See Table 25. Too low a barrel temperature should be avoided as this will tend to cause precure due to the development of large amounts of frictional heat. Such heat is likely to be unevenly distributed and so may result in highly stressed mouldings.

Many machines have two zones on the barrel; the temperature in each is regulated by circulating heated fluid through a jacket around the barrel. Low mould temperatures may result in porosity at the edges of components. High mould temperatures will give higher shrinkages.

23 Injection speed—mould filling speed: High injection speeds will cause large amounts of shear heat to be developed which will speed up the cure but may introduce problems, e.g. precure, air-trapping, etc. Therefore inject as

TABLE 25

Zone No.	*Location*	*Temperatures °C UF*		*MF*		*Temperatures °F UF*		*MF*	
		From	*To*	*From*	*To*	*From*	*To*	*From*	*To*
1	Barrel rear	20	60	60	70	68	140	140	158
4	Barrel front	70	90	75	95	158	194	167	203
5	Nozzle	90	105	90	110	194	221	194	230
	Mould	140	160	145	200	284	320	293	392

fast as possible but consistent with avoiding such problems. Speeds are lower than for phenolic.

24 Injection pressure: The machine should be capable of giving the following: First stage: Up to 2000 bars; 200 $MN\,m^{-2}$; 29 000 psi. Second stage (dwell or follow-up pressure): up to 1000 bars; 100 $MN\,m^{-2}$; 14 500 psi. The amount of packing pressure, and the time for which it is applied, should be kept to a minimum so as to minimise overpacking; adjust the packing pressure until sinking in the gate region is eliminated. Packing time may be of the order of 10 s.

25 Screw rotational speed (rpm): 30–55 rpm. The drive motor must be capable of giving sufficient torque so as to prevent stalling if some precure occurs. The torque required is less than that for PFs. Match speed to cure time.

26 Back pressure: Up to 70 bars; 7 $MN\,m^{-2}$; 1000 psi. High back pressures may cause premature setting; keep back pressure as low as possible—just sufficient to give good shot weight control.

27 Shutting down: Reduce barrel temperature, shut hopper slide, withdraw injection unit and purge the barrel clean; remove the nozzle and repurge. Scrape nozzle and front of barrel clean.

28 Reprocessing: Not normally recommended.

29 Finishing: Components can be finished by hand filling, buffing, etc. Mouldings can be decorated by painting, transfers or hot foiling; very durable finishes may be obtained during the moulding stage by using printed foils.

30 Other comments: Injection moulding tends to produce components which are highly stressed (cf. compression). The dry stoving test is used to determine whether samples are highly stressed, e.g. mouldings should withstand 2 days at 60 °C without cracking. MFs will withstand more severe heating tests than UFs.

31 Typical components: MF materials have greater resistance to high temperatures and humidities than UF materials; however, because of their higher cost MFs are probably not so widely used as UFs. Common uses of UF materials are for electrical components, e.g. sockets, plug-tops, covers.

MF is used in more demanding situations, e.g. where better dry heat resistance is required. MF is used in nontracking electrical components which can withstand high temperatures and thermal stress, for example, the ship building and domestic appliance industries. Cutlery handles and other eating utensils are also made from MF.

27 PHENOLICS

1 Common name: Phenolics or phenol-formaldehyde.

2 Abbreviation: PF.

3 Systematic chemical name: Not known; one suggestion is phenol-methanal.

4 Some suppliers:	*5 Trade names or trade marks:*
Bakelite GMBH	Bakelite
Bayerisches Kunstoffwerk GMBH	Chemoplast
Dynamit Nobel	Troliton
Perstorp	Nestorite
Raschig	Resinol
Sud-West-Chemie	Supraplast
N.V. Vynckier	Vyncolite

6 Material properties: PFs are a class of thermosets which are based on PF resins (Novolak), fillers (such as woodflour), lubricants, hardeners, etc. They are supplied in powder form and the mouldings have a useful combination of low cost, ease and versatility of moulding, temperature resistance, solvent and chemical resistance. The electrical insulation properties are good and the material has better water resistance than MF. One major drawback of PFs is their limited colour range; black and brown are the most common colours. Mouldings are hard, stiff and exhibit low elongations. The mouldings have good creep resistance. Moulding properties are very dependent on the filler used.

7 Ease of flow: Available in a range of flows, e.g. soft, medium and stiff. These are stiff, hard materials which require high pressures to get smooth mould filling. PFs have more stable melt rheology than aminoplastics, i.e. they are not so temperature dependent.

8 Shrinkage: About 0·010 in/in for woodflour-filled phenolics (approximately 1 %).

9 Resistant to the following: Dilute acids (not all PF grades), alcohols, aromatic and chlorinated hydrocarbons, greases and oils. Mineral filled grades are also resistant to detergents.

10 Not resistant to: Strong oxidising acids and alkalis; some components (depending on grade) may be attacked by dilute acids and alkalis, ketones and detergents. Sunlight will cause general darkening.

11 Material detection or identification: Mouldings are usually dark coloured (e.g. brown or black) and are difficult to cut. This material burns with difficulty (those filled with glass or asbestos are virtually incombustible) and cracks when burnt, giving off a phenolic odour (carbolic acid). As the material density is 1·3–1·8 g cm^{-3} (dependent on filler type, concentration, etc.) it sinks in water. When scraped, filed or sawn the materials smell strongly of phenol and the mouldings will only soften slightly at high temperatures, e.g. above 130 °C.

12 Colouring: Usually carried out at the compounding or manufacturing stage. Because of their complex chemical structure only dark colours are available—usually dark browns, greens and blacks.

13 Material and component handling: As long storage will affect flow it is important that a first-in first-out storage system be adopted so as to obtain a constant turnover of materials.

Components moulded from PFs are not susceptible to scuffing and so individual packaging is not normally necessary.

14 Mould and gate considerations: Because of the abrasive nature of thermosetting materials, moulds should be hardened, polished and chrome plated. The mould should be constructed so that as wear occurs replacement is possible (e.g. the gates should be carried on pads which are replaceable). Electrically heated moulds are commonly used—allow approximately 1 kW of installed power for every 50 kg of mould weight and insulate the mould as much as possible. Adequate venting is essential. Circular runners give less pressure drop than trapezoidal or semi-circular runners but are more difficult to machine. Pinpoint gates of less than 0·7 mm are not usually used as the temperature rise and abrasive affect is excessive; high impact grades will require larger gates—mould venting is essential. Runnerless moulding, although difficult to perform, can show substantial material savings. Melt temperature control may need to be very accurate (e.g. ±2 °C) and accurate control over nozzle temperature is essential, e.g. using water-cooled nozzles. Live sprue moulding is a useful compromise; the runners are cured and ejected but not the sprue. This saves some material and is relatively cheap to do, the big advantage is that the mould opening movement is reduced and time is saved. Faster curing (stiffer flow) grades can also be used.

15 Flow path: wall thickness ratio: Information not available.

16 Projected area: Approximately 2·5 tsi (37 $MN\,m^{-2}$). Thermoset mouldings are usually of heavy cross-section and this helps to reduce clamp pressure requirements. However, the viscosity drop which occurs during mould filling off-sets this. Cavity pressure control and programmed injection speed should be of help in reducing clamp pressure requirements.

17 Cylinder equipment: A check ring assembly is not normally fitted and the compression ratio of the screw is often 1:1 or even 1:0·8. A negative compression ratio allows for ease of removal if precuring occurs. Screws may be cored for heating and cooling.

18 Screw cushion: As small as possible, e.g. 2 mm. May be eliminated if a process control system is fitted. As thermosets increase in volume during curing then the increase in pressure may be sensed and used to control the shot size setting (and cure time calculations) so that the screw may bottom. By allowing for this expansion, overpacking may be reduced and material usage reduced.

19 Shot capacity: Only about 80% of the machines theoretical shot capacity is usually available as material 'flow-back' down the screw occurs during injection—this is minimised by fast injection speeds and low rear zone temperatures.

20 Melt temperature—as measured in the nozzle or by an air-shot technique: Work done on the material by the screw will raise the material temperature by approximately 15 °C (depends on back pressure, screw speed, etc.). A GP phenolic will typically have a 'melt' temperature of 105–130 °C (220–260 °F). Below 105 °C it will be too viscous whereas above 130 °C precure may occur. (GP means general purpose—such grades may be woodflour filled.)

21 Barrel residence time: Keep as low as possible; do not allow the machine to stand for long periods with the heat on and do not excessively under-utilise the machine. The material should be formulated so that short stoppages, e.g. 5 min, can be tolerated. Use sprue-break to stop heat transfer from the mould.

22 Temperature settings: Please note that it is the melt temperature which is important; those in Table 26 are only suggested, initial settings. The temperature of the hydraulic oil and of the material in the hopper should not vary excessively.

The barrel temperature is normally maintained by circulating heated

TABLE 26

Zone No.	Location	Temperature °C From	Temperature °C To	Temperature °F From	Temperature °F To
1	Barrel rear	50	70	122	158
4	Barrel front	70	85	178	183
5	Nozzle	80	110	176	230
	Mould	150	210	302	410

water through jackets or coils attached to the outside of the barrel. Cure time is very dependent on temperature. For example a component with a wall thickness of 3·2 mm ($\frac{1}{8}$ in) will require a cure time of 28 s at a mould temperature of 175 °C; at 210 °C it will be 12 s.

23 Injection speed—mould filling speed: In general this should be as high as possible. When the material enters the cavity its temperature will be of the order of 150 °C (302 °F) because of the friction generated. However, very high speeds are not possible (e.g. 5–20 s is usual) as gas marking and/or flashing may occur. Often the machines are not powerful enough to give high speeds, i.e. the line pressures are insufficient.

24 Injection pressure: The machine should be capable of giving the following: First stage: Up to 2000 bars; 200 MN m^{-2}; 29 000 psi. Second stage (dwell or follow-up pressure): up to 1000 bars; 100 MN m^{-2}; 14 500 psi.

25 Screw rotational speed (rpm): 60–90. The screw-drive motor must be capable of generating sufficient torque. Normally the torque required is not very high.

26 Back pressure: Up to 70 bars; 7 MN m^{-2}; 1000 psi. Excessive back pressure may cause premature setting to occur in the barrel and will accelerate wear.

27 Shutting down: Reduce the cylinder temperatures to 55 °C (130 °F) and start to purge the barrel. Stop the material feeding into the barrel and purge the barrel clean. Turn off the heaters. When restarting allow the barrel to heat soak for about 15 min at the running temperature before making a few air purging shots; if necessary remove the nozzle and scrape out cured material.

28 Reprocessing: Not normally recommended although if finely ground (e.g. below 200 mesh) from 5 to 20 % of regrind can be incorporated in some grades without a serious loss of mechanical properties. GP grades are not drastically affected, fibre-filled grades may be seriously affected by regrind incorporation.

29 Finishing: PFs are only available in dark colours—this may be changed by stoving or painting. Hot foiling can be performed but it is difficult to do well. Flash is commonly removed by finishing and/or hand filling; holes or threads can be incorporated by drilling and tapping.

30 Other comments: PFs are rated as being relatively easy to mould compared to aminoplastics.

31 Typical components: Iron, saucepan and pressure cooker handles are commonly seen components; the material is chosen for these applications because it is heat resistant and durable; it is also a low cost material. Dashboard ashtrays, coffee percolator bases and electrical connectors are other obvious applications. The maximum service temperature depends on the filler used; for the highest heat resistance, mineral-filled grades should be used. Slate, asbestos and other minerals give heat resistant materials; the best electrical resistance is obtained with ground mica. Glass or other fibrous fillers give components with the highest impact strength. If the mouldings contain organic fillers then under hot, wet conditions, fungus or moulds may grow on the surface.

28 DOUGH MOULDING COMPOUND

1 Common name: Dough moulding compound.

2 Abbreviation: DMC.

3 Systematic chemical name: Not known.

4 Some suppliers:	*5 Trade names or trade marks:*
BIP	Beetle
Freeman Chemicals	Freemix

6 Material properties: These materials are based on unsaturated polyester resins—the resins are blended with a fibrous reinforcement (usually glass), a filler, pigments, styrene and a catalyst system. The mixture has a dough-like consistency (hence the name) and in order to ease handling the composition is often supplied in a rope-like form. A wide range of formulations is possible but in general mouldings are very strong and stiff. The compounds may be specially formulated to maximise a particular property, e.g. flexural strength, resistance, low coefficient of thermal expansion, flame retardance, etc. However, in general the impact strength of injection moulded DMC is lower than that of compression moulded DMC.

7 Ease of flow: Can be very easy flowing and fast curing, e.g. 20 s for a 3-mm section (mould temperature 150 °C).

8 Shrinkage: Can be very low, GP grades are only 0·2 %. Low profile grades are available (obtained by adding a thermoplastic to DMC) which have little or no shrinkage, e.g. 0·0055 in/in (below 0·05 %). Coefficient of thermal expansion is also low, e.g. similar to that of steel.

9 Resistant to the following: Water, alcohols, aliphatic hydrocarbons, greases and oils. The mechanical properties are usually retained on weathering but the surface usually deteriorates. Mouldings can withstand exposure to high temperatures (e.g. 160 °C) for long periods.

10 Not resistant to: Ketones and chlorinated hydrocarbons; the resistance to aromatic hydrocarbons, acids and alkalis is also not very good. Resistance to such chemicals is much worse at elevated temperatures.

11 Material detection or identification: These materials have a very high density, e.g. 1·7–2·1 $g\,cm^{-3}$ (dependent on grade) and the mouldings feel stiff, hard and heavy; they are difficult to cut and will sink rapidly in water. When heated in a flame a lot of soot is formed and styrene may be smelt. Some grades burn easily whereas others are of a reduced flammability rating. Inorganic residues (e.g. glass) may be left after burning.

12 Colouring: The natural colour depends on the filler system employed. When whiting is used as the filler then beige is the natural colour. A wide colour range is available but the material must be purchased already coloured as there is no sensible way of adding colour during the moulding operation.

13 Material and component handling: Feeding of such materials may be difficult—the material may be forced into the barrel by means of pneumatic pistons. It is normally supplied as a slightly tacky, dough-like mass in rope form—avoid contact with the skin and breathing the vapour. Store in sealed containers away from heat as styrene is inflammable. Adapt a first-in first-out storage system as the shelf-life of the material is limited, e.g. 3 months at room temperature. Storage-deteriorated materials will not have the expected gloss or strength; increasing the cure time will not help.

14 Mould and gate considerations: Fibre orientation is affected by the type and position of gate used—use fairly large gates to minimise fibre orientation and fibre degradation. Runners should also be as large as possible and abrupt changes in direction should be avoided. It must be accepted that there will be some fibre degradation as the material is forced through the feed system and that injection mouldings will be anisotropic. If the melt is divided by a core then poor weld strength can be a problem. Injection–compression moulding reduces flow orientation; warm runner moulding reduces waste and can reduce cycle times—this is because curing of the sprue is often the rate-determining step as sprue thickness can be large.

15 Flow path: wall thickness ratio: Not normally quoted, however mould filling is seldom a problem with DMC.

16 Projected area: DMC components are often of heavy cross-section and thus for a given shot weight, clamping requirements should be relatively low. However, because of the easy flow nature of the material, clamping pressures of the order of 2–2·5 tsi (30–37 MN m^{-2}) are commonly employed.

17 Cylinder equipment: Screws with little or no compression ratio are employed. The material will need to be forced or 'stuffed' on to the screw during screw rotation by a piston. The conventional conical hopper is replaced by a cylinder which contains the material; the ram operates inside the cylinder. To prevent material from flowing back down the screw during injection, back-flow valves may be fitted to the screw; a valve is also fitted to the nozzle. Both the screw and the barrel should ideally be constructed from wear-resistant materials.

18 Screw cushion: Material is so easy flowing that if a screw cushion is set then the screw may 'punch' through. This will damage the fibres and reduce material viscosity thus altering material properties.

19 Shot capacity: DMC compounds are very easy flowing and therefore require low injection pressures. Cylinders of large shot volume may thus be fitted to comparatively small machines and this allows the economic production of components with a large shot weight but of heavy cross-section.

20 Melt temperature—as measured in the nozzle or by an air-shot technique: DMCs are plastic at room temperature—they do not, therefore, need to be heated before flow can occur. However, barrel heating is normally employed as this reduces fibre damage and speeds up cure. Temperatures of up to 60 °C are used.

21 Barrel residence time: Material can reside in the barrel for long periods, e.g. at 60 °C it is still relatively stable. It is good practice not to allow the material to stand and 'cook' in the barrel.

22 Temperature settings: Please note that it is the melt temperature

TABLE 27

Zone No.	*Location*	*Temperatures °C* *From*	*To*	*Temperatures °F* *From*	*To*
1	Barrel rear	20	40	68	104
4	Barrel front	40	60	104	140
	Mould	140	160	284	320

which is important; those in Table 27 are only suggested, initial settings. The temperature of the hydraulic oil and of the material in the hopper should not vary excessively.

23 Injection speed—mould filling speed: High injection speeds can give good surface finish; because of the ease of flow of these materials such high speeds can be obtained with relatively low pressures. The rate of filling may have to be reduced if air trapping is a problem (it can cause porosity).

24 Injection pressure: The machine should be capable of giving the following: First stage: up to 500 bars; 50 MN m^{-2}; 7000 psi. Second stage (dwell or follow-up pressure): up to 400 bars; 40 MN m^{-2}; 5800 psi.

25 Screw rotational speed (rpm): 30–55 rpm. Screw rotational speed and back pressure should be as low as possible so as to minimise fibre damage.

26 Back pressure: 50 bars; 5 MN m^{-2}; 700 psi. High back pressures will cause fibre breakdown—keep the back pressure as low as possible.

27 Shutting down: Not essential to thoroughly clean the barrel out even for overnight stops. Cool the barrel and purge clean.

28 Reprocessing: Not recommended—the material would be expensive to reclaim and could only be used to replace the cheap, inorganic filler.

29 Finishing: During finishing operations (e.g. machining) there is the possibility of dust generation. Adequately ventilate the area and avoid ingestion of the dust by using masks. As gates are large, and the material is tough, pneumatic side cutters may need to be employed for cropping. Tungsten tipped drills should be employed for drilling operations.

Can be painted without prior abrasion of the surface—degreasing may be necessary. Stoving (of paints) may also be employed, e.g. 30 min at 160 °C.

30 Other comments: Styrene vapour may be smelt during processing—keep the moulding area well ventilated so that the styrene concentration in the working atmosphere does not exceed the threshold limit value (TLV), e.g. 100 ppm.

31 Typical components: DMC has high strength, excellent electrical properties and very good dimensional stability; it has better heat resistance and lower creep than most thermoplastics. Grades are available which permit thick sections to be moulded without internal cracks or voids. It is used to mould diesel engine head covers, drill housings, machine guards, appliance covers, heavy duty contactors and circuit breakers. In this last application resistance to repeated arcing is important. Because DMC compounds are inherently fast curing, injection moulding does not offer very dramatic reductions in cure time (compared to compression moulding); however, injection moulding saves weighing and dispensing time.

29 Elastomers

1 INTRODUCTION

The rubber injection moulding industry now makes extensive use of the so-called two-stage injection moulding machines. These are machines in which the rubber is softened by means of a screw in one barrel and then transferred to another; at this point, the rubber has a temperature of approximately 110 °C. The rubber is then forced into the closed heated mould which, for example, is held at 185 °C. At these high temperatures, the setting, or vulcanisation, is extremely rapid and once it has been taken to the desired degree, the rubber moulding is ejected from the machine.

There are many different materials which can be classified as rubbers and for the purpose of this discussion, they will be divided into, general purpose rubbers, oil resistant types, and special purpose (SP) rubbers. There are approximately a dozen or so elastomers which are used widely commercially. These materials offer a very wide range of properties and it is possible to vary the characteristics of any elastomer by use of a range of compounding ingredients. Compound flow behaviour and final product properties can therefore vary enormously, even if the same rubber is used, and for this reason only general guidelines, suitable for the more widely used materials, will be given.

2 GENERAL PURPOSE RUBBERS

2.1 Names, Abbreviations and Suppliers

2.1.1 Natural rubber

1 *Common name:* Natural rubber.
2 *Abbreviation:* NR.
3 *Systematic chemical name:* Natural poly(1-methyl-1-butenylene).

4 *Some suppliers:*	5 *Trade names or trade marks:*
Guthrie Estates	Dynat
Harrison Crossfield	Harub
Hecht Heyworth Alcan	Technorub
Lewis and Peat	Nirub and Suprarub
Mardec	Marub

This material is most commonly sold as a specified type or grade, e.g. pale crepe; ribbed smoked sheet RSS; Standard Malaysian Rubber SMR, etc.

2.1.2 Styrene-butadiene rubber

1 Common name: Styrene-butadiene rubber.

2 Abbreviation: SBR.

3 Systematic chemical name: Not known. The following has been suggested: poly(1-butenylene-*co*-1-phenylethylene).

4 Some suppliers:	*5 Trade names or trade marks:*
Anic	Europrene Sol R
ISR	Intol
Phillips	Solprene
Polysar	Krylene
Shell	Cariflex
SIR	Sirel

2.1.3 Butyl rubber

1 Common name: Butyl rubber.

2 Abbreviation: IIR.

3 Systematic chemical name: Not known. The following has been suggested: poly(1,1-dimethylethylene-*co*-1-methyl-1-butenylene).

4 Some suppliers:	*5 Trade names or trade marks:*
Esso	Esso butyl
Polysar	Polysar butyl

2.1.4 Butadiene rubber

1 Common name: Butadiene rubber or *cis*-polybutadiene.

2 Abbreviation: BR.

3 Systematic chemical name: Poly(1-butenylene).

4 Some suppliers:	*5 Trade names or trade marks:*
Anic	Europrene
Firestone	Diene
Phillips	Solprene
ISR	Intene
Shell	Cariflex BR

2.1.5 Cis-*polyisoprene*

1 Common name: Isoprene rubber or *cis*-polyisoprene.

2 Abbreviation: IR.

3 Systematic chemical name: Poly(1-methyl-1-butenylene).

4 Some suppliers:	*5 Trade names or trade marks:*
Anic	Europrene
Goodyear	Natsyn
Shell	Cariflex IR

2.1.6 Ethylene-propylene rubber

1 Common name: Ethylene propylene rubber.

2 Abbreviation: EPM or EPDM or EP(D)M.

3 Systematic chemical name: Not known. The following has been suggested for EPM poly(methylene-*co*-propylene).

4 Some suppliers:	*5 Trade names or trade marks:*
Du Pont	Nordel
Esso	Vistalon
Goodrich	Epcar
Hüls	Buna
ISR	Intolan
Montedison	Dutral

2.2 Material Properties

2.2.1 NR and SBR

The total world production of all rubbers is now of the order of 11 million tons (Mt); the two most widely used rubbers are SBR (4 mt) and NR (3 Mt). These materials are therefore the work horses of the rubber industry and are used when a high level of resistance to oxidation, ozone and petroleum oil is *not* required. Both materials cost roughly the same and the maximum service temperature of moulded components is of the order of 70 °C. Both are available in a range of grades (range of molecular weights, viscosities, etc.) and all grades of SBR are compatible with NR in all proportions; mixtures are quite often used in order to get desired properties.

There are, however, differences between them; for example, if natural rubber is stretched, then crystallinity occurs and this reinforces the rubber so that it resists deformation. Because of this increased resistance to deformation the fillers used with natural rubber can be of the nonreinforcing variety (such fillers are cheap). SBR does not crystallise when stretched and so reinforcing (expensive) fillers must be used for many applications. NR has better hot tear resistance than SBR and this property is useful if there are difficulties in moulding. SBR is slower curing than natural rubber and although the cure rates can be matched, this means that more (expensive) accelerator must be used.

However, SBR has better abrasion resistance and the moulds should stay cleaner because there is less mould fouling with compounds based on this polymer; the ageing resistance of the mouldings is also slightly better.

The synthetic polymer can also be made in very high molecular weight grades (much higher than those found in NR) and these grades can then be extended with oil. Oil-extended SBR is relatively cheap and this technique helps to offset the cost of expensive reinforcing fillers (e.g. carbon black).

2.2.2 Butyl rubber

Butyl rubber is a copolymer of isobutylene and isoprene and because of this it is referred to as IIR. A copolymer is made because polyisobutylene will not vulcanise on its own; the amount of isoprene is relatively small (e.g. up to 3 %) and, for example, by varying the amount of this monomer, one can produce a range of grades. Increasing the isoprene content decreases the resistance to ageing but increases the rate of cure.

The isoprene is added in order to make the final polymer capable of being vulcanised by sulphur but it is found that if butyl is mixed with other unsaturated rubbers, then the sulphur will react preferentially with the added rubber and thus upset the cure reactions. For this reason, butyl rubber is incompatible with many other rubbers, e.g. NR, SBR, NBR and CR.

Butyl rubber is noted for its high impermeability to gases, its excellent ageing resistance and good chemical resistance. Despite its chemical resistance, butyl rubber is not classified as a solvent-resistant polymer, although it is better than NR and SBR in this respect. The material is also noted for being relatively slow curing and mouldings produced from this polymer exhibit low resilience (e.g. at room temperature a butyl ball will not rebound very far when dropped).

2.2.3 Butadiene rubber

This polymer is also known as *cis*-polybutadiene (BR) and it is the only rubber which has better resilience than natural rubber. Like other butadiene polymers, it has poor gum strengths (see SBR) and so it is essential to compound this polymer with reinforcing fillers. By using reinforcing blacks it is possible to produce mouldings which have excellent abrasion resistance. Another noteworthy feature of this polymer is its ability to accept large quantities of oil and black—such additions help, of course, to reduce compound costs. Both BR and SBR resist breakdown by mechanical shear. Another peculiarity of BR is that, because of its chemical structure, it requires less sulphur than other diene rubbers to achieve an optimum crosslink density. Because of this material's reasonable resistance to ozone and its good oxidation and chemical stability, it has been adopted as a replacement for natural rubber for some applications, e.g. engine mounts.

2.2.4 Polyisoprene rubber (IR)

This material, like natural rubber, is a polymer of isoprene and in order to differentiate it from that material it is sometimes known as synthetic polyisoprene. Like NR, polyisoprene has proved to be the only synthetic rubber which exhibits self-reinforcement at high elongations. However, in order to show these properties, the material must be made so that its chemical structure is very regular, i.e. it has a high *cis* content (e.g. 96 %). Both straight and oil-extended varieties are available and by varying parameters such as *cis* content, molecular weight, etc., it is possible to produce a very wide range of grades. It would be expected that lowering the molecular weight and/or the *cis* content would produce grades which exhibit improved flow during injection moulding.

Despite its chemical similarity to natural rubber, the material does exhibit differences. For example, it cures at a slower rate than natural rubber and lightly filled compounds have poorer hot tear strength. In general, processing is possible at lower temperatures and it is usually found that there is less heat generation when fillers are mixed with this material. As IR is compatible with NR in all proportions, it can be usefully used in blends with NR where scorch (premature vulcanisation) is a problem.

2.2.5 Ethylene propylene rubbers

These materials are sometimes referred to as polyolefin rubbers and are produced in two main types; the saturated copolymers (which are referred to as EPM or EPR) and the terpolymers (which are referred to as EPDM). Where one does not wish to differentiate between the two types of rubber, then occasionally the abbreviation EP(D)M is used.

The copolymer can only be vulcanised with an organic peroxide but the terpolymer can be vulcanised with sulphur.

As these polymers have very poor gum strengths, it is necessary to compound them with reinforcing fillers. In some ways these materials resemble butyl rubber, i.e. they have very good chemical stability and are noted for their good ageing resistance.

They also have a high resistance to breakdown during mechanical operations which generate shear. However, they are easier processing than butyl rubbers and have better resilience at ordinary temperatures. A major application for both types of ethylene–propylene rubbers is in blends with thermoplastics materials (such as polypropylene); such blends are widely used to produce items such as car bumpers.

3 OIL-RESISTANT RUBBERS

3.1 Names, Abbreviations and Suppliers

3.1.1 Nitrile rubber

1 Common name: Nitrile rubber or acrylonitrile butadiene rubber or nitrile butadiene rubber.

2 Abbreviation: NBR.

3 Systematic chemical name: Not known. The following has been suggested: poly(1-butenylene-*co*-1-cyanoethylene).

4 Some suppliers:	*5 Trade names or trade marks:*
Anic	Europrene
Bayer	Perbunan
Goodrich	Hycar
Goodyear	Chemigum
ISR	ISR Nitrile
Polysar	Krynac
SIR	Sirban

3.1.2 Neoprene

1 Common name: Neoprene or polychloroprene.

2 Abbreviation: CR.

3 Systematic chemical name: Poly(1-chloro-1-butenylene).

4 Some suppliers:	*5 Trade names or trade marks:*
Bayer	Baypren
Denki Kagaku	Denka
Distugil	Butachlor
Du Pont	Neoprene
Toyo Soda	Skyprene

3.2 Material Properties

3.2.1 Acrylonitrile-butadiene rubber (nitrile rubber)

An oil-resistant rubber can be made by copolymerising butadiene with acrylonitrile, the material is known as acrylonitrile-butadiene rubber, nitrile rubber or NBR. By varying the ratio of acrylonitrile to butadiene, it is possible to produce a range of grades in which improved oil resistance is given by those grades which have the highest acrylonitrile content (referred to as nitrile content). Unfortunately, high acrylonitrile content usually means more difficult processing and the finished mouldings are harder and exhibit lower resilience.

It is also found that as the proportion of acrylonitrile is increased, the low-temperature flexibility of the resultant polymer is decreased progressively. This means that a compromise between oil resistance and low-temperature properties is necessary.

Unfilled vulcanisates have very low tensile strengths and it is therefore necessary to compound the material with a suitable reinforcing filler (e.g. a carbon black of a suitable type).

In general, NBR is compounded along lines similar to those practised with the other copolymer of butadiene, SBR. Like SBR, it is found that there is only a small change in plasticity with mastication and the polymer does not crystallise on stretching. It has a slower rate of cure (that cannot be so easily varied by the choice of accelerator) and mouldings exhibit better ageing than SBR. The material has good resistance to aliphatic-type oils and solvents and exhibits reasonable resistance to aromatic-type oils and solvents. (The oil resistance of a high nitrile NBR (45%) would be better than a low nitrile NBR (20%).)

Interesting properties are obtainable from blends of NBR and PVC; the major attraction of these blends is that they can be processed just like thermoplastics and the mouldings have good abrasion and oil resistance. Such mouldings do not suffer from plasticiser loss in the same way that plasticised PVC does, i.e. PVC plasticised with more conventional liquid plasticisers.

3.2.2 Polychloroprene

This material (CR) was the world's first commercial synthetic rubber and is probably better known as neoprene. Mouldings made from this material are highly resistant to oxidation and can be compounded (e.g. with carbon black and antioxidants) so that they exhibit good long-term outdoor weathering properties. The material is also resistant to ozone cracking and CR mouldings have a high level of resistance to flex-cracking. Mouldings also exhibit good resistance to aliphatic (paraffinic) oils and solvents and, because the material contains a large amount of chlorine in its structure, it has very good flame resistance.

However, mouldings made from this material lose their flexibility as the temperature is lowered and it is also found that the raw polymer, and compounds based on this polymer, harden during storage due to crystallisation.

The polymer can be softened by mastication but the rate of change is slower than that found with natural rubber. The material has a tendency to stick to metal surfaces (this is characteristic of chlorinated rubbers) and in general compounds based on this polymer tend to be scorchy during processing.

4 SPECIAL PURPOSE (SP) RUBBERS

Many types of special purpose rubber are possible but three have been chosen to illustrate the range of materials available.

4.1 Names, Abbreviations and Suppliers

4.1.1 Thiokol rubbers

1 Common name: Thioplasts or thiokol rubbers.

2 Abbreviation: T.

3 Systematic chemical name: Not known.

4 Some suppliers:	*5 Trade names or trade marks:*
Thiokol Corp.	Thiokol

4.1.2 Silicones

1 Common name: Silicones.

2 Abbreviations: Depends on type—all contain 'Q' (which identifies the material as a silcone) and other letters. These other letters identify side group substituents, e.g. MQ, rubbers based on polydimethyl siloxanes; VMQ, general purpose rubbers which contain vinyl groups; PMQ/PVMQ, materials which contain phenyl groups; FVMQ, rubbers which also contain the element fluorine.

3 Systematic chemical name: Not known.

4 Some suppliers:	*5 Trade names or trade marks:*
Dow Corning	Silastic
ICI	Silcoset

4.1.3 Urethanes

1 Common name: Urethane.

2 Abbreviation: AU or EU.

3 Systematic chemical name: Not known.

4 Some suppliers:	*5 Trade names or trade marks:*
Bayer	Urepan
Du Pont	Adiprene
Thiokol	Elastothane

4.2 Material Properties

4.2.1 Thioplasts or thiokol rubbers (T)

In these materials the polymer chains consist of organic sections and sections which contain more than one sulphur atom—these groups are referred to as polysulphide and this structure confers some unusual properties on the final mouldings. The materials have excellent resistance to oils and solvents and in this respect they are better than nitrile rubber and neoprene. The materials also have good chemical stability, resist degradation by oxygen and ozone, and are fairly impermeable to gases (in this respect they are better than most other rubbers but they are not as good as butyl rubber). However, the materials are not very strong and have poor resilience; the feature which most people remember them for is their disagreeable odour.

4.2.2 Silicones (e.g. MQ)

In this class of materials the main polymer chains consist of alternating atoms of silicon and oxygen; this structure is responsible for the thermal stability of the resultant polymers. The outstanding characteristic of this type of rubber is the wide temperature range of use, e.g. from −100 to +300 °C. Compounds based on silicone polymers do not (i) appear to be affected by atmospheric exposure, (ii) show ozone cracking and (iii) taste or smell. They are chemically inert and have excellent electrical insulation properties which are maintained over a wide temperature range. However, the materials do not have a good oil and solvent resistance and high shrinkage occurs when the materials are set or vulcanised. Their poor strength at room temperature is offset by the fact that they maintain this strength up to very high temperatures. For good resistance to swelling, caused by solvents or fuel, specify a fluorosilicone material in preference to other types.

Reinforcing fillers that are used with silicone rubbers are not based on carbon blacks but are usually based on silica which may or may not have been treated (e.g. in order to improve the storage life of the mixed compound).

Traditionally silicone rubber is supplied already compounded; now a range of base rubbers and modifiers are available so that in-house compounding is easier. Silicones are usually cured with peroxides and the resultant decomposition products may need to be removed by postcuring, e.g. by heating the mouldings in an air-circulating oven at 200 °C for 12 h. Such a treatment will often improve physical properties, e.g. compression set.

Liquid silicone rubbers (LSR) are available which are very easy flowing and yet they have similar properties to conventional peroxide-cured silicones. They are supplied as two component parts which when blended will cure in a few seconds (for small parts) at high mould temperatures, e.g. 225 °C.

The components are pumped through a static mixing device and then the mix is fed to a modified moulding machine, for example, dynamic seals must be fitted to the machine so as to contain the liquid material. Low injection pressure and clamping forces are desirable characteristics of such machines (Cush and Winnan, 1981).

4.2.3 Urethanes

This group of rubbers is based on the reaction of polyols with diisocyanates; if polyols containing ether groups are used then the product will be a polyether polyurethane (EU type) and if the polyol used contains ester groups then the product will be a polyester polyurethane (AU type).

By varying the molecular weight it is possible to produce materials which range from liquids to solids. If the molecular weight is low then the liquid can be chain extended and crosslinked by the addition of liquid polyols or diamines; such liquids may be cast into a mould and allowed to set. Very high strengths are possible, without the addition of reinforcing fillers, even for quite soft materials. Higher molecular weight polymers are soft gums which can be mixed and moulded on conventional rubber processing equipment; the crosslinking agents may be diamines, sulphur, etc. Such materials seem to possess the disadvantages of 'traditional' rubbers without any of the major advantages (e.g. cost) as they must be vulcanised and they are expensive. It is also possible to synthesise elastomers which are composed of high molecular weight linear molecules which remain thermoplastic on moulding and which can be reprocessed if required. Such materials are referred to as thermoplastic rubbers or thermoplastic elastomers and they are available in a very wide range of grades.

Polyurethane components are noted for their good ageing, resistance to oils and solvents, their high strength and excellent abrasion resistance. However, their acid and alkaline resistance is not outstanding and the compression set resistance of the thermoplastic materials is poor.

MATERIAL RECOMMENDATIONS (6–31) FOR ELASTOMERS

6 *Material properties:* See Sections 2.2, 3.2 and 4.2.

7 *Ease of flow:* Natural rubber is a stiffer flow material at equal filler

loadings than the synthetic rubbers and its viscosity is more temperature sensitive. If a compound's viscosity is too high then reformulate (make it safer) so that it can be run at a higher temperature, i.e. do not try and use oil extenders, etc. as these will result in lower melt temperatures.

8 Shrinkage: Because of the higher mould temperatures employed, shrinkage may be higher than that found in compression moulding. It will probably be different because of the more complex flow patterns found in injection moulds, i.e. it will be nonuniform. If this is excessive then it may be reduced by using a slow curing compound and/or a lower mould temperature; this should give the material more time to relax or randomise. Allowances should be made in mould manufacture for final sizing of cavities and pins to be made, i.e. after initial moulding and measurement.

9 Resistant to: See Table 28.

10 Not resistant to: See Table 28.

11 Material detection or identification: Simple tests are not so useful for rubber materials as they are for plastics. Most rubber compounds are black and for many applications a compound with the desired properties is obtained by using a blend of elastomers. Elemental analysis (i.e. to determine what elements are present) and/or infra-red spectroscopy (i.e. so as to determine what organic groups are present) is usually necessary if base polymer identification is to be successful.

12 Colouring: Although coloured compounds are possible nearly all rubber components are black; this is because of the reinforcement possible with carbon black. Reinforcing fillers, such as carbon black, cause greater temperature rises than nonreinforcing fillers during plasticisation (thermal softening) and injection.

13 Material and component handling: Compounded material is usually fed to the machine (prepared either using an internal mixer or a continuous compounder) in the form of a strip. The technique of feeding the material in the form of a powder dry blend (similar to PVC technology) is being investigated as it can give good results and saves on compounding costs.

Components are usually removed from the mould by hand, e.g. because of the flexibility of the runners and of the moulding and because of part complexity.

Design the compound for scorch safety and ensure that the mixing procedure is uniform as mixing time and temperature influence compound flow.

14 Mould and gate considerations: The mould must be designed to take account of the material's characteristics and the operating procedures employed, e.g. high operating temperatures, the fluidity of many rubber

TABLE 28

POLYMER SELECTION GUIDE (COURTESY OF DU PONT)

Properties (Chemical name)	*Natural rubber (polyisoprene)*	*SBR or GR-S or Buna S (butadiene styrene)*	*Butyl (isobutylene isoprene)*	*Nitrile (Buna N) (butadiene acrylonitrile)*	*Neoprene (chloroprene)*	*Silicone (polysiloxane polymer)*
Material designation (ASTM D-2000 SAE J200 classification)	AA	AA	AA	BF, BG, BK, CH	BC, BE	GE
Tensile strength (MPa)						
Pure gum	Over 20·67	Below 6·89	Over 10·34	Below 6·89	Over 20·67	Below 10·34
Black loaded stocks	Over 20·67	Over 13·78	Over 13·78	Over 13·78	Over 20·67	Over 10·34
Hardness range (Durom A)	30–90	40–90	40–75	40–95	40–95	40–85
Specific gravity (base material)	0·93	0·94	0·92	1·00	1·23	1·14–2·05
Adhesion to metals	Excellent	Excellent	Good	Excellent	Excellent	Excellent
Adhesion to fabrics	Excellent	Good	Good	Good	Excellent	Excellent
Tear resistance	Very good	Fair	Good	Fair	Good	Poor
Abrasion resistance	Excellent	Good to excellent	Good	Good	Excellent	Poor
Compression set	Good	Good	Fair	Good	Fair to good	Fair
Rebound						
Cold	Excellent	Good	Poor	Good	Good	Excellent
Hot	Excellent	Good	Very good	Good	Very good	Excellent
Dielectric strength	Excellent	Excellent	Excellent	Poor	Very good	Good
Electrical insulation	Good to excellent	Good to excellent	Good to excellent	Poor	Fair to good	Excellent
Permeability to gases	Fairly low	Fairly low	Very low	Low	Low	Fairly low
Acid resistance						
Dilute	Fair to good	Fair to good	Excellent	Good	Excellent	Excellent
Concentrated	Fair to good	Fair to good	Good	Good	Good	Fair
Solvent resistance						
Aliphatic hydrocarbons	Poor	Poor	Poor	Excellent	Fair to good	Poor
Aromatic hydrocarbons	Poor	Poor	Poor	Good	Fair	Poor
Oxygenated (ketones, etc.)	Good	Good	Good	Poor	Poor to fair	Fair
Lacquer solvents	Poor	Poor	Fair to good	Fair	Poor	Poor
Resistance to:						
Swelling in lubricating oil	Poor	Poor	Poor	Very good	Good	Fair
Oil and gasoline	Poor	Poor	Poor	Excellent	Good	Fair
Animal and vegetable oils	Poor to good	Poor to good	Very good	Very good	Good	Good to excellent
Water absorption	Very good	Good to very good	Very good	Good	Good	Excellent
Oxidation	Good	Fair	Excellent	Good	Very good	Excellent
Ozone	Poor	Poor	Excellent	Fair	Very good	Excellent
Sunlight ageing	Poor	Poor	Very good	Poor	Very good	Excellent
Heat ageing	Fair	Fair to good	Very good	Good	Good	Outstanding
Low temperature	Very good	Very good	Good	Fair to good	Good	Outstanding
Flame	Poor	Poor	Poor	Poor	Good	Fair to good

TABLE 2[illegible]—Continued

Properties (Chemical name)	Hypalon® (chlorosulphonated polyethylene)	Nordel® (EPDM rubber) (ethylene propylene polymer)	Epichloro-hydrin rubber	Chlorinated polyethylene	Viton® (fluoroelastomer)	Adiprene® (urethane rubber)	Hytrel® (polyester elastomer)
Material designation (ASTM D-2000 SAE J200 Classification)	CE	CA	CH	BC, BE, CE	HK	BG	Not applicable
Tensile strength (MPa)							
Pure gum	Over 10·34	Below 6·89	Below 6·89	Over 10·34	8·27	20·67–55·12	25·49–39·27
Black loaded stocks	Over 17·23	Over 13·78	17·23	Over 17·23	10·34–20·67	20·67–55·12	25·49–39·27
Hardness range (Durom A)	40–95	30–90	40–90	60–90	50–90	80A–75D	92A–72D
Specific gravity (base material)	1·12	0·86	1·36–1·37	1·16–1·32	1·8	1·02–1·20	1·17–1·25
Adhesion to metals	Excellent	Fair	Fair to good	Fair to good	Fair to good	Excellent	Good
Adhesion to fabrics	Good	Good	Fair to good	Fair	Good	Very good	Good
Tear resistance	Fair	Fair	Fair to good	Fair	Fair to good	Excellent	Outstanding
Abrasion resistance	Excellent	Good to excellent	Fair to good	Good	Good	Outstanding	Excellent
Compression set	Fair	Good	Poor	Good	Excellent	Fair	Fair
Rebound							
Cold	Fair	Excellent	Good	Fair	Fair	Good	Good
Hot	Good	Excellent	Good	Good	Good	Good	Good
Dielectric strength	Very good	Outstanding	Good	Excellent	Very good	Excellent	Good
Electrical insulation	Good	Outstanding	Good	Good	Good	Good	Fair to good
Permeability to gases	Low to very low	Fairly low	Low to fairly low	Very low	Very low	Fairly low	Very low
Acid resistance							
Dilute	Excellent	Excellent	Fair to good	Excellent	Good to excellent	Fair	Good
Concentrated	Good	Good	Fair	Good	Excellent	Poor	Poor
Solvent resistance							
Aliphatic hydrocarbons	Fair to good	Poor	Excellent	Good	Excellent	Fair to good	Excellent
Aromatic hydrocarbons	Fair	Poor	Good	Fair	Excellent	Poor	Good
Oxygenated (ketones, etc.)	Poor to fair	Excellent	Poor	Fair	Poor	Poor	Good
Lacquer solvents	Poor	Fair to good	Fair	Fair	Poor to fair	Poor	Good
Resistance to:							
Swelling in lubricating oil	Good	Poor	Excellent	Good	Excellent	Good	Very good
Oil and gasoline	Good	Poor	Excellent	Good	Excellent	Good	Very good
Animal and vegetable oils	Good	Good to excellent	Excellent	Good	Excellent	Good	Good
Water absorption	Good	Excellent	Good	Very good	Very good	Very good	Very good
Oxidation	Excellent	Excellent	Good	Excellent	Outstanding	Excellent	Excellent
Ozone	Outstanding	Outstanding	Excellent	Excellent	Outstanding	Excellent	Excellent
Sunlight ageing	Outstanding	Outstanding	Good	Outstanding	Outstanding	Very good	Very good
Heat ageing	Very good	Excellent	Very good	Very good	Outstanding	Fair to good	Very good
Low temperature	Good	Excellent	Good to very good	Good	Fair to good	Excellent	Outstanding
Flame	Good	Poor to good	Poor to fair	Good	Excellent	Fair to good	Fair to good

compounds, the pressure absorbing nature of the material, the tendency to evolve fumes at high temperatures, etc. The steel used to make the mould must be of very good quality (to withstand the pressures, erosion, corrosion, etc.) and types used include prehardened, hardened and stainless. Chrome plating has been used but may be eroded by some fillers. The steel must machine well and take a good polish, e.g. runners polished to a 16–20 μ-in finish. A well balanced runner system must be used and the use of long narrow runners, excessively restricted gates and abrupt changes in direction are to be avoided as these increase the risk of scorch and lengthen mould filling times. Good mould venting is essential; ejection is usually done using pins but 'air assist' is also used. The ejector pins should be hardened and may be of the valve type to prevent flash penetration.

15 Flow path: wall thickness ratio: The mould is hotter than the compound and this means that long flow paths into thin sections are possible. This means that moulds must be well constructed; some compounds can penetrate 5 μm cracks. To ensure consistent flow properties the rubber must be of consistent quality and must be available in a narrow, specified viscosity range. Viscosity and scorch measurements should be made on each batch of compound before any attempt is made to mould that batch.

16 Projected area: The clamping pressures required per square inch of projected area are relatively high (compared to compression moulding) and are of the order of 2 tsi (30 MN m^{-2}); the figure used depends on the ease of flow of the component, depth of draw of the component, the amount of flash that can be tolerated, etc.

17 Cylinder equipment: To stop the injection pressure being transmitted to the screw's thrust bearings it is important that there is a back-flow valve between the plunger and the screw. The nozzle diameter is very important and should match a particular compound and mould—it is usually about 3 mm. As the diameter is decreased, heat build-up and dispersion (e.g. of filler and sulphur) increases. If the compound temperature can be built-up in the nozzle then the compound temperature can be reduced so as to decrease scorch tendency. If the nozzle diameter is too small then it will restrict mould filling and cause scorch; if this happens increase the diameter, e.g. in 0·5 mm steps, until scorch is eliminated.

18 Plunger cushion: Usually as small as possible (e.g. 1–2 mm) so that barrel residence time is kept as low as possible.

19 Shot capacity: As little as 5 % of the machine's rated capacity may be run, although it is better to match the injection unit to the mould if at all possible.

20 Melt temperature: As high as possible consistent with safe (e.g. scorch-free) working (e.g. 100 °C). Higher temperatures may be employed if a screw delay facility is available. If there are signs of scorch at a high temperature (e.g. specks of overheated material in the moulding, slow or hesitant screw and/or plunger movements, etc.), reduce barrel temperatures in 5 °C steps and/or reduce the screw speed until the signs of scorch disappear.

21 Barrel residence time: All compounds will cure in the barrels if left to stand for too long—as the temperature increases the barrel residence time decreases. If there is any break in moulding drop the temperatures and purge the barrels clean.

22 Temperature settings: See Table 29. The mould temperature should be hot enough to give the desired speed of vulcanisation but not so hot as to cause scorch, reversion and/or nonuniformly cured mouldings. The mould is usually heated by electrical resistance elements—these may be located in the mould itself or in the machine platens. Both the injection unit and the plasticising unit have a heating/cooling system. Around each barrel zone there is an outer jacket through which oil can be circulated; resistance elements (cuff heaters) are clamped around this outer jacket. If the temperature exceeds what is desired or if the machine needs to be crash-cooled then the oil will remove the excess heat. The liaison block (which contains the nozzle and links the injection unit to the extruder unit) is also oil heated.

TABLE 29

Location	*Temperatures °C*		*Temperatures °F*	
	From	*To*	*From*	*To*
Plasticising unit	80	110	176	230
Injection unit	90	125	194	257
Nozzle (liaison block)	90	120	194	248
Mould	170	200	338	392

23 Injection speed—mould filling speed: Use a safe (i.e. scorch resistant) mix, adjust the machine controls (injection speed, back pressure, etc.) so as to obtain the highest possible injection temperature and mould filling speed consistent with freedom from scorch. The temperature of the material may be raised by up to 40 °C if high speeds and a small nozzle are employed.

24 Injection pressure: The machine should be capable of giving the following: First stage: up to 1500 bars; 150 MN m^{-2}; 21 400 psi. If possible do not use a compound which demands the maximum pressure as any small variation will make moulding difficult and may demand compound changes.

Second stage (dwell or follow-up pressure): up to 1200 bars; 120 MN m^{-2}; 17 400 psi.

25 Screw rotational speed (rpm): Screw speeds of up to 200 rpm are employed—because of the low temperatures set on the 'extruder' barrel (up to 120 °C), the use of a high screw speed (e.g. 150 rpm) provides a convenient way of generating heat.

26 Back pressure: Up to 250 bars; 25 MN m^{-2}; 3620 psi. Back pressure is used to generate a temperature rise and is also used to exclude air from the compound.

27 Shutting down: If the stoppage is going to be a long one then cool the barrel as quickly as possible (crash cool) and purge out; if the stoppage is only going to be for a short time then reduce the temperature settings (each 10 °C drop will double the resistance to scorching) by about 20 °C.

28 Reprocessing: For noncritical applications some companies regrind the waste and compound the cured, granulated material in with another batch. Such reprocessing is not normally employed, however, as either the compound properties suffer or regranulation is difficult and expensive. The use of cold runner moulds can help to reduce the amount of scrap produced (manifold temperature 125 °C).

29 Finishing: Because the material is soft, trimming is relatively easy and many gates may be removed by simply 'twisting'. With some mouldings, it is difficult to avoid the formation of flash and even though this is much less than that produced by compression moulding, it still must be removed, e.g. by knife or scalpel. Because it is thin and flexible it may present problems if it is removed by barrelling—this is why many moulders reduce the temperature during barrelling (cryogenic tumbling).

30 Other comments: The scorch resistance (premature vulcanisation) of a mix is controlled mainly by the accelerator system used; more than one accelerator may be used to improve speed of cure. If a mix is too scorchy, a retarder may be used.

31 Typical components: Both mouldings and rubber-to-metal bonded products are produced by injection moulding. The mouldings are made in this way for a number of reasons, e.g. because they are cheaper, little or no finishing is necessary, etc. Rubber-to-metal products benefit because fresh, clean rubber is presented to the metal surface and this results in a stronger, more consistent product (e.g. a typical product would be an engine mount).

The automotive industry uses rubber injection moulding to produce windscreen wiper blades, seals, bushes, engine mounts, tyre valves, etc. In the pharmaceutical industry this process is used to produce bottle stoppers, teats, pipettes, etc.

The shoe industry uses injection machines to produce shoe sole units and to mould the shoe soles on to the uppers. Another large market is the manufacture of seals, e.g. for oil and water pipes. In short it is difficult to find an industry that does not use injection moulded rubber components.

REFERENCES

Cush, R. J. and Winnan, H. W. Silicone rubbers, *Developments in Rubber Technology—2*. Eds Whelan, A. and Lee, K. S. Applied Science Publishers Ltd, London, p. 203, 1981.

APPENDIX I

Euromap Documents Concerned with Injection Moulding

Euromap is the European committee of machinery manufacturers for the plastics and rubber industries. Its membership is drawn from representatives of associations, which cover machinery manufacturers, in Belgium, West Germany, France, Italy, The Netherlands, Spain, Switzerland and the UK.

Euromap's technical committee has compiled a series of recommendations relating to special features of primary processing equipment. These recommendations are designed to standardise and simplify the wealth of technical data for the benefit of machinery manufacturers and processors.

At present the documents cover injection moulding machines, extruders and blow moulding machines. In the UK such Euromap documents are available from The British Plastics Federation, 5 Belgrave Square, London, SW1X 8PH.

Euromap 1: Description of Injection Moulding Machines

Provides guidelines on technical information to be included in a description of an injection moulding machine and draws attention to the details that should be taken into account when evaluating a machine.

Euromap 2: Mould Fixing Dimension for Injection Moulding Machines

Specific information relating to mould mounting is provided on fixing holes and slots, symbols, dimensions and correlation of threads or slots to the lines.

Euromap 3: Injection Moulding Machine Clamping Units, Essential Dimensions

Descriptions of injection moulding machines should contain a sketch of the clamping unit showing all important dimensions. This recommendation lists the minimum number of dimensions.

Euromap 4: Determination of the Available Injection Power in Injection Moulding Machines

This recommendation provides conditions for measuring, according to a uniform method, the available injection power in injection moulding machines hydraulically driven by means of a piston or a screw and obtaining comparable results.

Euromap 5: Procedures for the Determination of Important Production Data for an Injection Moulding Machine

Procedures are described for the determination of plasticising flow, rate of injection and specific energy consumption, by means of any injection moulded part. Conditions of manufacture must be clearly set out in the manufacturing report and this ensures that results can be reproduced.

Euromap 6: Determination of the Duration of the No-load Cycle of Injection Moulding Machines

This recommendation lays down requirements for determining, by a uniform method, the duration of the no-load cycle of injection moulding machines and for obtaining comparable results.

Euromap 7: Determination of the Maximum Clamping Force of Injection Moulding Machines

Two methods are set out for determining uniformly the clamping force of injection moulding machines and for obtaining comparable results.

Euromap 8: Determination of the Nominal Opening Force of Injection Moulding Machines

The recommendation establishes the requirement that the opening force of injection moulding machines be determined by clearly defined methods.

Euromap 9: Testing the Parallelism of the Clamping Plates of Injection Moulding Machines

A uniform method for determining deviations from the parallel of the clamping plates of injection moulding machines so that comparable results can be obtained.

Euromap 10: The Testing of Injection Moulding Machines—General Rules for Testing

These rules, which are applicable to injection moulding machines processing both plastics and rubber, establish a uniform basis for preparing and carrying out the tests and include directions relating to the conditions under which the tests are to be performed. The appendices contain a specimen test report and declaration of acceptance.

APPENDIX II

Troubleshooting—An Instant Checklist of Likely Remedies for the Most Common Faults on Injection Machines

POSSIBLE REMEDIES / POSSIBLE FAULTS	Material dribbles out between sprue and nozzle	The mould has not been completely filled	Screw does not return	Sink marks in moulded parts	Burning in the moulded parts	Moulded parts show matt areas	Flash on moulded parts	Inferior surface finish (gloss)	Flaking	Moulded parts sticking in cavities	Tearing off of the sprue	Warping of the moulded parts	Discolouration at the sprue	Flow seams	Brittleness of the moulded parts	Wavy condition of the surface	Worm lines in the moulded part	Material temperature too high (burning)	Streak formations and marks in moulded parts	Voids, ie air bubbles in moulded parts
Increase injection pressure		●												●		●				
Decrease injection pressure							●					●								
Increase cylinder temperature		●				●		●	●			●		●	●	●	●		●	
Decrease cylinder temperature					●		●			●			●			●		●	●	●
Increase holding pressure time				●												●				●
Decrease holding pressure time							●			●	●	●			●			●		
Increase nozzle temperature		●		●		●			●								●			●
Nozzle seal or nozzle blocked		●																		
Nozzle seal is jammed	●	●																		
Increase screw rotation speed			●			●		●	●						●		●			
Decrease screw rotation speed																		●	●	
Nozzle seal or nozzle loose												●							●	
Wrong type of nozzle seal		●										●							●	
Inject with rotating screw		●				●		●	●			●		●	●					
Increase moulding locking force							●													
Injection pressure starts too early	●																			
Decrease injection speed				●	●	●	●	●	●	●		●	●		●	●	●		●	●

Increase injection speed		●												●		●				
Increase screw back pressure		●				●		●	●			●		●	●		●		●	
Decrease screw back pressure			●	●														●		
Increase nozzle orifice		●		●	●			●	●				●	●		●	●			
Increase mould temperature		●				●		●	●	●		●	●	●	●					●
Decrease mould temperature				●		●	●	●		●	●	●			●	●		●		●
Polish mould and round off edges						●		●		●										
Mould requires maintenance							●			●										
Polish sprues, runners and channels						●					●									
Sprues and runners should be increased		●		●	●	●		●	●				●	●	●		●			●
Introduce mould venting		●			●									●						
Introduce cold slug pocket						●							●		●		●			
Pre-dry material			●			●		●	●					●	●		●		●	
Material is contaminated								●	●											
Hopper is empty or throat blocked			●																	
Metering is insufficient		●		●										●		●				●
Use mould release agent										●										
Check on nozzle setting	●																			
Check on nozzle/sprue bush radiuses	●									●	●									
Temperature too high; increase nozzle contact time	●																			
Decrease cylinder temp at intake zone (except nylon)			●																	
Bridging in hopper/vent attachment			●																●	
Check for equal filling (multi-cavity tools)				●																
Introduce air ejector										●										
Increase cooling and pause times											●	●				●			●	
Decrease cooling and pause times																		●		

Index

Aachen mixing system, 66–7
Accelerators, 22
Accessibility, 106
Accidents, 102
Accumulators, 87, 94
Acetals, 316–19
Acrylic, 264–7
Acrylonitrile-butadiene rubber, 372
Acrylonitrile-butadiene styrene (ABS), 260–4
  electroplating, 122
  electroplating tests, 123–5
Activation, 126
Activators, 22
Additives, 21, 222
Adhesive(s)
  bonding, 158, 160–6
    design considerations, 165
  cyanoacrylate, 164–5
  design considerations, 165
  epoxide, 164
  hot-melt, 163
  pressure-sensitive, 161
  reactive, 163
Adiabatic machines, 67–8
Adjustable length scotch, 106
Air shot technique, 229
Aminoplastics, 354–9
Amorphous materials, 169
Anisotropy, 180, 183
Annealing, 177–9
ASTM D 647, 224
ASTM D 1130-63, 223
ASTM D 1897-72, 223
Auger granulation, 12
Automatic systems, 249

Back pressure, 227, 228
Barrelling, 115–16
Base coat applications, 131–2
Birefringence, 180–2
Blending, 16, 25, 27
Blowing agents, 22
Bolts, 154
Bonding cements, 159
BPF Code of Practice, 102–7
British Standard 5378, 104
Buffing, 117
Burns, 102
Butadiene rubber, 368, 370
Butyl rubber, 368, 370

Cavity pressure control (CPC), 236
Celluloid Corporation, 45
Cellulosics, 275-80
Charpy impact testing, 209, 211
Clamp speeds, 92
Clamp stroke, 92
Clamping, 91, 118
  force, 91
  pressure, 227
  unit, 91–4
Cleaning for electroplating, 125
Codes of Practice, 102–7, 223–4
Colourants, 22

Colouring systems, 32–5
  colour additions, 35
  costs, 34
  dry colouring, 32
  liquid colouring, 33
  masterbatch colouring, 33
Component finishing and decoration, 114–67
Component testing, 240–4
Constant load testing, 219
Contamination
  hydraulic system, of, 96
  regrind, of, 13
Control location, 97–8
Control system, 96–7
Coolers, 96
Cooling
  capacities, 116
  cycle, 179
  rate, 178, 185
  time, 226, 232, 233
Core-pulling systems, 101
Correlation coefficients, 186, 187
Costs
  acquisition and support, 16
  colouring systems, 34
  materials, 17
  reclaiming, 18
Counter, operating hours, 100
Cracks, 166, 195
Creep curves, 221
Creep tests, 219
Crosslinking systems, 23
Crystallinity, 169–71, 179, 184–5, 222
Crystallisation rate, 186
Cutting, 119
  tools. *See* Machining
Cyanoacrylate adhesives, 164–5
Cycle time, 230

Deflashing, 115–16
Deformation recovery, 118
Delivery, 112
Density effects, 178–9
Density measurement, 217–18
Differential shrinkage, 177
Dimensional stability, 168
Direct screw transfer, 75
Dough moulding compound, 363–6
Dry-offset printing, 140
Drying and dryers, 18–21
  hopper dryers, 19–21
  oven dryers, 18
  vented injection units, 21
Durometer hardness, 219
Dwell pressure, 236
Dwell time, 225, 231
Dynamic load application, 222

Ejection system, 11, 93–4, 100
Elastomers, 367–83
  material recommendations, 367–83
Electrical equipment, 102, 107
Electrical power outlets, 101
Electrically conductive mouldings, 143
Electroless copper plating, 127
Electroless metal deposition, 126
Electroless nickel plating, 127, 129
Electroplating, 122–9
  ABS, 122–5
  alternative systems, 127
  applications, 129
  metals used, 122
  process details, 125–9
  production of mouldings, 123
Environmental stress-cracking resistance (ESC), 171
Epoxide adhesives, 164
EPROM bit processor, 98
Ethylene propylene rubber, 369, 371
Euromap, 37, 85, 384–6
Experiment design, 198–203
Experimental card layout, 200
Experimental procedure, 201
Extrusion moulding, 69

Fail safe, 107
Fault finding, 98, 387
Feed separation, 12
Feed system, 9, 15, 71
Fibre reinforced thermoplastics, 344–9
Fillers, 23, 65
  fibrous, 174–5, 211

Filling speed, 191
First-stage pressure, 227
Fixed length scotch, 106
Flash removal, 114
Flexural strength, 209
Flow
  moulding, 73
  path, wall thickness ratio (FR), 217
  properties, 213–17
  tendency, 241
Fountain effect, 174
Freeze time, 232–3
Frozen-in-strain, 169
Frozen-in-stress, 169

Gate
  changing, 195
  freeze time, 232
  position, 193, 197
  size, 193, 197, 233
Glass fibres, 23, 175, 211
Glass transition temperature, 176, 186
Granulated material, 14
Granulating equipment, 10, 13
Guards, 103–6

Hand trimming, 114
Hardening, 160
Hardness, 218–19
Hazards, 104
Health and Safety at Work Act 1974, 103
Heat distortion temperature, 214
Heating system, 89
High density polyethylene, 297–301
High shear rheometer, 215
Horizontal machines, 59
Hot foil marking, 134–7
  application areas, 136
  application methods, 135
  carrier film or foil, 134
  foil composition, 134–5
  moulded-in foils, 136
Hydraulic components, 95
Hydraulic control, 95
Hydraulic ejectors, 100
Hydraulic equipment, 102
Hydraulic interlock, 106
Hydraulic line pressure, 187
Hydraulic system, 86, 94–6
  contamination of, 96
Hydraulic valves, 103
Hygroscopic materials, 18

Impact modifiers, 24
Impact testing, 170, 211, 239, 242–3
Injection
  blow moulding, 77–8
  capacity, 80–1
  line pressure, 95, 226
  moulding process, 1
  pressure, 87, 189, 226
  rate, 85–7, 191
  speed, 189
  thermoforming, 76–7
  time, 230
  transfer moulding, 76
  unit, 80–91
  velocity, 85–7
Injection–compression moulding, 74, 75
In-line screw machine, 50, 55, 73
  alternative arrangements, 59–61
  components, 58
  horizontal, 59–61
  lay-out, 58
  vertical, 59–61
Inspection, 105, 239–40, 248, 249
Instrumented falling weight test, 213
Insulated runner, 71
Intensifiers, 94
Interlocking principles, 106
Intrusion moulding, 73
Isochronous stress–strain curves, 221
Isometric stress–log time curves, 221
Isoprene rubber, 368
Izod impact strength, 211, 238–9
Izod impact tests, 170, 239

Joining, mechanical, 150–60

Labelling, 141
Laser cutting, 120
L:D ratio, 50, 81, 82
L:D screws, 83
Legislation, 102
Letterflex printing, 140
Level allocation, 200
Level of use, 14–15
Limit switch, 104, 106
Line pressure, 95, 226
Long-term testing, 219–22
Low density polyethylene, 292–6
Lubricants, 24

Machine
  construction, 105
  costs, 112
  installation, 107
  operation, 234
  purchase, 108–13
  purging, 233–4
  selection, 108
  settings, 203, 222, 225, 229
  specifications, 80–113
  survey, 112
  types, 37
Machining, 117–21
Maintenance, 111
Material(s), 1–36, 109
  blending, 16
  bulk containers or bins, 3
  bulk storage, 5
  compounding, 28–32
  considerations, 90
  costs, 17
  handling, 5–9
    automated systems, 9
    bulk containers, 3, 7
    sacks, 2, 5
    vacuum hoppers, 9
  hygroscopic, 18
  packed in sacks or bags, 2
  packed in tins or cans, 3
  properties, 117–18, 151, 369–76
  reclamation, 9–18
    automatic systems, 11

Material(s)—*contd.*
  reclamation—*contd.*
    imperfect mouldings, sprues and runners, 14
    manually operated system, 11
  recommendations, 251–383
    elastomers, 376–83
  temperature requirements, 5
  testing, 240–4
Maximum daylight, 93
Mechanical joining, 150–60
Mechanical properties, 168–70, 186–7
Mechanical restraints, 106
Melt
  compaction, 225
  flow index, 214
  temperature, 185, 186, 187, 189, 191, 200, 228, 229
Metal inserting, 145, 157–60
Metallisation, 121
Microprocessors, 97, 98, 249
Mixing, 21
  equipment types, 30
  high-speed mixers, 26
  hoppers, 28
  machine, on, 27
Moisture content, 241
Molecular orientation. *See* Orientation effects
Monitoring, 247
Moulded-in inserts, 157–60
Moulding
  conditions, 185–93, 198–203
  procedure report, 235
  techniques, 73–8
  trials, 109
Mould(s), 68–73, 110
  closed time, 226
  cold-runner, 71
  family, 72
  filling, 172–5, 225
  hot runner, 71
  insert, 224
  multicavity, 228
  open time, 226, 230
  opening force, 92
  runnerless, 71
  sprueless, 71

Mould(s)—*contd.*
stack, 72
thickness, 92
three-plate, 71
two-plate, 69–71
undercut, 72
unscrewing, 73
variable gate, 195
vertical flash, 72
Multibarrel machines, 63–4
Multiple-colour mouldings, 63
Multistation machines, 62–4, 74
Multistep machines, 236–8
Multistep setting, 236, 238

Natural rubber, 367, 369
Neoprene, 372
Nitrile rubber, 372
No-load cycle, 99
Normal distribution curve, 244–5
Notch sensitivity, 239
Nozzle contact force, 88
Nuts, 154
Nylon, 174, 309–16

Oil
cooling, 96
temperature regulation, 96, 100
Oil-resistant rubbers, 372–4
Orientation effects, 169, 176–7, 179–86, 195, 198
Orientation stress relaxation, 182
O-rings, 74
Oven heating, 242

Packing, 175–7
Painting, 141–3
curing or setting, 143
process details, 142
reasons for, 141–2
Permeability to gases and liquids, 171
Phenolics, 359–63
Photoelastic inspection, 242
Physical properties, 168, 182–3
'Piggy-back' machines, 46
Piston machines, 41
Plasticisation
early systems, 44
terminology, 44
Plasticised polyvinyl chloride, 272–5
Plasticisers, 24
Plasticising capacity, 82–5
Platens, 93
Plunger machines, 41
conflicting requirements, 42
development, 38
disadvantages, 41–3
injection unit requirements, 41
operating sequence, 38–41
Pneumatic conveying system, 14
Polishing, 117
Polybutylene terephthalate, 305–9
Polycarbonate, 280–4
Polychloroprene, 373
Polyethersulphone, 284–9
cis-Polyisoprene, 368
Polyisoprene rubber, 371
Polymer morphology, 169
Polymer selection guide, 378
Polyphenylene oxide (modified), 289–92
Polyphenylene sulphide, 323–7
Polypropylene, 187, 189–91, 301–5
Polystyrene, 186, 232, 251–4
Polyvinylidene fluoride, 319–23
Postmoulded inserts, 160
Power failure, 107
Preplasticisation
in-line systems, 51–64
two-stage, 44–51
attractions of, 49–51
operating sequence, 46
Pressure control, 95, 101
Pressure effects, 175–7, 183, 187, 189, 199, 226–7
Printing, 139–41
dry-offset, 140
letterflex, 140
silk-screen, 139
wood, 140
Processing conditions, 168–9
Production capability, 244–7
Progressive scotch, 106

Protective agents, 23
Pump drive, 94
Pump size, 87

Quality
  assurance, 243
  control, 240, 243, 247
  limits, 248
  requirements, 169, 203–4

Ram machines. *See* Plunger machines
Random sampling, 248–9
Range coefficient, 246–7
Reaction injection moulding, 78
Records, 238
Regrind material, 13–15
  contamination of, 13
Relaxation, 176–7, 182
Repeatability, 99
Replacement machines, 108
Rigidity, 171, 209
RIM process, 78
Risk involvement, 106
Riveting, 152–4
Rivets
  bifurcated, 153
  blind, 154
  tubular, 153
Robots, 107, 249
Rockwell hardness, 219
Rotor and ram system, 65
RSG process, 78
Rubber(s)
  injection moulding machine, 46
  reinforced polypropylene, 327–31
  rubber compounds, and, 1, 91
    general purpose, 367–71
    oil-resistant, 372–4
    preparation, 29–32
    special purpose, 374–6
Runner dimensions, 233

Safety checks, 105
Safety measures, 101–8
Sample gauging, 247
Sample production, 234
Sandwich moulding, 63–4
SBS block copolymers, 336–40
Screw
  advance, 225
  decompression, 89
  delay timer, 57
  forward time, 225
  recovery time, 226
  rotation, 88
  rotational speed, 199, 224, 228
  stroke, 82
Screw/plunger combinations, 46
Screwed assemblies, 154–60
Second-stage pressure, 227
Self-tapping screws, 155–7
Sensitisation, 126
Service, 110–12
Service engineers, 111
Setting operation, 105
Shear rate, 215
Shear stress, 215
Short-term testing, 207–19, 238
Shrinkage, 171, 176–7, 182, 187, 198, 218, 231
Shut-off nozzles, 61
Silicone mould release, 123
Silicones, 374
Silk-screen printing, 139
Simulated end-use testing, 243
Single-screw machine, 51–5, 59, 64–5
Single-stage machine, 65
Snap fits, 150
Society of the Plastics Industry Inc. (SPI) comparison test, 83, 229
Solvent
  cementing systems, 163
  systems, 161–3
  welding, 162
Spares availability, 110
Specification matching, 108–9
Speed controls, 95, 101
Speed nuts, 155
Spiral flow testing, 215–17
Spread clips, 155
Spreaders, 43–4
Spring clips, 155
Spruce break, 88

Staking, 146–7, 152
Standard deviation, 244
Standardisation, 207
Starting time, 225
Static electricity, 142
Statistical analysis, 202, 244
Stress cracking, 166
Stress relieving, 175–7
Stress/strain ratio, 209
Strip production, 30–2
Structural foam, 349–54
Styrene acrylonitrile, 254–7
Styrene-butadiene rubber, 368, 369
'Suck-back', 59
Surface
  conditioning, 126
  gloss, 171
  imperfections, 123

Television, 249
Temperature
  control, 89, 101, 228
  effects, 5, 89, 123, 201, 222, 228, 232, *see also* Melt temperature
  regulation, 96, 100
  settings, 228
Tensile creep modulus, 221
Tensile impact test, 213
Tensile modulus, 171
Tensile strength, 209
Tensile testing, 208
Test specimens, 223–38
Testing, 205, 238–49
  long-term, 219–22
  short-term, 207–19, 238
  simulated end-use, 243
  time, 222
Thermal conductivity effects, 42
Thermal expansion coefficient, 151, 154
Thermal properties, 213–17
Thermoplastic materials, 1, 2, 10, 120, 149, 166, 233
  delivery, 2–5
  machine differences, 57
  production, 28
Thermoplastic polyetherester, 340–4
Thermoplastic polyurethane, 331–6
Thermosetting materials, 1, 10, 91
  machine differences, 57
  moulding techniques, 56–7
Thiokol rubbers, 374, 375
Thioplasts, 375
Thread cutting, 120–1
Tie-bar clearance, 93
Tie-bar construction, 93
Timers, 101
Tolerance limits, 246
Torpedoes, 43–4
Toughened polystyrene, 257–60
Training courses, 111
Transparency, 171
Twin-screw system, 51
Two-stage machines, 65

Ultrasonic assembly, 144
Ultrasonic measurements, 183–4
Ultrasonic staking or riveting, 146–7
Ultrasonic welding, 144–7
Unplasticised polyvinyl chloride, 268–72
Urethanes, 374

Vacuum
  hoppers, 9
  metallising, 129–34
    areas of use, 134
    base coat applications, 131–2
    moulding preparation, 131
    moulding production, 124–31
    process details, 132
    top lacquer application, 133
  sputtering, 137–9
    areas of use, 138
    coating thickness, 138
    metals used, 137
    post-treatment, 138
    pretreatment, 138
    process details, 137
Vented injection unit, 21
Vicat softening point, 213
Viscosity, 215

Water concentration, 222
Weight tests, 241
Weissenberg effect, 68
Weld strength, 189–91
Welding, 143–50
  contact, 144
  friction, 149
  hot-plate, 147–9
  induction, 149–50
  remote, 144
Welding—*contd.*
  spin, 149
  ultrasonic, 144–7
Wetting, 160
Wood graining, 140

X-ray microradiographs, 184

Young's modulus, 209

GPSR Compliance

The European Union's (EU) General Product Safety Regulation (GPSR) is a set of rules that requires consumer products to be safe and our obligations to ensure this.

If you have any concerns about our products, you can contact us on ProductSafety@springernature.com

In case Publisher is established outside the EU, the EU authorized representative is:

Springer Nature Customer Service Center GmbH
Europaplatz 3
69115 Heidelberg, Germany

Batch number: 10607194

Printed by Printforce, the Netherlands